Agricultural Production: Technological Advances

Agricultural Production: Technological Advances

Edited by Dustin Knox

SYRAWOOD
PUBLISHING HOUSE

New York

Published by Syrawood Publishing House,
750 Third Avenue, 9th Floor,
New York, NY 10017, USA
www.syrawoodpublishinghouse.com

Agricultural Production: Technological Advances
Edited by Dustin Knox

© 2022 Syrawood Publishing House

International Standard Book Number: 978-1-64740-052-1 (Hardback)

Cataloging-in-Publication Data

Agricultural production : technological advances / edited by Dustin Knox.
    p. cm.
Includes bibliographical references and index.
ISBN 978-1-64740-052-1
1. Agricultural productivity. 2. Agricultural innovations. 3. Crop yields--Technological innovations.
4. Agriculture--Technology transfer. I. Knox, Dustin.

S494.5.I5 A37 2022

338.16--dc23

TABLE OF CONTENTS

Permissions

List of Contributors

Index

PREFACE

The art and science of cultivating livestock and plants is known as agriculture. Agricultural production has increased dramatically in the past few decades. This has been made possible due to the widespread use of machines such as tractors along with harvesters as well as use of agrochemicals like fertilizers and pesticides. It also makes use of various scientific fields and processes such as genetic engineering and mutation breeding to modify crops. Genetic material is altered by a technique known as recombinant DNA technology. Mutation breeding, also known as variation breeding, exposes the seeds to radiation and chemicals to produce mutant plants with desirable characteristics. Modern technological advances have led to increased durability, insect and virus resistance, nutritional content and herbicide tolerance in agricultural production. This book elucidates the concepts and innovative models around prospective developments with respect to agricultural production. The aim of this book is to present researches that have transformed this field and aided its advancement. It will provide comprehensive knowledge to the readers.

After months of intensive research and writing, this book is the end result of all who devoted their time and efforts in the initiation and progress of this book. It will surely be a source of reference in enhancing the required knowledge of the new developments in the area. During the course of developing this book, certain measures such as accuracy, authenticity and research focused analytical studies were given preference in order to produce a comprehensive book in the area of study.

This book would not have been possible without the efforts of the authors and the publisher. I extend my sincere thanks to them. Secondly, I express my gratitude to my family and well-wishers. And most importantly, I thank my students for constantly expressing their willingness and curiosity in enhancing their knowledge in the field, which encourages me to take up further research projects for the advancement of the area.

Editor

Determinants of technical efficiency of freshwater prawn farming in southwestern Bangladesh

Mst. Esmat Ara Begum [*], Stefanos A. Nastis , Evangelos Papanagiotou

Department of Agricultural Economics, School of Agriculture, Aristotle University of Thessaloniki, Greece

Abstract

This paper estimates a translog stochastic production function to examine the determinants of technical efficiency of freshwater prawn farming in Bangladesh. Primary data has been collected using random sampling from 90 farmers of three villages in southwestern Bangladesh. Prawn farming displayed much variability in technical efficiency ranging from 9.50 to 99.94 % with mean technical efficiency of 65 %, which suggested a substantial 35 % of potential output can be recovered by removing inefficiency. For a land scarce country like Bangladesh this gain could help increase income and ensure better livelihood for the farmers. Based on the translog production function specification, farmers could be made scale efficient by providing more input to produce more output. The results suggest that farmers' education and non-farm income significantly improve efficiency whilst farmers' training, farm distance from the water canal and involvement in fish farm associations reduces efficiency. Hence, the study proposes strategies such as less involvement in farming-related associations and raising the effective training facilities of the farmers as beneficial adjustments for reducing inefficiency. Moreover, the key policy implication of the analysis is that investment in primary education would greatly improve technical efficiency.

Keywords: Bangladesh, Prawn farming, Technical efficiency, Translog stochastic frontier production function

1 Introduction

Bangladesh with its vast and highly diverse aquatic resources and agro-climatic conditions is widely recognised as one of the suitable countries in the world for freshwater prawn (*Macrobrachium rosenbergii*) (locally known as *golda*) farming. A sub-tropical monsoonal climate, low laying agricultural land and a vast area of shallow water provide ideal conditions for freshwater prawn production (Ahmed *et al.*, 2008a). Within the last three decades, prawn farming has become one of the most important sectors in the agricultural based economy of Bangladesh, because it has created jobs,

earned foreign exchange and supplied additional protein to an undernourished population. Approximately 1.2 million people directly and an additional 4.8 million rural people indirectly earn income from prawn and shrimp production and its associated activities (USAID, 2006). In 2011–12, Bangladesh exported 48,007 tons of prawn and shrimp valued at US\$ 428.75 million, of which 25 % was prawn (FSYB, 2013).

The total area under prawn cultivation in Bangladesh is estimated to be around 50,000 ha (Khondaker, 2009). More than 71 % of prawn farms are located in southwest Bangladesh, particularly in the Bagerhat, Khulna and Satkhira districts (Ahmed *et al.*, 2008a). The families living in the densely populated southwest Bangladesh tend to be resource poor, income poor, and vulnerable to environment, climate and economic variability (Bundell & Maybin, 1996; Muir, 2003). Prawn farming therefore opens a new frontier for income and live-

[*] Corresponding author
Agricultural Economics Division, Bangladesh Agricultural Research Institute, Joydebpur, Gazipur-1701, Bangladesh
Email: esmatbau@yahoo.com

lihood for farmers and other farming related people of this region (DIFTA, 1993; Ahmed, 2001; Ito, 2004). The most spectacular boost of prawn farming has taken place in the Bagerhat district where a large number of farmers have converted their rice fields to profitable prawn farms (Ahmed et al., 2008b). The reasons behind the widespread adoption of prawn farming in southwest Bangladesh are the availability of wild postlarvae and low-lying rice fields, a warm climate, fertile soil, and cheap and abundant labour (Haroon, 1990; Ahmed et al., 2008a).

However, the average yield of prawn is low, at 467 kg ha^{-1} (Ahmed et al., 2010a) much lower than in other Asian countries[1]. The potential gain from closing this yield gap is high for Bangladesh. This gap indicates a difference in productivity between 'best practice' farm and another less efficient farm that operate with comparable resource constraints under similar circumstances (Wadud, 1999; Villano, 2005). The difference between the actual and potential output for prawn farming implies great opportunities for increasing income and foreign exchange through improvements in productivity. For a densely populated and resource scarce country such as Bangladesh, where opportunities to develop and adopt new technologies are rare, empirical investigations of technical efficiency and its determinants in prawn farming are a dire necessity. Such studies help to determine the level at which farmers are using existing technologies as well as explore the possibility of raising productivity by increasing efficiency for prawn farming.

In Bangladesh, a number of studies have been conducted on prawn farming, examining conversion of rice fields to prawn farms (Ahmed et al., 2010a), sustainability of freshwater prawn farming (Ahmed et al., 2010b), livelihood analysis of prawn farmers and associated groups (Ahmed et al., 2008a), prawn farming in gher systems (Ahmed et al., 2008b), history of prawn farming (Ahmed et al., 2008c), economic returns to prawn and shrimp farming (Islam et al., 2005), agrarian change and economic transformation (Ito, 2004), and prawn and shrimp marketing (Ahmed et al., 2009; Islam, 2008). To the best of our knowledge, only two studies have analysed the technical efficiency of prawn and gher farming in Bangladesh namely diversification economies and efficiencies of prawn-carp-rice farming (Rahman et al., 2011) and production efficiency of rice fish farming

(Ahmed et al., 2011), and two more studies in neighbouring countries, i.e. Devi (2004) in India and Thi et al. (2007) in Vietnam. Given this backdrop, the present study aims to estimate the determinants of technical efficiency and each factor's contribution to prawn farming inefficiency. The present study selects the appropriate functional form of the inefficiency component and a suitable production function model that fits the data best based on several empirical hypotheses. Our results reinforce existing theoretical arguments that education, non-farm income, training and tenure status may impact farm productivity and efficiency. An understanding of these relationships can provide policy makers with information about the nature of the problems facing prawn farms in Bangladesh and help design programs that improve efficiency.

2   Materials and methods

2.1   Study area

The study was undertaken in the Fakirhat upazila (lowest administrative unit) in Bagerhat district of southwest Bangladesh. Fakirhat was selected for this study as it is an important area for prawn culture because of the availability of natural- and hatchery-produced postlarvae, ponds and low-lying agricultural land.

2.2   Culture season and methods

The peak season of prawn farming in the study area is from May to January. Prawn postlarvae are stocked in May to June and are harvested primarily from December to January, a culture period of around nine months. Between January and April, some of the farmers grow HYV Boro rice on the land inside the gher[2], which is irrigated by water from the inside canals using either traditional methods (swing basket) and/or pumps. Farmers use lime to reduce soil acidity at the time of gher preparation and during the culture period. Farmers use a wide range of homemade and commercially available supplementary feeds to increase prawn production including snail meat, rice bran, wheat bran, oil cake, and pulses.

---

[1] Prawn yield in China 1500 kg ha^{-1} (Weimin & Xianping, 2002), India 600–1000 kg ha^{-1} (Raizada et al., 2005), Taiwan 1500 kg ha^{-1} (New, 1995), Thailand 2338 kg ha^{-1} (Vicki, 2007) and Vietnam 1000–1500 kg ha^{-1} (Ridmontri, 2002).

[2] The local term gher is an enclosure made for prawn cultivation by modifying rice fields through building higher dikes around the rice field and excavating a canal several feet deep inside the periphery to retain water during the dry season (Kendrick, 1994). During the rainy season the whole water body is used for the cultivation of prawn and other fishes, while only trenches are used for fish during dry season (Chapman & Abedin, 2002). The dikes are used for growing vegetables and fruits throughout the year.

The system is costly and labour intensive. Farmers repair the *gher* dikes and trenches almost annually, before releasing prawn postlarvae.

Almost all the freshwater prawn farming practice is intensive, while a few farmers (5 %) still practice extensive farming. The intensive production system is characterised by relatively high stocking densities and high inputs, such as industrial manufactured pellet feeds, chemicals and drugs that increase the nutrients and organic matter load to the ecosystem (Shang *et al.*, 1998). In contrast, the extensive system typically uses slightly modified versions of traditional methods and is called low density system (10,000–17,000 post larvae ha^{-1}) and low input system. The extensive system relies mainly on the natural productivity of the pond, but organic and inorganic fertilisers are occasionally used to promote growth of natural foods (Shang *et al.*, 1998).

2.3 Analytical framework

The seminal paper of Farrell (1957) developed several approaches to efficiency and productivity analysis. Among these, the stochastic frontier production function (Aigner *et al.*, 1977; Meeusen & van den Broeck, 1977) and Data Envelopment Analysis (DEA) (Charnes *et al.*, 1978) are the two principal methods for efficiency measurement. In developing countries' agriculture where data are heavily influenced by measurement errors and the effects of weather conditions, diseases, etc, the stochastic frontier analysis has more advantages compared with data envelopment analysis (Färe *et al.*, 1985; Kirkly *et al.*, 1995, 1998; Jaforullah & Devlin, 1996; Coelli *et al.*, 1998; Dey, 2000; Dey *et al.*, 2005). This also applies to the applications of frontier techniques to prawn farming, hence a stochastic frontier production function is specified.

The stochastic frontier production function for the cross section data can be defined as follows:

$$Y_i = f(X_i; \beta) + V_i - U_i \qquad (1)$$

where Y_i denotes the production for the ith farm ($i = 1, 2, \ldots, n$); X_i is a $1 \times k$ vector of functions of inputs quantities used by the ith firm; β is a $k \times 1$ vector of unknown parameters to be estimated; the V_i's are random variables which are assumed to be independently and identically distributed N(0, σ_v^2) and are distributed independently of the technical inefficiencies U_i's; and the U_i's are non-negative random variables associated with technical inefficiency in production, which are assumed to be independently distributed as truncations of the N($Z_i\delta, \sigma_u^2$) distribution.

Following Battese & Coelli (1995), U_i's can be represented as:

$$U_i = Z_i \delta + W_i \qquad (2)$$

where Z_i is a $1 \times p$ vector of variables which may influence efficiency of a farm; δ is a $p \times 1$ vector of parameters to be estimated; and W_i's are the random variables defined by the truncation of the normal distribution with mean 0 and variance σ_u^2, such that the point of truncation is $-Z_i\delta$, i.e., $W_i \geq Z_i\delta$. These assumptions are consistent with U_i being a non-negative truncation of the N($Z_i\delta, \sigma_u^2$) distribution (Battese & Coelli, 1995).

The technical efficiency of production for the ith farm (TE_i) is defined as:

$$TE_i = \exp(-U_i) = \frac{Y_i}{f(X_i; \beta) \exp(V_i)} \qquad (3)$$

The prediction of the technical efficiencies is based on conditional expectation of expression in equation 3, given the model's assumptions.

2.4 The data

This study is based on farm level cross sectional data for the crop year 2011 collected from three villages (Faltita Baniyakhali, Saittala and Bailtali[3]) of Fakirhat upazila. For the sampling method, a database of prawn farmers was collected from upazila and district Fisheries Offices. A total of 90 *gher* farmers were randomly selected. The survey was conducted for a period of 3 months from August to October 2011. Questionnaire interviews with *gher* farmers were preceded by preparation and testing of the questionnaire and the use of enumerators to fill in the questionnaire.

2.5 Limitation of the Study

A caveat of the paper that needs to be mentioned is the sample size. Due to cost constraints, the researchers have been able to collect at present a small sample of *gher* farms. As mentioned, only a few attempts have been undertaken to determine prawn farm efficiency in Bangladesh. Thus, the results should be interpreted in this light.

2.6 The empirical model

The Cobb-Douglas (CD) and the transcendental logarithmic (TL) are the two most popular functional forms in the stochastic frontier analysis economics literature.

---

[3] Total number of *ghers* of these three villages is 1565 which represents more than 10 % of *ghers* in Fakirhat upazila (Haque & Saifuzzaman, 2002).

Various studies have been conducted using a CD production function due to its linearity in logarithms; however its elasticity is constant and the elasticity of substitution is unity. The TL is more flexible in that it imposes few assumptions on the function and its elasticities, but it is more difficult to estimate due to the large number of parameters and the attendant problem of multicollinearity among the regressors (Irz & McKenzie, 2003). We first specified a translog (TL) stochastic frontier model in this study that is then tested against a Cobb-Douglas (CD) to determine which functional specification best fits the data on the prawn farming.

The translog (TL) functional form employed to estimate the stochastic production frontier is specified as:

$$\ln Y_i = \beta_0 + \sum_{j=1}^{7} \beta_j \ln \overline{X}_{ji} + \frac{1}{2} \sum_{j=1}^{7} \sum_{k=1}^{7} \beta_{jk} \ln \overline{X}_{ji} \ln \overline{X}_{ki} + V_i - U_i \tag{4}$$

where subscript i refers to the ith farm in the sample; ln represents the natural logarithm; Y output variable and $\overline{X}$'s are means of input variables (creating variables by deducting each observation from its sample mean, i.e., $\overline{X}_1 = X_1 - \overline{x}_1$) as defined in Table 1. V_i's are iid $N(0, \sigma_v^2)$ random variables; U_i's are independently distributed $(|N(Z_i\delta, \sigma_u^2)|)$.

Following Battese & Coelli (1995), it is further assumed that the technical inefficiency distribution parameter, U_i is a function of various operational and farm specific variables hypothesized to influence technical inefficiencies as:

$$U_i = \delta_0 + \sum_{k=1}^{9} \delta_k Z_{ki} + W_i \tag{5}$$

where Z's are various farm specific variables, as defined in Table 3; δ's are unknown parameters to be estimated; and W_i is a random variable as defined in equation 3. These farm specific variables (age, education, training, involvement in fish farm association, non farm income, family size, distance from the canal, water quality and lease area) may affect efficiency. Choice of these variables is based on the existing literature, and the justification for their inclusion is briefly discussed.

Use of the education level and training of farmers as a technical efficiency shifter is fairly common (Wang et al., 1996; Wadud & White, 2000; Asadullah & Rahman, 2009; Haque, 2011; Rahman et al., 2011). The education and training variables are also used as a surrogate for a number of factors. At the technical level, access to information as well as capacity to understand the technical aspects related to production are expected to improve with education and training, thereby, influen-

cing technical efficiency. The justification for including age is straightforward as older, and hence more experienced, farmers are more likely to be more efficient in decisions regarding the use and allocation of scarce inputs (Liewelyn & Williams, 1996; Coelli & Fleming, 2004).

Large families seem to have the tendency to adopt earlier new technologies, as has been found in Bangladesh by Hossain et al. (1990). The number of family members is incorporated to test whether it influences technical efficiency as proposed by Haque (2011). The proportion of lease area is included in this study as the number of farmers who cultivate prawn solely on leased land was limited. The prawn farms were often a mixture of own land and leased land. Thus, in this case we have considered lease area as a continuous variable. Involvement in fish farm associations exposes them to information, technology and other facilities. Thus, the association of the prawn farmers with fisheries-related activities is expected to increase technical efficiency. Non-farm income is included as an indication of the farmers' economic condition. Closer distance of the pond to the main water channel has advantages as the water contains more postlarvae, natural foods and minerals. It is expected that the prawn farmers who are having more non-farm income would be able to invest more in the farms and hence would be able to achieve better technical efficiency.

At the point of approximation (i.e., sample mean), the translog production function should be well behaved, satisfying all regularity conditions namely positive and diminishing marginal products (the first order parameters are all between zero and one, while the bordered Hessian matrix of the first and second order partial derivatives is negative semi-definite).

We assume perfect competition in the prawn industry and monotonicity condition necessitates positive marginal product which can be derived from the translog production in equation 4. More specifically,

$$MP = f_i = \frac{\partial Y}{\partial X_i} = \frac{\partial \ln Y}{\partial \ln X_i} \times \frac{Y}{X_i} \tag{6}$$

$$f_i = \frac{Y}{X_i}\left(\beta_i + \sum_{j=1}^{n} \beta_{ij} \ln X_j\right) > 0 \tag{7}$$

Diminishing marginal productivity is

$$f_{ii} = \frac{Y}{X_i^2}\left[\beta_{ii} + \left(\beta_i - 1 + \sum_{j=1}^{n} \beta_{ij} \ln X_j\right)\right.$$
$$\left.\left(\beta_i + \sum_{j=1}^{n} \beta_{ij} \ln X_j\right)\right] < 0 \tag{8}$$

where f_i is marginal product of inputs, f_{ii} is the diminishing marginal product of input, Y is the production of prawn, i and j are the inputs, $X_i X_j$ are the input levels, β_i is the estimated coefficient of the X term and β_{ij} is the estimated coefficient of the X^2 term.

It should be noted that the technical inefficiency model in equation 5 can only be estimated if the technical inefficiency effects, U_i's are stochastic and have particular distributional properties (Coelli & Battese, 1996). Therefore, it is of interest to test the null hypothesis that the technical inefficiency effects are absent: $\gamma = \delta_0 = \delta_E = \delta_A = \delta_T = \delta_I = \delta_N = \delta_F = \delta_D = \delta_W = \delta_L = 0$ (where subscripts represent education, age, training, involvement in fish farm associations, non-farm income, family size, distance of the *gher* from the water channel, water quality and lease area, respectively). The stochastic frontier model reduces to a traditional average function in which the explanatory variables in the technical inefficiency model are included in the production function. Failure to reject the null hypothesis $H_0 : \gamma = 0$ implies the existence of a stochastic frontier. Similarly, $\gamma = 1$ implies that all the deviations from the frontier are due to the technical inefficiency (Coelli *et al.*, 1998). These and related null hypotheses can be tested using the generalized likelihood-ratio statistic, λ, given by:

$$\lambda = -2 \left[\ln\{L(H_0)\} - \ln\{L(H_1)\} \right] \qquad (9)$$

where $L(H_0)$ and $L(H_1)$ denote the values of the likelihood function under null (H_0) and alternative (H_1) hypotheses, respectively. If the given null hypothesis is true, λ has approximately χ^2-distribution or mixed χ^2-distribution when the null hypothesis involves $\lambda = 0$ (Coelli, 1995b,a).

Given the model specifications, the technical efficiency index for the ith farm in the sample (TE_i), defined as the ratio of observed output to the corresponding frontier output, is given by

$$TE_i = \exp(-U_i) \qquad (10)$$

The prediction of technical efficiencies is based on the conditional expectation of expression in equation 10, given the values of $V_i - U_i$ evaluated at the maximum likelihood estimates of the parameter of the stochastic production frontier model (Battese & Coelli, 1995). The frontier production for the ith farm can be computed as the actual production divided by the technical efficiency

Table 1: *Description of output, input and farm specific variables*

Variable	Description	Unit
Y	Total production of prawn of the sampled farmers during the year	kg
Variables in the production frontier		
X_G	*Gher* size	Hectare
$X_L{}^*$	Quantity of labour employed per hectare per year	Man-days
X_{Fi}	Quantity of fingerlings stocked in pond per hectare per year	Number
X_{Fd}	Quantity of feeds (pulses, oilcake and wheat bran) applied per hectare per year	kg
X_{Li}	Quantity of lime applied per hectare per year	kg
X_{Fe}	Quantity of fertiliser used per hectare per year	kg
X_C	Amount of cost incurred for other inputs per ha per year	US$
Variables in the inefficiency function		
Z_E	Education (years of schooling) of the farmers	Year
Z_A	Age of the farmers	Year
Z_T	Training received by the farmers (1 if received, 0 otherwise)	1, 0
Z_I	Involvement of the farmers in the fish farm association (1 if involve, 0 otherwise)	1, 0
Z_N	Share of non-farm income to total income of the farmers	Percent
Z_F	Family size of the farmers	Persons
Z_D	Distance of the pond from the channel/creek (1 if less than 500 metres, 0 otherwise)	1, 0
Z_W	Water quality of the ponds (1 if good enough, 0 otherwise)	1, 0
Z_L	Proportion of lease area to total prawn farm area	Percent

* 1 man day equals 8 working hours/day

estimate. The parameters for the stochastic production frontier function in equation 4 and those of the technical inefficiency model in equation 5 are estimated simultaneously using the maximum likelihood (ML) estimation method, using the computer programme, FRONTIER 4.1 (Coelli, 1994), which gives the variance parameters of the likelihood function in terms of $\sigma^2 = \sigma_v^2 + \sigma_u^2$ and $\gamma = \sigma_u^2/\sigma^2$.

3   Results

3.1   Sample characteristics

Of the farms analysed, the average *gher* size is 1.96 ha ranging from 0.20 ha to 6.32 ha. About 30 % of operations have a *gher* size of less than 1 ha. Stocking density (number of fingerling released per ha) is 30,721 pieces on average. The supplementary feed application for prawn production is mostly a mixture of wheat bran, mustard oil cake and pulses. The average feed application is 3280 kg ha^{-1}. The mean yield of prawn is 589 kg ha^{-1} ranging from a minimum of 534.4 kg ha^{-1} to as high as 673.6 kg ha^{-1}. The mean of liming for pond preparation is reported 242.23 kg ha^{-1}. All the sample farmers apply fertiliser for pond preparation and water treatment ranging from 79 kg ha^{-1} to 888 kg ha^{-1}. The mean non-farm annual income of the prawn farmers is US$ 1995.78. The average labour use is 248.17 man-days ha^{-1} ranging from 118.78 man-days ha^{-1} to 576.49 man-days ha^{-1} (Table 2).

A likelihood ratio test was conducted to test the null hypothesis that the translog stochastic frontier production function can be reduced to a Cobb-Douglas production function. The test statistic $H_0: \beta_{jk} = 0, H_1: \beta_{jk} \neq 0$, has a likelihood ratio value of 165.21, which implies a rejection of the null hypothesis at the 5 % significance level. In other words, the translog production function is more suitable to the prawn farm survey data that adequately captures the production behaviour.

Before proceeding with any explanation to the analysis, monotonicity test is performed and results are presented in Table 4. The results show that the marginal products of the inputs are positive and diminishing marginal products are negative.

The maximum-likelihood estimates of the parameters for the stochastic frontier production model and those for the technical inefficiency model of prawn farming in Bangladesh are shown in Table 3. Most of the slope coefficients of inputs on the first order terms were positive, with the exception of the other cost coefficient. However, this negative coefficient was insignificant. The coefficients associated with *gher* size, labour and fingerling were highly significant while the coefficient for feed was significant at the 5 % level. Other independent variables such as lime and fertiliser have pos-

Table 2: *Summary statistics*

Name of variable	Mean	Standard deviation	Minimum	Maximum
Prawn production (kg ha^{-1})	589.1	28.54	534.4	673.6
Gher size (ha)	1.96	1.37	0.20	6.32
Labour (man-days ha^{-1})	248.17	79.84	118.75	576.49
Fingerlings (number ha^{-1})	30 720.6	2181.88	24 754.58	35 085.23
Feed (kg ha^{-1})	3280.09	529.37	2400	5200
Lime (kg ha^{-1})	242.23	34.12	155	301
Fertiliser (kg ha^{-1})	149.21	86.93	79	888
Other cost (US$) *	53.51	28.70	24.23	199.34
Education of head of household((years of schooling)	11.3	2.1	5	16
Age of head of household(years)	42.9	6.3	29	55
Total non-farm income (US$)	1995.78	1538.68	142.60	7130.13
Family size (persons)	5.16	1.11	3	8
Proportion of leased area to total operational area under prawn culture (%)	7.29	15.57	0	80.77

* US$ 1 ≈ 84.15 Bangladeshi Taka in 11/02/2012.

itive coefficients but are insignificant under the translog production function. The coefficients on second order terms are also significantly different from zero, thereby confirming nonlinearities in the production process, and hence, justify the use of translog production function specification.

The γ- parameter associated with the variances in the stochastic production frontier is estimated to be close to 1 (Table 3). Although the γ- parameter cannot be interpreted as the proportion of the total variance explained by technical inefficiency effects, the result indicates that technical inefficiency effects do make a significant contribution to the level and variation of prawn farming in Bangladesh.

3.2 Factors explaining inefficiency

The results indicate that the farm specific variables included in the technical inefficiency model contribute significantly, both as a group and several of them individually, to the explanation of the technical inefficiencies (Table 3). The parameter estimates showed that factors such as training, involvement in fish farm association, family size, distance and water quality were positively related to inefficiency while education, age, non-farm income and lease area were negatively related to inefficiency.

It is expected that involvement in fish farm associations should be positively related to technical efficiency. It is assumed that prawn farmers who belong to fish farm associations are likely to benefit from better access to inputs and to information on improved farming practices. Being a member in farmers' association may lead to sharing of information on farming technologies, which tends to influence the production practices of members through peer learning. However, our estimate showed a negative statistically significant relationship between membership in fish farm associations and technical efficiency.

3.3 Hypotheses tests for γ and δ parameters

Generalized likelihood ratio tests of various null hypotheses involving the restrictions on the variance parameter, γ, in the stochastic production frontier and δ coefficients in the technical inefficiency model are presented in Table 5. The first hypothesis is tested for the presence of inefficiencies in the model. Thus, γ is defined between zero and one, where if $\gamma = 0$, technical inefficiency is not present, and where $\gamma = 1$, there is no random noise. The test of significance of the inefficiencies in the model (H_0: $\gamma = \mu = 0$) was rejected at the 5 % significance level, indicating that the maximum

likelihood estimation is a significant improvement over an Ordinary Least Squares (OLS) specification and inefficiencies are present in the model. The calculated value of the test statistic is 51.10, which is greater than the critical value (Table 5).

The second null hypothesis, H_0: $\delta_0 = \delta_{Ed} = \delta_{Tr} = \ldots = \delta_{Pl} = 0$, specifies that technical inefficiency follows a half-normal distribution with zero mean originally proposed by Aigner et al. (1977). This null hypothesis is rejected at 5 % significance level suggesting that, given the stochastic frontier with the model for technical inefficiency effects, the standard stochastic error component model is not appropriate for the half-normal distribution. The third null hypothesis, H_0: $\delta_{Ed} = \delta_{Tr} = \ldots = \delta_{Pl} = 0$, implies that technical inefficiency effects follow a standard truncated normal distribution (Stevenson, 1980) as the null hypothesis is rejected at the 5 % level of significance. This indicates that the farm-specific variables involved in the technical inefficiency model contribute significantly as a group to the explanation of the technical inefficiency effects in prawn production although, based on asymptotic t ratios, some slope coefficients are not significant individually (Table 5).

The confidence intervals of inefficiency parameters show the effect size of individual parameters has on technical efficiency (Table 6).

3.4 Technical efficiency distribution

The technical efficiency (TE) scores range from 9.50 to 99.94%, with a mean score of 65 % (Table 7). The implication is that, on average, 35 % of the potential output can be recovered by eliminating technical inefficiency, which is substantial and could improve the competitiveness of the Bangladesh prawn farming. The indices of TE indicate that if the average farmer of the sample could achieve the TE level of its most efficient counterpart, then average farmers could increase their output by 34.67 % [1-(65/99.94)]. Similarly, the most technically inefficient farmer could increase the production by 90.45 % [1-(9.50/99.50)] if he/she could increase the level of TE to his/her most efficient counterpart. For a land-scarce country like Bangladesh, this gain in production will increase income and ensure better livelihood for the farmers.

The distribution of the efficiency score is quite similar at the higher and lower end of the efficiency spectrum. About 32 % of the farmers are producing at an efficiency level of less than 50 % while 27 % of the farmers are producing at an efficiency level of 90 % and above, which is encouraging.

Table 3: *Maximum likelihood estimates of the stochastic production frontier and inefficiency model*

Variable	Parameter	Coefficient	Standard error
Production frontier			
Constant	β_0	7.43187***	0.01624
ln *Gher* size	β_1	0.45219***	0.15078
ln Labour	β_2	0.37631***	0.13786
ln Fingerling	β_3	0.23425***	0.04939
ln Feed	β_4	0.12426**	0.05826
ln Lime	β_5	0.13062	0.10032
ln Fertiliser	β_6	0.03970	0.08494
ln Other cost	β_7	−0.07229	0.04835
ln *Gher* size × ln *Gher* size	β_{11}	0.78245***	0.14488
ln Labour × ln Labour	β_{22}	−0.19381***	0.05796
ln Fingerling × ln Fingerling	β_{33}	−0.07449***	0.00591
ln Feed × ln Feed	β_{44}	−0.04022**	0.02029
ln Lime × ln Lime	β_{55}	−0.12309**	0.05977
ln Fertiliser × ln Fertiliser	β_{66}	−0.03109	0.04922
ln Other cost × ln Other cost	β_{77}	−0.11126	0.07506
ln *Gher* size × ln Labour	β_{12}	0.08047*	0.04595
ln *Gher* size × ln Fingerling	β_{13}	0.00566	0.01457
ln *Gher* size × ln Feed	β_{14}	−0.05822***	0.02153
ln *Gher* size × ln Lime	β_{15}	0.02421	0.02908
ln *Gher* size × ln Fertiliser	β_{16}	−0.08853***	0.03137
ln *Gher* size × ln Other cost	β_{17}	0.29947***	0.03752
ln Labour × ln Fingerling	β_{23}	−0.00864**	0.00413
ln Labour × ln Feed	β_{24}	0.00342	0.00489
ln Labour × ln Lime	β_{25}	−0.01136	0.01093
ln Labour × ln Fertiliser	β_{26}	−0.01937	0.01345
ln Labour × ln Other cost	β_{27}	0.03941***	0.01478
ln Fingerling × ln Feed	β_{34}	−0.00308***	0.00273
ln Fingerling × ln Lime	β_{35}	0.02108***	0.00414
ln Fingerling × ln Fertiliser	β_{36}	0.00499	0.00454
ln Fingerling × ln Other cost	β_{37}	−0.04288***	0.00965
ln Feed × ln Lime	β_{45}	−0.00019	0.00953
ln Feed × ln Fertiliser	β_{46}	−0.00544	0.00505
ln Feed × ln Other cost	β_{47}	0.06487***	0.00489
ln Lime × ln Fertiliser	β_{56}	−0.01061**	0.00433
ln Lime × ln Other cost	β_{57}	0.07898***	0.01014
ln Fertiliser × ln Other cost	β_{67}	0.05461***	0.01784
Inefficiency function			
Constant	δ_0	1.67062*	0.89052
Education	δ_1	−0.13651***	0.03562
Training	δ_2	1.01326***	0.24709
Age	δ_3	−0.01661	0.01636
Involvement of fish farm association	δ_4	0.59012*	0.31291
Non-farm income	δ_5	−0.01471***	0.00541
Family size	δ_6	0.00039	0.07865
Distance	δ_7	0.72474**	0.30490
Water quality	δ_8	0.01216	0.22742
Lease area	δ_9	−0.01418	0.01039
Variance parameters			
Sigma-squared	σ^2	0.28584***	0.08179
Gamma	γ	0.99999***	0.00002
Log likelihood	−14.52533		
Mean TE index	65.0%		

Note: *, ** and *** statistically significant at 10%, 5% and 1% level respectively,
Number of observations: 90

Table 4: *Estimated output elasticities, marginal products and diminishing marginal products for prawn farming*

Inputs	Output Elasticities	Marginal products	Diminishing marginal products
Land	0.603	355.064	−319.857
Labour	0.0007	0.002	−0.00016
Fingerlings	0.058	1.310	−0.00013
Feed	0.027	0.005	−0.0000007
Lime	0.047	0.114	−0.00002
Fertiliser	0.005	0.029	−0.00002
Other cost	0.095	2.046	−0.00171

Table 5: *Generalised likelihood ratio tests of hypotheses of parameters*

Test of null hypotheses (H_0)	Log-likelihood value of the reduced model	Test statistic (λ)	DF	Critical $\chi 2$ value at 95 %	Conclusion
1. No inefficiency effects (H_0: $\gamma = \delta_0 = \delta_{Ed} = \ldots = \delta_{Fs} = 0$)	−40.08	51.10	11	19.045	Reject H_0
2. Technical inefficiency effects have a half normal distribution with mean zero (H_0: $\delta_0 = \delta_{Ed} = \ldots = \delta_{Fs}$)	−40.32	51.58	10	17.670	Reject H_0
3. No effects of inefficiency factors included in the inefficiency model (H_0: $\delta_{Ed} = \ldots = \delta_{Fs} = 0$)	−40.32	51.58	9	16.274	Reject H_0

Note: The value of the log-likelihood function under the specification of alternative hypothesis (unrestricted/full model) is 53.89. The correct value for the null hypothesis of no inefficiency effects are obtained from Kodde & Palm (1986).

Table 6: *Confidence intervals of inefficiency parameters on technical efficiency*

Inefficiency Parameters	Coefficient	Std.err.	z	P > z	95 % Confidence intervals	
					Upper bound	Lower bound
Education	−0.350	0.190	−2.850	0.015	−0.722	−0.821
Training	1.416	2.537	4.560	0.017	0.657	0.788
Age	−0.007	0.074	−0.100	0.920	−0.153	−0.138
Involvement of fish farm association	1.669	0.864	1.930	0.053	0.635	0.762
Non-farm income	−0.034	0.029	−2.990	0.013	0.590	0.622
Family size	0.145	0.407	0.360	0.722	0.653	0.943
Distance	1.286	1.009	3.180	0.040	0.690	0.753
Water quality	0.449	0.739	0.610	0.543	0.500	0.898
Lease area	−0.135	0.195	−0.690	0.487	−0.517	−0.746
Constant	−0.033	2.921	−2.690	0.090	−13.598	13.532

Table 7: *Distribution of technical efficiency scores*

Variables	Estimates
Efficiency levels (Percent)	
≤ 50	32.22
$50 \leq 60$	14.44
$60 \leq 70$	14.44
$70 \leq 80$	5.56
$80 \leq 90$	5.56
$90 \leq 100$	27.78
Mean efficiency level	65
Minimum	9.50
Maximum	99.94
Number of observations	90

4   Discussion

4.1   Output coefficient

The coefficient of output with respect to *gher* size is the highest among all the inputs, which demonstrates the importance of scarce land in boosting prawn production in Bangladesh. The policy implication of this finding is that the government could encourage farmers to keep and increase their existing *gher* size. Coefficient of labour is the second highest, but excess use of the labour exerts negative impacts on output as observed from the second order of labour.

4.2   Parameters of the inefficiency function

Results indicate that education significantly improves technical efficiency, consistent with Asadullah & Rahman (2009) and Sharif & Dar (1996) for Bangladeshi farms. Similar results have been reported in studies that have focused on the association between formal education and technical efficiency (Uaiene & Arndt, 2009; Bozoglu & Ceyhan, 2007). In general, more educated farmers are able to perceive, interpret and respond to new information and adopt improved technologies. The educated prawn farmers are expected to follow the prawn management practices properly, which might have led to higher efficiency for them. This result is consistent with the findings by Abdulai & Eberlin (2001), which established that an increase in formal education will augment the productivity of farmers since they will be better able to allocate family-supplied and purchased inputs, select and utilise the appropriate quantities of purchased inputs while applying available and acceptable techniques to achieve the portfolio of household pursuits such as income. The age coefficient is positive and insignificant with technical efficiency, which indicates that older farmers are more capable to take proper decisions regarding farm management practices as they have many years of practical experience. This conforms to the results obtained by Dey *et al.* (2000); Alam *et al.* (2011) and Rahman *et al.* (2011). The training coefficient is negatively significant with technical efficiency, which was unexpected but consistent with Bhattacharya (2008). This contradictory result may be due to lack of participation of the most successful farmers in training programs, and thus the real impact of training may be disguised. In general, the participants of training programs in Bangladesh are farmers who have good contact with NGOs, local extension officers, and other organisations. Small and medium-scale farmers have lack of such contacts and only large farmers have good relations with the aforementioned organisations. However, large-scale farmers are not actively participating in farming activities. Only their representatives take the responsibilities in farm operation. The training program might, also in addition, be inappropriate for the farmers that are participating. The prawn farmers might require a more hands-on training, rather than a governmental/NGOs/other organisations' lecture-based training program. Involvement in fish farm associations is negatively related to the technical efficiency and is significant at the 10 % level. Thus, we conclude that the association is not useful and not fit for the job. This result conforms to that obtained by Bhattacharya (2008) who found similar relations for shrimp farmers in India. Non-farm income is positively and significantly related with technical efficiency of prawn farmers. This indicates that higher non-farm income increases the technical efficiency of prawn farmers as they are able to invest the earned money in their farming activities. This result is consistent with Haque (2011). There are large numbers of farmers who have higher level of education and who have income from non-farm activities, especially working as government employees. The family members of the farmers also contribute to non-farm income as they work outside the farms, even abroad. As prawn farming activities incur high cost, non-farm income significantly contributes additional income. Family size is negative and insignificant with technical efficiency, consistent with Irz & McKenzie (2003), which indicates that those farmers that have large families are less efficient. This might be the result of large families having excess labour if all members stay on the farm which is often the case in Bangladesh. Distance from the water canal significantly degrades technical efficiency due to inferior water quality (less natural food, organic materials and postlarvae).

4.3 Technical efficiency

The mean technical efficiency of 65 % is quite similar to the estimates for agricultural farms (aquaculture and livestock/dairy farms) in Bangladesh (Bravo-Ureta *et al.*, 2007; Coelli *et al.*, 2002; Wadud & White, 2000). Rahman *et al.* (2011) found the technical efficiency of prawn farming to be 68 %. Technical efficiency of carp culture in other Asian countries, however, ranges from 42 % in all farm types in Malaysia (Iinuma *et al.*, 1999) as well as in extensive farms in Vietnam, to 93 % amongst intensive farms in China (Dey *et al.*, 2005). Other studies such as Alam *et al.* (2011) found the TE of tilapia for Bangladesh farmers at 78 %. Sharma & Leung (2000) estimated the TE of carp polyculture in Bangladesh to be 47.5 % for extensive farming and 73.8 % for semi-intensive farming. ICLARM (2001) found the TE of carp polyculture at 70 % while Arjumanara *et al.* (2004) estimated TE of 62 and 86 % for different groups of carp farmers in Bangladesh. The wide inefficiency spectrum in this study is therefore not surprising and is similar to those reported in literature.

The inefficiency effect is significant, and education, age, training, involvement in fish farm associations, family size, non-farm income, water quality, distance of the farm from the canal and lease area, as a group, are significant determinants of technical inefficiency. By operating at full technical efficiency levels, prawn production can be improved on average from the current level of 589 to 795 kg ha^{-1}. As a result farm income would increase on average Tk. 134377 (US$ 1655.91).

5 Conclusion and Policy Implications

This study uses a translog stochastic frontier production function on survey data to determine the technical efficiency and its determinants in prawn farming in Bangladesh. The production frontier involves seven variables, including *gher* size, labour, fingerling, feed, lime, fertiliser and other cost. Similarly, the technical inefficiency model includes nine farm-specific variables, namely education, training, age, involvement to fish farm association, non-farm income, family size, distance, water quality and land lease.

The level of technical efficiency of prawn farming is low at 65 % implying that a substantial 35 % of the potential output from the system can be recovered by eliminating inefficiency, given the existing technology and resource endowments. Our results confirmed that training, involvement in fish farm associations, family size, distance and water quality positively affected technical inefficiency whereas education, age, non-farm income

and lease area negatively affected technical inefficiency. In particular, policies leading to improving water quality through lime application, construct the ponds close to the water channel or digging supplementary channels for reducing farm distance, proper involvement in farming-related associations which ensure the information flow and technology change, encourage the family members involvement in the off farm activities, consider all farmers as participants of training programs (encourage small and medium farmers to participate in training programs) could be beneficial for reducing inefficiency in prawn farming in Bangladesh.

More investment in education in rural areas through private and public partnerships, initiating progress to encourage those at school-going age and 'food for education' programs may be harnessed as a central ingredient in the development strategies. Moreover, the farmer field schools (FFS) program, promoted by different development agencies may be rigorously implemented and practiced. This would help farmers through 'learning by doing' to improve their analytical and decision-making skills that contribute to adapting improved farming technologies. These measures in the long run may shift the farmers' production frontier upward, which may in turn, reduce technical inefficiency on the one hand and lead to increase income and standard of living on the other.

References

Abdulai, A. & Eberlin, R. (2001). Technical efficiency during economic reform in Nicaragua: evidence from farm household survey data. *Economic Systems*, 25 (2), 113–125.

Ahmed, N. (2001). *Socio-economic aspects of freshwater prawn culture development in Bangladesh*. Ph.D. thesis, Institute of Aquaculture, University of Stirling, Scotland, UK.

Ahmed, N., Allison, E. H. & Muir, J. F. (2008a). Using the sustainable livelihoods framework to identify constraints and opportunities to the development of freshwater prawn farming in southwest Bangladesh. *World Aquaculture Society*, 39, 598–611.

Ahmed, N., Allison, E. H. & Muir, J. F. (2010a). Ricefields to prawn farms: a blue revolution in southwest Bangladesh? *Aquaculture International*, 18, 555–574.

Ahmed, N., Brown, J. H. & Muir, J. F. (2008b). Freshwater prawn farming in gher systems in southwest Bangladesh. *Aquaculture Economics and Management*, 12, 207–223.

Ahmed, N., Demaine, H. & Muir, J. F. (2008c). Freshwater prawn farming in Bangladesh: history, present status and future prospects. *Aquaculture Research*, 39, 806−819.

Ahmed, N., Lecouffe, C., Allison, E. H. & Muir, J. F. (2009). The sustainable livelihoods approach to the development of freshwater prawn marketing systems in southwest Bangladesh. *Aquaculture Economics and Management*, 13, 246−269.

Ahmed, N., Stephen, T. & Garnett (2010b). Sustainability of freshwater prawn farming in rice fields in southwest Bangladesh. *Journal of Sustainable Agriculture*, 34, 659–679.

Ahmed, N., Zander, K. K. & Garbett, S. T. (2011). Socioeconomic aspects of rice-fish farming in Bangladesh: Opportunities, challenges and production efficiency. *The Australian Journal of Agricultural and Resource Economics*, 55, 199–219.

Aigner, D. J., Lovell, C. A. K. & Schimidt, P. (1977). Formulation and estimation of stochastic frontier production models. *Journal of Econometrics*, 6, 21–37.

Alam, M. F., Khan, M. A. & Huq, A. S. M. A. (2011). Technical efficiency in tilapia farming of Bangladesh: a stochastic frontier production approach. *Aquaculture International*, 1–16.

Arjumanara, L., Alam, M. F., Rahman, M. M. & Jabber, M. A. (2004). Yield gaps, production losses and technical efficiency of selected groups of fish farmers in Bangladesh. *Indian Journal of Agricultural Economics*, 59, 806–818.

Asadullah, M. N. & Rahman, S. (2009). Farm productivity and efficiency in rural Bangladesh: the role of education revisited. *Applied Economics*, 41, 17−33.

Battese, G. E. & Coelli, T. J. (1995). A model of technical inefficiency effects in a stochastic frontier production function for panel data. *Empirical Economics*, 20, 325−332.

Bhattacharya, P. (2008). Economics of shrimp farming: a comparative study of traditional vs scientific shrimp farming in West Bengal. Working paper No. 218, The Institute for Social and Economic Cahnge, Bangalore.

Bozoglu, M. & Ceyhan, N. (2007). Measuring the technical efficiency and exploring the inefficiency determinants of vegetables farms in Samsung Province. *Turkey Agricultural System*, 94, 649−656.

Bravo-Ureta, B. E., Solis, D., Lopez, V. H. M., Maripani, J. F., Thiam, A. & Rivas, T. (2007). Technical efficiency in farming: a meta regression analysis. *Journal of Productivity Analysis*, 27, 57−72.

Bundell, K. & Maybin, E. (1996). After the prawn rush: the human and environmental costs of commercial prawn farming. Christian Aid Report, London, UK.

Chapman, G. & Abedin, J. (2002). A description of the rice-prawn-fish systems of southwest Bangladesh. *In:* Edwards, P., Little, D. C. & Demaine, H. (eds.), *Rural Aquaculture*. pp. 111−116, CABI Publishing, New York.

Charnes, A., Cooper, W. W. & Rhodes, E. (1978). Measuring the efficiency of decision-making units. *European Journal of Operational Research*, 2, 429−444.

Coelli, T. J. (1994). A guide to FRONTIER version 4.1: a computer program for stochastic frontier production and cost function estimation. Department of Econometrics, University of New England, Armidale, Australia.

Coelli, T. J. (1995a). Estimators and hypothesis tests for a stochastic frontier function: a Monte Carlo analysis. *Journal of Productivity Analysis*, 6, 247–268.

Coelli, T. J. (1995b). Recent development in frontier modeling and efficiency measurement. *Australian Journal of Agricultural Economics*, 215–245.

Coelli, T. J. & Battese, G. E. (1996). Identification of factors which influence the technical efficiency of Indian farmers. *Australian Journal of Agricultural Economics*, 40, 103−128.

Coelli, T. J. & Fleming, E. (2004). Diversification economies and specialization efficiencies in a mixed food and coffee smallholder farming system in Papua New Guinea. *Agricultural Economics*, 31, 229−239.

Coelli, T. J., Rahman, S. & Thirtle, C. (2002). Technical, allocative, cost and scale efficiencies in Bangladesh rice cultivation: a non-parametric approach. *Journal of Agricultural Economics*, 53, 607−626.

Coelli, T. J., Rao, D. S. P. & Battese, G. E. (1998). *An Introduction to Efficiency and Productivity Analysis*. Kluwer, Boston.

Devi, K. U. (2004). Economic Analysis of prawn culture in coastal Andhra Pradesh. *Agricultural Economics Research Review*, 17, 284–285.

Dey, M. M. (2000). The impact of genetically improved farmed Nile tilapia in Asia. *Aquaculture Economics and Management*, 4, 107−124.

Dey, M. M., Paraguas, F. J., Bimbao, G. B. & Regaspi, P. B. (2000). Technical efficiency of Tilapia grow out pond operations in Philippines. *Aquaculture Economics and Management*, 4, 33–47.

Dey, M. M., Paraguas, F. J., Sricantuk, N., Xinhua, Y., Bhatta, R. & Dung, L. T. C. (2005). Technical efficiency of freshwater pond polyculture production in selected Asian countries: estimation and implication. *Aquaculture Economics and Management*, 9, 39—63.

DIFTA (1993). Sub-sector Study on the Freshwater Prawn Macrobrachiumrosenbergii in Bangladesh. Danish Institute for Fisheries Technology and Aquaculture (DIFTA), Hirtshals, Denmark.

Färe, R., Grosskopf, S. & Lovell, C. A. K. (1985). *The Measurement of Efficiency of Production.* Kluwer-Nijhoff Publishing, Boston.

Farrell, M. J. (1957). The measurement of productive efficiency. *Journal of the Royal Statistical Society, Series A (General)*, 120 (3), 253–281.

FSYB (2013). Fisheries Resources Survey System. Fisheries Statistical Yearbook of Bangladesh (FSYB), Department of Fisheries, Ministry of Fisheries and Livestock, Bangladesh.

Haque, S. (2011). *Efficiency and Institutional Issues of Shrimp Farming in Bangladesh.* Margraf Publishers, Kanalstr. 21, Weikersheim.

Haque, Z. & Saifuzzaman, M. (2002). Social and environmental effects of shrimp cultivation in Bangladesh: notes on study methods. *In:* Rahman, M. (ed.), *Globalization Environmental Crisis and Social Changes in Bangladesh.* Winnpig.

Haroon, A. K. Y. (1990). Freshwater prawn farming trails in Bangladesh. *Naga, the ICLARM Quarterly*, 13, 6–7.

Hossain, M., Quasem, M. A., Akash, M. M. & Jabber, M. A. (1990). *Differential impact of modern rice technology: the Bangladesh case.* Bangladesh Institute of Development Studies, Dhaka.

ICLARM (2001). Genetic improvement of carp species in Asia. Final report. Asian Development Bank regional technical assistance no 57111, WorldFish Center, Penang.

Iinuma, M., Sharma, K. R. & Leung, P. S. (1999). Technical efficiency of carp pond culture in peninsula Malaysia: an application of stochastic frontier and technical inefficiency model. *Aquaculture*, 175, 199—213.

Irz, X. & McKenzie, V. (2003). Profitability and technical efficiency of aquaculture systems in Pampanga, Philippines. *Aquaculture Economics and Management*, 7, 195–212.

Islam, M. S. (2008). From pond to plate: towards a twin-driven commodity chain in Bangladesh shrimp aquaculture. *Food Policy*, 33, 209—223.

Islam, M. S., Milstein, A., Wahab, M. A., Kamal, A. H. M. & Dewan, S. (2005). Production and economic return of shrimp aquaculture in coastal ponds of different sizes and with different management regimes. *Aquaculture International*, 13, 489—500.

Ito, S. (2004). Globalization and agrarian change: a case of freshwater prawn farming in Bangladesh. *Journal of International Development*, 16, 1003–1013.

Jaforullah, M. & Devlin, N. J. (1996). Technical efficiency in the New Zealand dairy industry: a frontier production function approach. *New Zealand Economic papers*, 30, 1—17.

Kendrick, A. (1994). *The gher revolution: the social impacts of technological change in freshwater prawn cultivation in southern Bangladesh.* Bangladesh Aquaculture and Fisheries Resource Unit, Dhaka.

Khondaker, H. R. (2009). Prawn hatchery development in Bangladesh: problems and potentials. *In:* Wahab, M. A. & R., H. M. A. (eds.), *Abstracts National Workshop on Freshwater Prawn Farming in Bangladesh: Technologies for Sustainable Production and Quality Control.* p. 11, Dhaka, Bangladesh.

Kirkly, J. E., Squires, D. & Strand, I. E. (1995). Assessing technical efficiency in commercial fisheries: the mid-Atlantic sea scallop fishery. *American Journal of Agricultural Economics*, 77, 686—697.

Kirkly, J. E., Squires, D. & Strand, I. E. (1998). Characterizing managerial skill and technical efficiency in fishery. *Journal of Productivity Analysis*, 9, 145—160.

Kodde, D. A. & Palm, E. C. (1986). Wald criteria for jointly testing equality and inequality restrictions. *Econometrica*, 54 (5), 1243–1248.

Liewelyn, R. V. & Williams, J. R. (1996). Nonparametric analysis of technical, pure technical, and scale efficiencies for food crop production in East java, Indonesia. *Agricultural Economics*, 15, 113—126.

Meeusen, W. & van den Broeck, J. (1977). Efficiency estimation from Cobb-Douglas production function with composed errors.

International Economic Review, 8, 435—444.

Muir, J. F. (2003). The future for fisheries: livelihoods, social development and environment, Fisheries Sector Review and Future Development Study. Commissioned with the association of the World Bank, DANIDA, USAID, FAO, DFID with the cooperation of the Bangladesh Ministry of Fisheries and Livestock and the Department of Fisheries, Dhaka. , p 81.

New, M. B. (1995). Status of freshwater prawn farming: a review. *Aquaculture Research*, 26, 1–44.

Rahman, S., Barmon, B. K. & Ahmed, N. (2011). Diversification economics and efficiencies in a 'blue-green revolution' combination: a case study of prawn-carp-rice farming in the '*gher*' system in Bangladesh. *Aquaculture International*, 19 (4), 665–682.

Raizada, S., Chadha, N. K., Javed, H., Ali, M., Singh, I. J., Kumar, S. & Kumar, A. (2005). Monoculture of giant freshwaterprawn, *Macrobrachium rosenbergii* in inland saline ecosystem. *Journal of Aquaculture in the Tropics*, 20, 45–56.

Ridmontri, C. (2002). Agifish doubles freshwater shrimp fry production. *Asian Aquaculture Magazine*, March/April, 18–20.

Shang, Y. C., Leung, P. S. & Ling, B.-H. (1998). Comparative economics of shrimp farming in Asia. *Aquaculture*, 164, 183—200.

Sharif, N. R. & Dar, A. (1996). An empirical study of the patterns and sources of technical inefficiency in traditional and HYV rice cultivation in Bangladesh. *Journal of Development Studies*, 32, 612—629.

Sharma, K. R. & Leung, P. S. (2000). Technical efficiency of carp pond culture in south Asia: an application of a stochastic meta-production frontier model. *Aquaculture Economics and management*, 4, 169—189.

Stevenson, R. E. (1980). Likelihood functions for generalized stochastic frontier estimation. *Journal of Econometrics*, 13, 57–66.

Thi, D. D., Tihomir, A. & Michael, H. (2007). Technical efficiency of prawn farms in the Mekong Delta, Vietnam. No. 10351, 2007 Conference (51st), February 13–16, 2007, Queenstown, New Zealand, Australian Agricultural and Resource Economics Society.

Uaiene, R. W. & Arndt, C. (2009). Farm household technical efficiency in Mozambique. Contributed paper prepared for presentation at the international Association of Agricultural Economists Conference. Beijing. China 16–22 August.

USAID (2006). A pro-poor analysis of the shrimp sector in Bangladesh. The United States Agency for International Development (USAID), Development and Training Services, Arlington, VA, p. 93.

Vicki, S. S. (2007). *Social, economic and production characteristics of freshwater prawn Macrobrachium rosenbergii culture in Thailand*. Master's thesis, School of Natural Resources and Environment, University of Michigan, USA.

Villano, A. R. (2005). Technical efficiency of rainfed rice farms in the Philippines: a stochastic frontier production function approach. Working Paper, School of Economics, University of New England, Armidale, NSW, 2351.

Wadud, A. (1999). *Farm efficiency in Bangladesh*. Ph.D. thesis, Department of Agricultural Economics and Food Marketing, University of Newcastle upon Tyne, UK.

Wadud, M. A. & White, B. (2000). Farm household efficiency in Bangladesh: a comparison of stochastic frontier and DEA methods. *Applied Economics*, 32, 1665—1673.

Wang, J., Cramer, G. L. & Wailes, E. J. (1996). Production efficiency of Chinese agriculture: evidence from rural household survey data. *Agricultural Economics*, 15, 17—28.

Weimin, M. & Xianping, G. (2002). Freshwater prawn culture in China: an overview. *Aquaculture Asia*, 7, 7–12.

2

Changes in biochemical characteristics and Na and K content of caper (*Capparis spinosa* L.) seedlings under water and salt stress

Hossein Sadeghi *, Laleh Rostami

Department of Natural Resources and Environmental Engineering, College of Agriculture, Shiraz University, 71441-65186, Shiraz, Iran

Abstract

In order to investigate the effect of water and salt stress on caper (*Capparis spinosa* L.) seedlings, a randomized complete block design with five replications was carried out in 2013 at Shiraz University, Iran. Water stress had three levels: 100 % (control), 75 %, and 50 % field capacity (FC), and five levels of salinity were applied: 0 (control), 4, 8, 12, and 18 dS m^{-1}. The results indicated that salinity had a significantly negative effect on chlorophyll content of caper seedlings, while drought increased this content. The carotenoid content in caper seedlings under water and salinity stress was significantly increased. Proline and total protein content increased also under both salinity and water stress. Antioxidant enzyme activity; superoxide dismutase (SOD), catalase (CAT), peroxidase (POD) and ascorbate peroxidase (APX) also increased in response of salinity and drought. Salinity stress significantly increased the content of Na$^+$ in cells but decreased K$^+$ content. It seems that caper seedlings could tolerate a salinity level up to 4–8 dS m^{-1} as well as water stress of 75 % FC, no significant differences were observed between these two salinity levels, the water stress level and the control. The interaction effect of water stress and salinity had a significant effect on biochemical characteristics of caper. The highest content of carotenoid, proline and total protein content were obtained in 50 % FC and 18 dS m^{-1}. The results of biochemical characteristics and leaf content of K$^+$ and Na$^+$ suggest that caper plant is a very tolerant species to salinity and drought stress which make it a suitable crop for most arid and semi-arid regions of Iran.

Keywords: caper, catalase, peroxidase, salt stress, superoxide dismutase

1 Introduction

Caper (*Capparis spinosa* L.), is a multi-purpose shrub native to the Mediterranean regions and (semi-)arid tropics (Legua *et al.*, 2013). Due to the recent severe droughts in Iran and most arid and semi-arid regions in the world, farmers have attempted to cultivate drought and saline resistant plants (such as caper) instead of plants with high water requirements (Sadeghi & Rostami, 2016). Cultivation of an alternative crop (such as caper) can increase income of poor and marginal land holding farmers in arid regions and can prevent them from rural to urban migration. Commercial cultivation of caper in Iran is still in its infancy with the possibility of future expansion owing to its economic importance which can contribute to the livelihoods of many small farmers due to its low cultivation requirements and its tolerance to adverse environmental conditions.

Caper has a deep root system, is resistant to drought conditions, and can tolerate temperatures exceeding 40 °C (Suleiman *et al.*, 2012). Because of its vegetative canopy, caper gives an excellent soil cover, thus preserving soil water (Saifi *et al.*, 2011; Rostami *et al.*, 2016). *C. spinosa* grows wild in different parts of Iran, especially in dry and arid regions, and has a variety of economic, ecological, and medicinal uses in Iran. This plant is further considered to be excellent for wind screens and sandy soil stabilisation, and its introduction

* Corresponding author

Email: sadeghih@shirazu.ac.ir;

in arid and semi-arid environments could help to prevent the disruption of the equilibrium of those fragile eco-systems (Sozzi, 2001). Recently, the cultivation of this plant has been initiated to reduce the negative effects of dust phenomenon in south and southwest of Iran.

Salinity and drought are the most common abiotic stresses that induce a significant reduction in photosynthesis, which depend on photosynthesizing tissue and photosynthetic pigments (Saed-Moocheshi et al., 2014a). During stress, active solute accumulation such as soluble carbohydrates, proteins, and free amino acids is claimed to be an effective stress-tolerance mechanism. Certain plant species adapt to high salt concentrations in soil by lowering their tissue osmotic potential and by accumulating these osmotic solutes (Saed-Moocheshi et al., 2014b). Salinity affects plant growth in two ways, by increasing osmotic pressure of the soil solution and/or by the specific effect of the salt ions, mainly Na^+ and Cl^-. The increased osmotic pressure of the soil solution resulting from increased salt content impairs the ability of plants to absorb water by lowering leaf water potential. Under osmotic stress plants need to maintain water potential below that of the soil and maintain turgor and water uptake for growth. This requires an increase in the osmoregulants, either by accumulation of inorganic solutes (e.g., K^+, Na^+ and Cl^-) or by synthesis of organic solutes (e.g., proline and glycine betaine).

The present study was performed to determine the changes in biochemical characteristics and leaf content of K^+ and Na^+ under drought and salinity stress in caper plant in order to evaluate the tolerance and adaptability of this plant under water and salt stress conditions.

2   Materials and methods

2.1   Experimental procedure

Seeds of the caper plant were gathered from Farashband belonging to Fars province of Iran., separated, washed with deionized water and sterilised with 70 % ethanol for five minutes. The seeds were placed in Petri dishes containing filter paper with 5 mL polyethylene glycol (PEG) 6000 for dormancy breaking and kept in a germinator at 4 °C for a period of four weeks. After germination, the seeds were transported to 5 L pots filled with soil. Ten germinated seeds were sown in each pot.

Treatments were arranged in a randomized complete block design with two factors, water and salinity stress, and five replications. The first factor water stress had three levels of 100 % (control), 75 %, and 50 % FC. For the determination of field capacity, pots with dry soil

were weighed, soaked, and after total drain of the water, weighed again. Maximum water holding capacity (approximately 20 %) was determined by the difference between dry and soaked soil weights. The determination of water refill for all field capacities was calculated in relation to this difference. Drought treatment levels were applied based on the weighting method by daily weighting of pots (Sadeghi & Rostami, 2016). The second factor salinity stress had five levels 0 (control), 4, 8, 12, and 18 dS m^{-1}. For salinity treatments, sodium chloride and calcium chloride with the same ration were applied. The plants were grown at day/night temperatures of $28/22 \pm 2$ °C. Directly after the sowing of germinated seeds in the pots, drought and salinity treatments started. After an experimental period of forty days, the leaves of all plants were separated from the plant, frozen in liquid nitrogen and transported to the laboratory for measurements.

Total chlorophyll, chlorophyll a, chlorophyll b, and carotenoid contents were determined for the samples according to the Arnon (1949) method. Subsequently, the content of pigments was determined based on the following standard formulas (Lichtenthaler & Buschmann, 2001):

Total chlorophyll (mg/mL) = $20.2(A_{645}) + 8.02(A_{663})$

Chlorophyll a (mg/mL) = $12.7(A663) - 2.69(A_{645})$

Chlorophyll b (mg/mL) = $22.9(A645) - 4.68(A_{663})$

Carotenoid (mg/mL)   = $(1000A_{470} - 3.27[\text{Chl } a] - 104[\text{Chl } b])/227$

where A is the recorded number in the spectrophotometer and Chl a and Chl b denote chlorophyll a and chlorophyll b content, respectively.

Free proline was extracted from fresh leaves according to the method described by Bates et al. (1973).

Frozen leaves were ground to fine powder with a mortar and pestle in liquid nitrogen and were extracted with ice-cold 0.1 M Tris-HCl buffer (pH 7.5) containing 5 % (w/v) sucrose and 0.1 % 2-mercaptoethanol (3 : 1 buffer volume / fresh weight).The homogenate was centrifuged at 12 000 × g for 20 minutes at 4 °C and the supernatant was used to measure protein content and enzyme activity.

The protein content was estimated according to the method of Bradford (1976), using bovine serum albumin (BSA) as a standard and observance of 595 nm.

Superoxide dismutase (SOD) inhibits the photochemical reduction of nitro-blue-tetrazolium (NBT) (Beauchamp & Fridovich, 1973), and this ability was

Table 1: *Analysis of variance (ANOVA) for measured traits.*

Source	Degree of freedom	Mean squares								
		Proline	Protein	Sodium content	Potassium content	Carotenoid	Ascorbate peroxidase	Superoxide dismutase	Catalase	peroxidase
Drought	2	409.1**	203.15**	12.08**	204.01**	0.17**	0.47**	0.6**	0.18**	2.51**
Salinity	4	50.26**	99.15**	1.06**	101.15**	0.11**	0.18**	0.44**	0.16**	0.19**
Drought * Salinity	8	11.84**	4.1**	0.05**	7.02**	0.01^{ns}	0.02^{ns}	0.03^{ns}	0.05^{ns}	0.03^{ns}
Error	60	2.2	3.25	0.10	11.02	0.03	0.002	0.02	0.03	0.09
Coefficient of variation		8.14	10.12	14.23	13.06	15.19	11.24	12.61	16.06	14.06

[1] **, *, and ^{ns} are representation of significant in 1 % level, significant in 5 % level, and not significant, respectively.

used to determine its activity. For SOD assay, the reaction mixture contained 50 mM K-phosphate buffer (pH 7.8), 13 mM methionine, 75 μM NBT, 0.1 μM EDTA, 4 μM riboflavin, and extracted enzyme. The reaction started by adding riboflavin, after which the tubes were placed under two 15 W fluorescent lamps for 15 minutes. A complete reaction mixture-lacking enzyme, which gave the maximal colour, was considered as control. A non-irradiated complete reaction mixture was used as a blank. One unit of SOD activity was defined as the amount of enzyme required to cause 50 % inhibition of the reduction of NBT as monitored at 560 nm (Giannopolitis & Ries, 1977).

Peroxidase (POD) activity was assayed (Polle *et al.*, 1994) at 436 nm by its ability to convert guaiacol to tetraguaiacol ($\varepsilon = 26.6$ mM cm^{-1}). The reaction mixture contained 100 mM K-phosphate buffer (pH 7.0), 20.1 mM guaiacol, 10 mM H$_2$O$_2$, and enzyme extract. The increase in absorbance was recorded by adding H$_2$O$_2$ at 436 nm for 5 minutes. The activity of catalase (CAT) was determined by monitoring the disappearance of H$_2$O$_2$ at 240 nm ($\varepsilon = 40$ mM cm^{-1}). The reaction mixture contained 50 mM K-phosphate buffer (pH 7.0), 33 mM H$_2$O$_2$, and enzyme extract.

For measuring ascorbate peroxidase (APX) activity, 50 mm sodium phosphate buffer (pH = 6), 0.1 mM EDTA, 0.1 mM H$_2$O$_2$, and 0.5 mM ascorbate were mixed and 0.2 mL enzyme extract was added. After that, the absorption of the light was measured at 290 nm wave length and the enzyme activity was estimated (Nakano & Asada, 1981).

For the determination of leaf sodium (Na) and potassium (K) contents the samples were dry ashed at 550 °C, then 2 mol HCl solution was used for extraction (Chapman & Pratt, 1961). Subsequently, Na and K content were determined by atomic absorption by spectrophotometer (Varian model Spectera AA 220, Australia).

Univariate normality test was carried out on residuals of the ANOVA model for all measured traits for test-ing hypothesis related to normal distribution of the data using SAS 9.3 software. The main effects of factors and their interactions were tested using analysis of variance (ANOVA) by GLM procedure of SAS. Least significant difference (LSD) was used for mean comparison of main treatment factors and their interactions at the significant level of 5 %.

3 Results

The growth parameters for this study are presented by Sadeghi & Rostami (2016). The results of analysis of variance (Table 1) showed significant effect of salinity and drought stress on proline, protein, sodium and potassium content, as well as on carotenoid and antioxidant enzyme activity. The interaction effect of drought by salinity was only significant for proline, protein, sodium and potassium content. The results showed that the content of chlorophyll *a*, chlorophyll *b*, and total chlorophyll decreased with increase in salinity (Table 2). The decrease rate of chlorophyll *a* was higher than chlorophyll *b*. Chlorophyll *a* content was not affected by 4 dS m^{-1}, and 8 dS m^{-1} treatments, but it was significantly reduced at 12 dS m^{-1} and 18 dS m^{-1}. Salinity level of 18 dS m^{-1} was the only level which showed significant difference in chlorophyll *b* from other salinity levels. Total chlorophyll content showed similar results to chlorophyll *a*. Carotenoid content increased with an increase in salinity level. The highest content of carotenoid was observed at 18 dS m^{-1} salinity level, while the lowest content was obtained in the control. In contrast to salinity, the content of chlorophyll *a* and total chlorophyll increased with increase in severity of water stress (Table 2). However, chlorophyll *b* was not affected by different water stress levels. Carotenoid content increased with drought. The highest content of carotenoid was obtained in 50 % FC and the lowest content in 100 % FC irrigation.

Table 2: *Pigment content of caper seedlings under different irrigation and salinity levels at 40 days after seeding.*

	Chlorophyll a (mg/mL)	Chlorophyll b (mg/mL)	Total chlorophyll (mg/mL)	Carotenoid (mg/mL)
FC (%)				
100	11.20^b	3.10^a	14.30^b	1.87^c
75	12.50^a	2.90^a	15.40^a	2.45^b
50	12.70^a	3.10^a	15.80^a	2.90^a
Salinity levels ($dS\ m^{-1}$)				
0	12.34^a	3.40^a	15.74^a	1.56^d
4	12.00^{ab}	3.30^a	15.30^{ab}	1.90^c
8	12.40^a	3.20^a	15.60^a	2.11^c
12	11.70^b	3.00^{ab}	14.70^b	2.45^b
18	11.01^c	2.80^b	13.81^c	2.99^a

Means with the same letters in each column are not significantly different (least significant difference at 5 % level of probability). FC: field capacity

The contents of free proline and total protein increased under salinity and water stress treatments (Figs. 1 and 2). Highest contents of proline and total protein were observed under the highest level of salinity ($18\,dS\,m^{-1}$). The 12 and $18\,dS\,m^{-1}$ salinity levels showed a significant difference compared to the other levels (Fig. 1). Similar to salinity stress, water stress also caused an increase in proline and protein content (Fig. 2). Highest amounts of free proline and total protein were measured for 50 % FC, while the lowest amount was recorded for 100 % FC.

3.1 Antioxidant enzymes activity

Superoxide dismutase (SOD), catalase (CAT), peroxidase (POD), and ascorbate peroxidase (APX) activities increased with increase in salinity (Fig. 3). The rate of increase in response to salinity was highest for SOD, while lowest for APX. Control, 4, 8, and $12\,dS\,m^{-1}$ salinity levels showed no significant difference for APX, but $18\,dS\,m^{-1}$ treatment had the highest APX activity. Also, no significant differences in POD and SOD were observed between control and $4\,dS\,m^{-1}$ treatment and in CAT among the control, 4, and $8\,dS\,m^{-1}$ treatments. These results indicate the salinity threshold and its tolerance to salinity levels of about $4\text{-}8\,dS\,m^{-1}$ because these levels showed no significant difference for antioxidant enzyme activity in this experiment. Meanwhile, drought increases the activity of enzymatic antioxidant in caper plant as well (Fig. 4). Highest activity of enzymatic antioxidant was obtained in 50 % FC and the lowest in the 100 % FC.

The analysis of Na^+ and K^+ showed that salinity increased the sodium ion content of caper plant leaves, while it decreased its potassium content (Fig. 5). Lowest content of Na^+ was obtained in control, which showed no significant difference with 4 and $8\,dS\,m^{-1}$ salinity levels. Highest content of Na^+ was observed in $18\,dS\,m^{-1}$ salinity level. Conversely, lowest content of K^+ was observed in $12\,dS\,m^{-1}$ salinity level, while no significant differences with 4 and $8\,dS\,m^{-1}$ salinity levels occurred. The ratio of Na^+ / K^+ showed no significant difference between control, 4, and $8\,dS\,m^{-1}$ salinity levels, while this ratio was highest at $18\,dS\,m^{-1}$ level and significant different with the other levels. FC of 100 % showed the lowest content of Na^+ and also Na^+ / K^+ ratio, while it showed the highest content of K^+. On the other side, 50 % FC had the highest Na^+ and Na^+ / K^+, but the lowest K^+ in caper plant leaves (Fig. 5).

4 Discussion

The results of this study showed that the content of chlorophyll a, chlorophyll b, and total chlorophyll decreased under salinity stress, while, carotenoid content increased. Water stress increased the chlorophyll contents measured. Azooz et al., (2011) have argued that the reduction in photosynthetic pigment contents under salinity stress is related to pigment destruction and the instability of pigment complex. This occurrence is probably related to the interference of salt ions with the chlorophyll structural component, and protein synthesis, rather than the interruption of chlorophyll (Jaleel et al., 2008). Moreover, drought increased chlorophyll content of caper plant, which may be due to an increase in the concentration of chlorophyll. Similar results were obtained in sour orange (García-Sánchez et al., 2002), olive (Mousavi et al., 2008), and maize (Saed-Moocheshi et al., 2014a) under salinity stress.

The contents of free proline and total protein were increased under both salinity and drought stress conditions. Similar results were also reported for sugar beet and wild beet (Bor et al., 2003), rice (Türkan & Demiral, 2009), maize (Saed-Moocheshi et al., 2014c), and broad bean (Azooz et al., 2011). Ascorbate peroxidase is the key enzyme for scavenging hydrogen peroxide in chloroplast and cytosol of plant cells (Amako et al., 1994). Numbers of different reports have shown an enhanced expression of APX in plants in response to different abiotic stress such as drought and salinity (Saed-Moocheshi et al., 2014c). Over expression of APX in tobacco chloroplasts enhanced plant tolerance to salt and water deficit (Hebelstrup & Møller, 2015). As we have

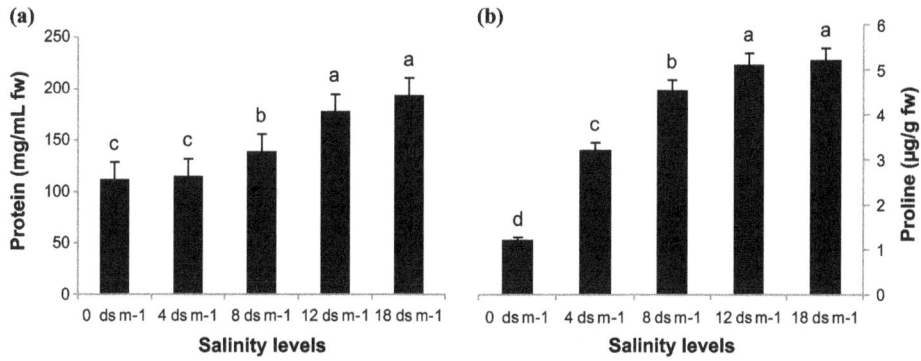

Fig. 1: *Proline and total protein content under different salinity levels. Means with the same letter(s) are not significantly different (5 % level).*

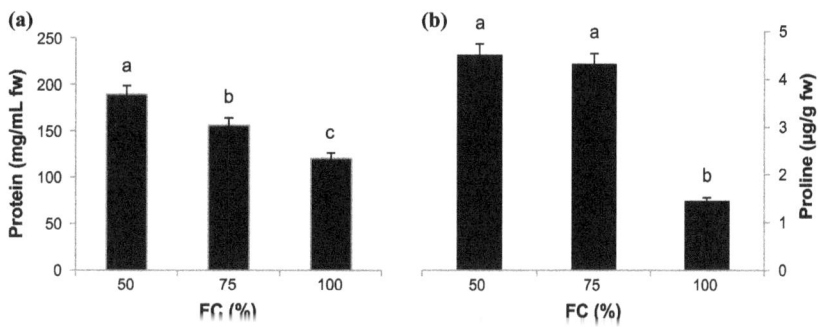

Fig. 2: *Proline and total protein content under different FC (%). Means with the same letter are not significantly different (5 % level).*

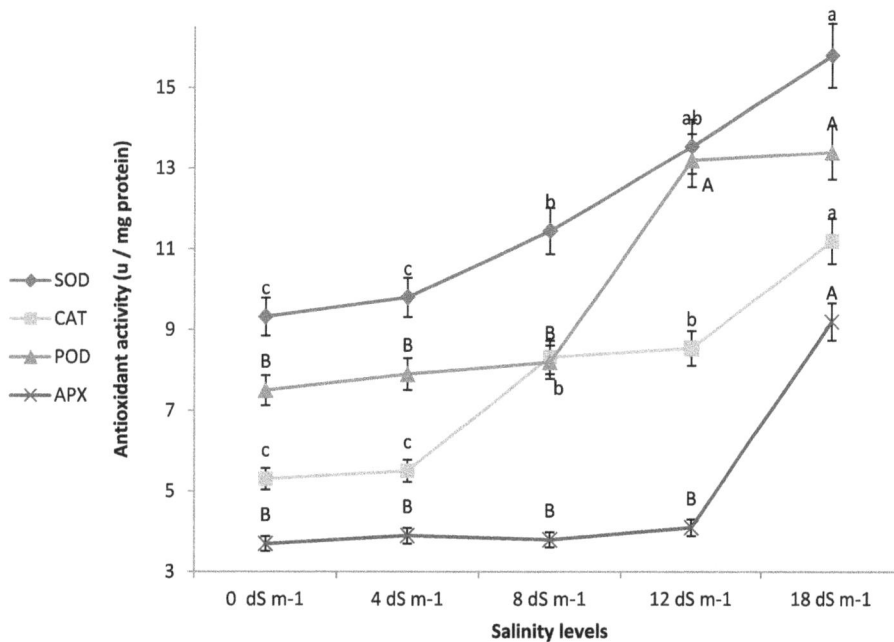

Fig. 3: *Antioxidant enzymes' activity under different salinity levels. Means with the same letter(s) are not significantly different (5 % level). SOD: Superoxide dismutase, CAT: catalase, POD: peroxidase , and APX: ascorbate peroxidase.*

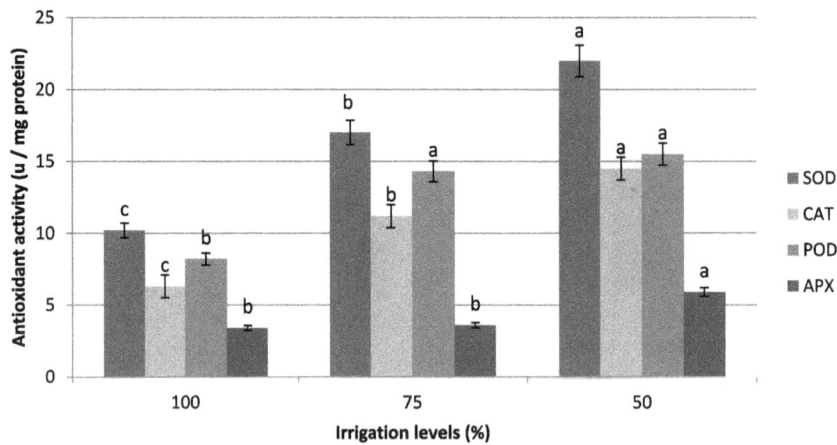

Fig. 4: *Antioxidant enzymes activity of caper plant under different field capacity (FC %). Means with the same letter are not significantly different (5 % level). SOD: Superoxide dismutase, CAT: catalase, POD: peroxidase , and APX: ascorbate peroxidase.*

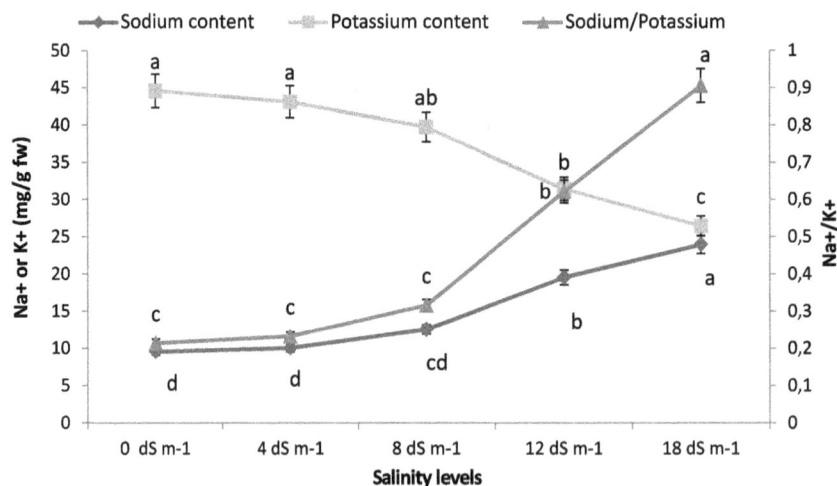

Fig. 5: *Sodium and potassium content in additon to their ration (Na+ / K+) under different salinity levels. Means with the same letter(s) are not significantly different (5 % level).*

mentioned above, abiotic and environmental stress (such as salinity and drought) increase the amount of react-ive oxygen species (ROS) which can damage other vital molecules and metabolites such as DNA, pigments, proteins, lipids (Hebelstrup & Møller, 2015). This mechanism (Antioxidant enzymes activity) can reduce ROS and in this way protect cells from further damage. The plant growth parameters showed a significant negative effect of salinity and drought stress on plant height, leaf num-ber, leaf length, root length, shoot fresh weight, shoot dry weight, root fresh weight and root dry weight (Sade-ghi & Rostami, 2016). The interaction effect of drought by salinity was only significant for plant height and shoot dry weight. The highest values for plant height, leaf number, leaf length and root length were observed in 100 % field capacity. Under both stress factors while increasing the severity of stress, the values of the traits decreased.

In the present study, the content of Na^+ increased un-der salinity in capper plant leaves, while K^+ content de-creased. Salinity, which is usually caused by the effect of NaCl in the soil, causes an imbalance in the ionic equilibrium in the soil solution and thereby decreases the absorption of the mineral elements and it also de-creases the content of K^+ in plants (Aroca *et al.*, 2007). Higher Na^+ in soil causes disequilibrium in nutrient ions in plants. The capacity of plants to absorb less Na^+ and more K^+ and thus maintaining a high cytosolic K^+ / Na^+ ratio is likely to be one of the key determinants of plant salt tolerance. In this study, caper plant showed no sig-nificant increase in the amount of Na^+ in response to salinity levels of 4 and $8\,dS\,m^{-1}$ and it could therefore be considered as a plant tolerant to salinity.

5 Conclusions

Cultivation of crops such as caper which is resilient to dry and saline conditions can minimise the production and income loss in drought periods, increase the production in areas with uncertain rainfall and on land with saline soils. The results indicated that antioxidant enzyme activity increased in response to salinity and drought in caper plant. Salinity had a negative effect on chlorophyll content of caper, while drought caused an increase in the content of chlorophyll pigments. Total protein content of caper plant under both salinity and water stress was increased.

References

Amako, K., Chen, G.-X. & Asada, K. (1994). Separate assays specific for ascorbate peroxidase and guaiacol peroxidase and for the chloroplastic and cytosolic isozymes of ascorbate peroxidase in plants. *Plant and Cell Physiology*, 35, 497–504.

Arnon, D. (1949). Copper enzyme in isolated chloroplast and chlorophyll expressed in terms of mg per gram. *Plant Physiology*, 24, 1–15.

Aroca, R., Porcel, R. & Ruiz-Lozano, J. M. (2007). How does arbuscular mycorrhizal symbiosis regulate root hydraulic properties and plasma membrane aquaporins in *Phaseolus vulgaris* under drought, cold or salinity stresses? *New Phytologist*, 173, 808–816.

Azooz, M. M., Youssef, A. M. & Ahmad, P. (2011). Evaluation of salicylic acid (SA) application on growth, osmotic solutes and antioxidant enzyme activities on broad bean seedlings grown under diluted seawater. *International Journal of Plant Physiology and Biochemistry*, 3, 253–264.

Bates, L., Waldren, R. & Teare, I. (1973). Rapid determination of free proline for water-stress studies. *Plant and Soil*, 39, 205–207.

Beauchamp, C. & Fridovich, I. (1973). Isozymes of superoxide dismutase from wheat germ. *Biochimica et Biophysica Acta (BBA) – Protein Structure*, 317, 50–64.

Bor, M., Özdemir, F. & Türkan, I. (2003). The effect of salt stress on lipid peroxidation and antioxidants in leaves of sugar beet *Beta vulgaris* L. and wild beet *Beta maritima* L. *Plant Science*, 164, 77–84.

Bradford, M. M. (1976). A rapid and sensitive method for the quantitation of microgram quantities of protein utilizing the principle of protein-dye binding. *Analytical biochemistry*, 72, 248–254.

Chapman, H. D. & Pratt, P. F. (1961). Ammonium vandate-molybdate method for determination of phosphorus. *In:* Chapman, H. D. & Pratt, P. F. (eds.), *Methods of Analysis for Soils, Plants and Water*. Vol. 1, pp. 184–203, University of California, Berkeley, USA.

García-Sánchez, F., Martínez, V., Jifon, J., Syvertsen, J. & Grosser, J. (2002). Salinity reduces growth, gas exchange, chlorophyll and nutrient concentrations in diploid sour orange and related allotetraploid somatic hybrids. *Journal of Horticultural Science & Biotechnology*, 77, 379–386.

Giannopolitis, C. N. & Ries, S. K. (1977). Superoxide dismutases I. Occurrence in higher plants. *Plant Physiology*, 59, 309–314.

Hebelstrup, K. H. & Møller, I. M. (2015). Mitochondrial Signaling in Plants Under Hypoxia: Use of Reactive Oxygen Species (ROS) and Reactive Nitrogen Species (RNS). *In:* Gupta, K. J. & Igamberdiev, A. U. (eds.), *Reactive Oxygen and Nitrogen Species Signaling and Communication in Plants*. pp. 63–77, Springer International Publishing.

Jaleel, C. A., Kishorekumar, A., Manivannan, P., Saankar, B., Gomathinayagam, M. & Panneerselvam, R. (2008). Salt stress mitigation by calcium chloride in *Phyllanthus amarus*. *Acta Botanica Croatica*, 67, 53–62.

Legua, P., Martínez, J., Melgarejo, P., Martínez, R. & Hernández, F. (2013). Phenological growth stages of caper plant (*Capparis spinosa* L.) according to the Biologische Bundesanstalt, Bundessortenamt and CHemical scale. *Annals of Applied Biology*, 163, 135–141.

Lichtenthaler, H. K. & Buschmann, C. (2001). Chlorophylls and Carotenoids: Measurement and Characterization by UV-VIS Spectroscopy. *In:* Current Protocols in Food Analytical Chemistry. F:F4:F4.3. John Wiley & Sons, Inc.

Mousavi, A., Lessani, H., Babalar, M., Talaei, A. & Fallahi, E. (2008). Influence of salinity on chlorophyll, leaf water potential, total soluble sugars, and mineral nutrients in two young olive cultivars. *Journal of Plant Nutrition*, 31, 1906–1916.

Nakano, Y. & Asada, K. (1981). Hydrogen peroxide is scavenged by ascorbate-specific peroxidase in spinach chloroplasts. *Plant and cell physiology*, 22, 867–880.

Polle, A., Otter, T. & Seifert, F. (1994). Apoplastic peroxidases and lignification in needles of Norway spruce (*Picea abies* L.). *Plant Physiology*, 106, 53–60.

Rostami, L., Sadeghi, H. & Hosseini, S. (2016). Response of caper plant to drought and different ratios of calcium and sodium chloride. *Planta Daninha*, 34, 259–266.

Sadeghi, H. & Rostami, L. (2016). Evaluating the physiological and hormonal responses of caper plant (*Capparis spinosa*) subjected to drought and salinity. *Desert*, 21, 49–55.

Saed-Moocheshi, A., Pakniyat, H., Pirasteh-Anosheh, H. & Azooz, M. (2014a). Role of ROS as signaling molecules in plants. *In:* Ahmad, P. (ed.), *Oxidative Damage to Plants – Antioxidant Networks and Signaling*. Ch. 20, pp. 585–620, Academic Press, Elsevier Inc., USA.

Saed-Moocheshi, A., Shekoofa, A. & Pessarakli, M. (2014b). Reactive oxygen species (ROS) generation and detoxifying in plants. *Journal of Plant Nutrition*, 37, 1573–1585.

Saed-Moocheshi, A., Shekoofa, A., Sadeghi, H. & Pessarakli, M. (2014c). Drought and Salt Stress Mitigation by Seed Priming with KNO_3 and Urea in Various Maize Hybrids: An Experimental Approach Based on Enhancing Antioxidant Responses. *Journal of Plant Nutrition*, 37, 674–689.

Saifi, N., Ibijbijen, J. & Echchgadda, D. (2011). Genetic diversity of caper plant (*Capparis* ssp.) from North Morocco. *Journal of Food, Agriculture & Environment*, 9, 299–304.

Sozzi, G. O. (2001). Caper bush: Botany and Horticulture. *Horticultural Reviews*, 27, 125–128.

Suleiman, M. K., Bhat, N. R., Jacob, S. & Thomas, R. R. (2012). Effect of Rooting Hormones (IBA and NAA) on Rooting of Semi Hardwood Cuttings of *Capparis spinosa*. *Journal of Agriculture and Biodiversity Research*, 1, 135–139.

Türkan, I. & Demiral, T. (2009). Recent developments in understanding salinity tolerance. *Environmental and Experimental Botany*, 67, 2–9.

3

Climate change adaptation strategies of maize producers of the Central Rift Valley of Ethiopia

Musa Hasen Ahmed [*]

School of Agricultural Economics and Agribusiness, Haramaya University, Dire Dawa, Ethiopia

Abstract

The impacts of climate change are considered to be strong in countries located in tropical Africa that depend on agriculture for their food, income and livelihood. Therefore, a better understanding of the local dimensions of adaptation strategies is essential to develop appropriate measures that will mitigate adverse consequences. Hence, this study was conducted to identify the most commonly used adaptation strategies that farm households practice among a set of options to withstand the effects of climate change and to identify factors that affect the choice of climate change adaptation strategies in the Central Rift Valley of Ethiopia. To address this objective, Multivariate Probit model was used. The results of the model indicated that the likelihood of households to adapt improved varieties of crops, adjust planting date, crop diversification and soil conservation practices were 58.73 %, 57.72 %, 35.61 % and 41.15 %, respectively. The Simulated Maximum Likelihood estimation of the Multivariate Probit model results suggested that there was positive and significant interdependence between household decisions to adapt crop diversification and using improved varieties of crops; and between adjusting planting date and using improved varieties of crops. The results also showed that there was a negative and significant relationship between household decisions to adapt crop diversification and soil conservation practices. The paper also recommended household, socioeconomic, institutional and plot characteristics that facilitate and impede the probability of choosing those adaptation strategies.

Keywords: climate change, adaptation strategies, multivariate, maize, Ethiopia

1 Introduction

The main challenge for agriculture in the twenty-first century is the need to nourish increasing numbers of people while conserving the ongoing soil degradation and water depletion in the face of limited resources and growing pressures associated with an increasing global population and changing diets (Tubiello, 2012). Climate change is already putting extra pressure on agriculture and its effects are expected to become more vital in the future (Apata *et al.*, 2009; Lobell *et al.*, 2011b; Rosenzweig *et al.*, 2014). Despite technological advancements that have already been reached, the agricultural system is still highly dependent on the climatic condition in many areas of the world (Müller *et al.*, 2011). For example, Parry (2007) has indicated that, by 2020, agricultural production would decline by 50 % in some countries with rain-fed agriculture due to climate change.

Climate change affects agriculture directly and indirectly. Directly, it affects by influencing the weather variables such as rainfall, temperature, solar radiation, wind speed and humidity (Sowunmi, 2010; Pryor *et al.*, 2014; Arimi, 2014). Indirectly, it affects through disease and pest outbreak as well as favoring the development of climate related diseases like malaria that affect the workforce (Newton *et al.*, 2011).

Studies on global climate change indicated that developing nations are expected to withstand the worst of

[*] School of Agricultural Economics and Agribusiness,
P O Box 95, Haramaya University, Dire Dawa, Ethiopia
Email: musa.hassen@haramaya.edu.et

the associated damages (Ericksen *et al.*, 2011; Skoufias *et al.*, 2011). Africa will be one of the continents that will be hard hit by the impact of climate change, though the continent represents only 3.6 percent of emissions (Parry, 2007; Alemneh, 2011). Sub-Saharan Africa (SSA) is arguably the most vulnerable region to many unpleasant effects of climate change due to a very high dependence on rain-fed agriculture (Cooper *et al.*, 2008). Thus, the impacts of climate change are likely to fall unreasonably on poorer nations and on poorer households. Ethiopia is among the most vulnerable countries in SSA due to its great reliance on climate vulnerable economy (Conway & Schipper, 2011).

Ethiopian economy is an agrarian economy as agriculture comprises about 41.3 % of GDP, generates 90 % of foreign exchange earnings, and employs more than 80 % of the population (MoFED, 2012). Currently, however, the performance of this sector is seriously eroded due to climate change induced problems. It is estimated that in Ethiopia, one drought event in 12 years lowers GDP by 7 to 10 % and increases poverty by 12 to 14 % (Makombe *et al.*, 2011). The projected reduction in the Ethiopian agricultural productivity due to climate change can reduce average income by 30 percent over the next 50 years (Gebreegziabher *et al.*, 2011). Climate change can also have a significant impact on the urban dweller in terms of higher food prices, limited job opportunities in the agro-processing industries and expensive imported food items due to foreign exchange shortages (Aragie, 2013). In addition to this, it can cause a decline in biodiversity, increases in human and livestock health problems, rural-urban migration and dependency on external supports (Daniel, 2008). Therefore, adaptation strategies that minimize the negative outcomes associated with changing climate are urgently needed. Because a society with high adaptive capacity will be less susceptible in the future than other communities to the potentially detrimental and often unpredictable effects of climate change (Petheram *et al.*, 2010). However, most of the farmers in the country have low access to education, information, technology, and basic social and support services, and, as a result, have low adaptive capacity to deal with the consequences of climate variability and change (The World Bank Group, 2010).

The Central Rift Valley of Ethiopia (CRV) where this study was conducted is evidently the hardest hit region of the country in terms of drought (Emana *et al.*, 2010). In CRV, fluctuations in precipitation and temperature rates are directly affecting the production and productivity of the agricultural systems (Deschênes & Greenstone, 2007). In the area, poverty and degradation of natural resources are absolutely intertwined. On the one hand, the harsh poverty forces people to exhaust natural resources in their fight for survival, on the other hand, degraded natural resources worsen poverty (Jansen *et al.*, 2007). In general, the land, water resources and ecosystem of the valley have been affected by rapid population growth, deforestation, overgrazing and soil erosion (Bekele & Drake, 2003; Legesse *et al.*, 2004). This diverse climate change in the study area influences the livelihood activities of the farming community. However, farmers in the study area have been responding to climate change through various adaptation strategies. Nevertheless, there was no empirical data that substantiate or support the relationship between existing adaptation strategies practiced by the farmers in the area. The purpose of this study was, therefore, to identify the nature of the relationship that exist between the adaptation strategies that have been used by maize producing farmers of the CRV. Alongside this, the paper also analyzed the factors that jointly facilitate and impede the probability of choosing a particular adaptation strategy.

2   Hypothesis of the research

H1: Climate change adaptation strategies of smallholder maize producers of the Central Rift Valley of Ethiopia are interdependent and farmers adopt them as complements, substitutes or supplements.

H2: The decision by the maize producers of the Central Rift Valley of Ethiopia to adapt a climate change adaptation strategy is influenced significantly by the plot, household, socioeconomic, institutional and environmental characteristics.

3   Maize production in Ethiopia

Maize is the most widely distributed cereal crops in the world. According to The World Bank Group (2011), in developed countries 70 % of maize is destined for feed, 3 % is consumed directly by humans and the remaining is used for bio-fuels, industrial products and seed. While in SSA outside of South Africa, 77 % of maize is used as food and only 12 % serves as a feed. Maize covers 25 million ha in SSA, largely by smallholder farmers that produced 38 million tons in 2008, primarily for food. Despite the importance of maize in SSA, yields remain low (Shiferaw *et al.*, 2011). While maize yields in the top five maize producing countries in the world (USA, China, Brazil, Mexico and Indonesia) have increased three-fold since 1961 (from 1.84 t ha^{-1} to 6.10 t ha^{-1}), maize yields in SSA have stagnated at less than 2 t ha^{-1}, (Cairns *et al.*, 2013).

In Ethiopia, maize accounts for the largest share of production by volume and is produced by more farms than any other crop. CSA (2012a) indicated that about nine million smallholders were involved in maize production in the 2011/12 production season. The same source also indicated that maize covered about 2.05 million ha of land at the national level that is equivalent to 21.43 % of the total area covered by all cereals. Out of this area, 30.64 % of the land was covered by improved seed varieties and 23.26 % and 27.7 % of the land had utilized organic and inorganic fertilizer, respectively. The total output of maize in the same year at national level was 60.69 million qt that is 32.27 % of the total cereal production in the same year.

Maize is more susceptible to climate change compared to other crops (Schlenker & Lobell, 2010). About 40 % of Africa's maize producing areas face irregular drought stress in which yield losses are 10–25 % and 25 % of the maize crop suffers recurrent drought, with losses of up to half the harvest (CIMMYT, 2013). The findings of Slingo *et al.* (2005) indicated that maize crops, tend to have the highest water requirement when the maximum leaf area index combines with the highest evaporative demand. Thus, maize crop is very susceptible to water shortfall during its critical period for two reasons: high water requirement in terms of evapotranspiration and high physiological sensitivity when determining its principal yield components such as the number of ears per plant and number of kernels per ear (Omoyo *et al.*, 2015). Specifically, in Africa, under non-drought conditions 65 % of the area that is under maize cultivation would experience yield losses from a uniform 1°C warming. Under drought conditions, this figure will increase to 100 %, with 75 % of this area suffering yield losses of at least 20 % (Lobell *et al.*, 2011a).

4 Methodology

4.1 Description of study areas

This study was undertaken in the Central Rift Valley Ethiopia (CRV), explicitly in Arsi-Negele district. Geographically, it is situated at 7° 09′–7°41′ N and 38° 25′–38° 54′ E. The study area covers three agro-ecological zones (low, mid and high land) based on temperature, rainfall, altitude and vegetation (ICRA, 2002). The temperature of the area ranges from 16°C to 25°C and annual rainfall ranges between 500–1150 mm. The topography of the area is a gentle slope or flatter. Some parts of the highlands in the study area are covered by natural forest, bush and shrub. The main crops grown in the area include wheat, maize, *teff*, barley, sorghum, onion and potato. The rainfall of the area is a bimodal, with short rain occurring from February to April and the main rain from June to October. The short rain allows farmers to grow potato early and later replace by small cereals specifically wheat. Livestock are an important component of the farming system and a source of intermediate products in the district. The area is intensively cultivated and private grazing land is unavailable. Communal pasture and straw from crops are the main source of feed for livestock production. According to CSA (2012b), the district has a total population of 303,223 of which 150,245 are male and 152,978 are females.

4.2 Data sources and collection methods

A combination of purposive and random sampling techniques was employed to draw sample respondents for this study. Firstly, Arsi Negele district was selected since it is one of the major maize producing areas in CRV. A two-stage random sampling technique was then applied to select sample households. In the first stage, three Kebeles namely, *Refu-Hargisa*, *Meko-Oda* and *Aliweyo* were selected from the district. In the second stage, 135 household heads were selected randomly using probability proportional to size. The data were collected by preparing and distributing semi-structured questionnaire. The schedule was first pre-tested; and based on the result of the pre-test some modifications were made on the questionnaire before the execution of the formal survey. Enumerators who are familiar with the study area, who can understand the local language and who have prior experience in data collection were recruited.

4.3 Methods of data analysis

Descriptive statistical tools and econometrics model were employed to analyze the collected data. Descriptive statistics such as mean, percentage and frequency were used to explain the different socioeconomic characteristics of the sample respondent households. In the econometrics part, Multivariate Probit model (MVP) was used to identify factors that are determining the choice of adaption strategies of farmers. This model was also used to examine the tradeoffs and complementarities that exist between the strategies that have been adopted by farmers. This technique simultaneously models the influence of the set of explanatory variables on

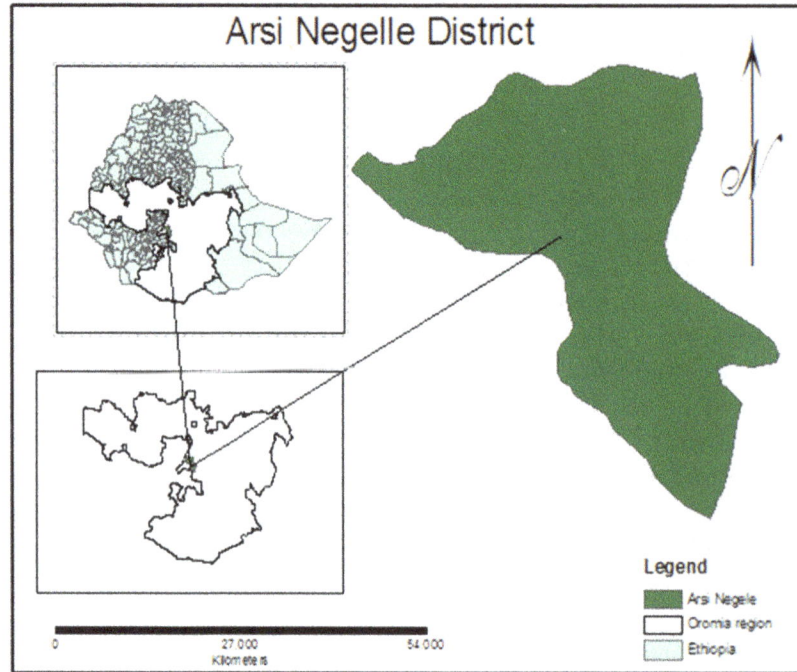

Fig. 1: *Location of the study area in Oromia Regional State of Ethiopia*

each of the different strategies while allowing for the potential correlation between unobserved disturbances, as well as the relationship between the strategies of different practices (Kassie *et al.*, 2009). The results on correlation coefficients of the error terms indicate that there is complementarity (positive correlation) and substitutability (negative correlation) between different adaptation options being used by farmers. Failure to capture unobserved factors and interrelationships among adaptation strategies will lead to bias and inefficient estimates (Greene, 2008).

Following Lin *et al.* (2005), the MVP model for this study is characterized by a set of m binary dependent variables Y_{hj} such that:

$$Y_{hj}^* = X_{hj}' \beta_j + u_{hj} \quad \text{and} \tag{1}$$

$$Y_{hj} = \begin{cases} 1 & if \ Y_{hj}^* > 0 \\ 0 & otherwise \end{cases} \tag{2}$$

Where $j = 1, 2 \ldots m$ denotes the type of adaptation strategy available; X_{hj}' is a vector of explanatory variables, β_j denotes the vector of parameter to be estimated, and u_{hj} are random error terms distributed as multivariate normal distribution with zero mean and unitary variance. It is assumed that a rational h^{th} farmer has a latent variable, Y_{hj}^* which captures the unobserved preferences or demand associated with the jth choice of adaptation strategy.

5 Results and discussion

5.1 Characteristics of the sample respondents

The mean age of the sample farmers was about 42 years with a range of 22 to 70 years (Table 1). On average, the sample respondents have cultivated maize for about 20 years. The family size of the sample respondents ranged from one to 13 with a mean of 5.73 persons per household. Concerning their literacy level, the mean educational level of sample respondents was 4.56. The minimum size of land cultivated by the respondents was 0.50 ha while the mean was 1.85 ha. On average, respondent farmers owned livestock of 8.46 TLU ranging from zero to 81.11 TLU. The survey result showed that 44 % of the sample farmers accessed credit from different sources.

All of the sample respondents reported that they received extension services though the frequency of contact differs. About 65 % of respondents have indicated that they had extensive contact on a weekly basis. While a nearly quarter of the sample respondents had contact with extension workers twice a month. Forty percent of respondents indicated that they have social responsibilities such as religious, administrative and/or community leadership roles. The mean distance from the nearest market to the homestead was 3.80 kilometres. On average, the plots are 0.84 kilometres far from homestead.

Table 1: *Summary of the explanatory variables hypothesized to affect adaptation to climate change*

Variable	Mean	Std. Dev.	Min	Max
Age	42.46	11.21	22	70
Education	4.56	4	0	12
Family Size	5.73	2.24	1	13
Experience	20.3	10.63	2	50
Off/nonfarm income	460.12	1592.55	0	10800
Social responsibility	0.40	0.49	0	1
Size of cultivated land	1.85	1.25	0.50	7
Tropical Livestock Unit (TLU)	8.46	11.57	0	81.11
Extension contact	38.04	13.68	12	48
Credit	0.44	0.5	0	1
Distance to market	3.80	1.94	0.10	9
Plot to home distance	0.84	0.97	0.01	5

Source: Survey results

5.2 Climate change adaptation strategies

The sampled households were asked if they have used adaptation strategies to reduce the impact of climate change. Accordingly, they reported that they were using different adaptation strategies to reduce the negative impact of climate change. These include, use of improved crop varieties, soil conservation techniques, crop diversification and adjusting planting dates. These strategies, however, are mostly used in combination with one another (Table 2).

Table 2: *Summary of adaptation strategies used by farmers*

Adaptation strategies	Number of respondents (n= 135)	Percent *
Use of improved crop varieties	80	59.26
Adjusting planting date	79	58.52
Crop diversification	48	35.56
Soil conservation practices	56	41.48

Source: Survey results
* Percentage cannot be added to 100 as a farmer can have more than one adaptation strategy

One of the most commonly used adaptation options used to cope with the adverse effect of climate change in the study area is using improved crop varieties. About 60 % of farmers used improved crop varieties (such as drought resistant and short maturing varieties of crops) as adaptation strategy to reduce the adverse effect of climate change. Around 58 % of sample households used adjusting planting date (from early planting to late planting or vice versa) as adaptation strategy to reduce the adverse effect of climate change on their farm.

Crop diversification is also a common practice in the study area. From the total sampled households, 35.56 % of them used crop diversification as adaptation strategy to reduce the adverse effect of climate change on farm. Out of the total sampled households, 41.48 % of them also used soil conservation as adaptation strategy to reduce the adverse effect of climate change on farm.

5.3 Relationship between the climate change adaptation strategies

The results of the correlation coefficients of the error terms are significant for any pairs of equations indicating that they are correlated (Table 3). The correlation coefficients are statistically different from zero in 3 of the 6 cases, confirming the appropriateness of the MVP specification. The result of the model shows that the likelihood of households to adapt improved varieties of crops, adjusting planting date, use of crop diversification and soil conservation practices were 58.73 %, 57.72 %, 35.61 % and 41.15 %, respectively. The result also shows that the joint probability of using all adaptation strategies was only 7.74 % and the joint probability of failure to adopt all of the adaptation strategies was 14.56 %.

Table 3: *Correlation matrix of the adaptation strategies from the MVP model*

	Crop Diversification	Adjusting planting date	Improved crop varieties	Soil conservation
Rho2	0.22 (0.156)			
Rho3	0.32 (0.148)[**]	0.38 (0.142)[***]		
Rho4	−0.34 (0.156)[**]	0.11 (0.159)	0.18 (0.164)	
Predicted probability	0.3561	0.5772	0.5873	0.4115
Joint probability (success)	0.0774			
Joint probability (failure)	0.1456			
Log likelihood	−278.86			
Likelihood ratio test of $Rho_{ij} = 0, P > \chi^2(6)$	0.0011			

Numbers in the parentheses are standard errors
[***] and [**] significant at 1 % and 5 % probability level, respectively.

The Simulated Maximum Likelihood estimation of the MVP results suggested that there was positive and significant interdependence between household decisions to adapt crop diversification and using improved varieties of crops; and between adjusting planting date and using improved crop varieties. The Simulated Maximum Likelihood estimation results also suggested that there was negative and significant interdependence between household decisions to adapt crop diversification and soil conservation practices.

5.4 Determinants of farmers' choice of adaptation strategies

Though farmers adopt a combination of strategies to reduce the impact of climate change, there would be a number of factors that can influence their decision to choose a particular strategy. This section has identified those variables, which determine the use of various adaptation strategies using MVP (Table 4). The selection of those explanatory variables for the model was done through literature review. The results of the model shows that the adaptation decisions of households to different strategies are quite distinct and largely factors governing the adaptation decision of each of them are also different indicating the heterogeneity in the adaptation strategies.

The square of age of the household head was found to have an inverse relationship with crop diversification strategy. Thus, middle age farmers are more interested in crop diversification strategy than the very young and

older ones. Since farming as any other professions need to accumulate knowledge, skill and physical capability, it is decisive in determining the right cropping portfolio. The knowledge, the skills as well as the physical capability of farmers are likely to increase as their age increases. However, this tends to decrease after a certain age level. Moreover, older farmers may be more interested in following traditional methods that are familiar to them rather than adapting new practices (Acquah, 2011; Quayum & Ali, 2012).

The educational level of household head was found to have positive and significant relation with the use of crop diversification and adjusting time adaptation strategies. This result showed that education increases the awareness of farmers about the consequence of climate change on productivity and the benefit of crop diversification and adjusting of planting time to reduce the impact of climate change. This finding is in line with the investigation of Deressa *et al.* (2008), Uddin *et al.* (2014) and Zuluaga *et al.* (2015).

The model result showed that family size has negative and significant impact on the likelihood of using improved crop varieties and soil conservation as adaptation strategy to reduce the negative impact of climate change. The possible reason for the inverse relationship between family size and using improved crop varieties is that increase in family size would increase expenditure for home consumption and creates financial constraints for other inputs such as improved crop varieties. This finding is in line with the investigation of Tazeze *et al.* (2012) and Zuluaga *et al.* (2015).

Table 4: *Multivariate probit simulation results for households' climate change adaptation decisions*

Variables	Crop Diversification		Adjusting planting date		Improved crop varieties		Soil Conservation	
	Coef.	Std. Err.	Coef.	Std. Err.	Coef.	Std. Err.	Coef.	Std. Err.
Age^2	0.0006**	0.0003	−0.0005	0.0003	−0.0002	0.0003	−0.0003	0.0003
Education	0.1645***	0.0353	0.0628*	0.0328	0.05068	0.03246	−0.0099	0.0306
Family size	−0.0792	0.0666	−0.0962	0.0651	−0.10692*	0.06244	−0.1284**	0.0627
Experience	−0.0162	0.0272	0.0646**	0.0287	0.02203	0.02748	0.0269	0.0267
Off/nonfarm income	0.0001	0.0001	−0.0001	0.0001	0.00002	0.00009	0.0001	0.0001
Social responsibility	−0.8033***	0.2892	−0.4880*	0.2701	−0.0264	0.25903	−0.6041**	0.266
Cultivated land	−0.27	0.18	−0.0447	0.1805	−0.2185	0.17194	0.1404	0.1596
Tropical Livestock Unit (TLU)	0.025	0.0184	0.012	0.0181	0.04358*	0.02386	0.0081	0.0172
Extension contact	0.0842	0.1296	0.5037***	0.1387	0.50489***	0.12766	0.1487	0.1252
Credit	0.5112	0.2738	0.5347**	0.2646	0.0387	0.25839	0.0349	0.2554
Distance to market	−0.104	0.068	−0.1733**	0.0737	−0.0337	0.06635	0.2401***	0.0638
Plot to home distance	−0.0689	0.1334	−0.5577***	0.1862	−0.31852**	0.15236	−0.1851	0.1664
_cons	−1.1518	0.6432	−0.3176	0.6873	−0.4767	0.62627	−0.8194	0.6182

Log likelihood = −272.76792; Prob > χ^2 = 0.0000

***, **, and * significant at 1 %, 5 %, and 10 % probability level, respectively.

The negative relation between family size and soil conservation adaptation strategy may be due to the fact that households with larger family size may be forced to switch part of the labor force to off-farm or non-farm activities in an effort to earn income in order to ease the consumption pressure imposed by a large family. This finding is in line with the finding of Legesse *et al.* (2013).

An increase in the experience of a household head has positive relationship with adjusting planting dates. This is because as the farming experience of the household head increases, the farmer is expected to acquire more experience in weather forecasting. Therefore, they can easily adjust themselves to climate change stresses. This result is consistent with Hassan & Nhemachena (2008).

The results also indicated that frequency of extension visit to the households has a positive and significant impact on adjusting planting date and use of improve crop varieties. An extension service is an important source of information on climate change as well as agricultural production and management practices in the country (Nhemachena & Hassan, 2007). This implies that farmers with more access to information and technical assistance on agricultural activities have more awareness about the consequence of climate change. This result is consistent with Nhemachena & Hassan (2007), Deressa *et al.* (2009), Deressa *et al.* (2011), Di Falco *et al.* (2011), Fosu-Mensah *et al.* (2012) and Zuluaga *et al.* (2015).

Distance from the market center is associated with soil conservation and use of adjusting planting date as an adaptation strategy with opposite signs. The negative relation between market distance and adjusting planting dates as adaptation strategy is plausible because the market serves as a means of exchanging information with other farmers (Maddison, 2007). Better access to market enable farmers to obtain information on climate change and other important inputs they may need if they are to change their practices to cope with predicted changes in future climate. When farmers are far from the market, the transaction cost for acquiring input will be high, and this will in turn, reduce the relative advantage of adapting new technologies/ strategy. This finding is in line with the investigation of Tazeze *et al.* (2012).

The positive relationship between soil conservation strategy and distance to the market implies farmers who are far from the market choose soil conservation strategy and this may be due to the fact that as the farmer become far from the market, getting inputs such as improved seed from the market become costly for them. This will force them to choose labor-intensive adaptation measures such as soil and water conservation. The other reason for this relationship may be since they are far away from the market, they will have less opportunities to participate in off-farm or non-farm activities, which will in turn allow them to invest in labor-intensive adaptation measures such as soil and water conservation.

Distance between the farm where maize was cultivated and the residence of the respondents was found to have a negative relation with adjusting planting date and using improved crop varieties as adaptation strategies to mitigate the impact of climate change. This is reasonable because as the farm becomes far from the homestead it will receive less attention as the farmer requires longer time to visit the farm and manage it properly.

Livestock ownership is an important variable affecting adaption decision at the farm level. The ownership of livestock of the households has a positive and significant impact on use of improved crop varieties as adaptation strategy. The possible reason could be if the farmer possesses more number of livestock will have better capacity to purchase agricultural inputs as income obtained from livestock serves for investment on crop production. The result is in line with the finding of Deressa et al. (2011) and Okonya et al. (2013).

The results also indicated that access to credit has a positive and significant impact on the likelihood of adjusting planting date. With more financial and other resources at their disposal, farmers are able to change their management practices in response to changing climatic and other factors and are better able to make use of all the available information they might have on changing conditions both climatic and other socioeconomic factors. The finding is in line with the investigation of Nhemachena & Hassan (2007), Deressa et al. (2011), Di Falco et al. (2011), Fosu-Mensah et al. (2012), Temesgen et al. (2014) and Zuluaga et al. (2015).

Meanwhile, social responsibility was found to have an inverse relation with crop diversification, adjusting cropping date and soil conservation. This is plausible because household who spent more time on social responsibility may not carry out major farm activities on time.

6 Conclusion and policy implications

The study attempted to identify factors affecting the choice of climate change adaptation strategies used by maize producing farmers of CRV based on data collected from 135 sampled households. Adaptation strategies used by farmers in the study area include adjusting planting date, use of soil conservation techniques, use of improved crop varieties and crop diversification.

This study examined determinants of household level climate change adaptation strategies using MVP model. The model allows the simultaneous identification of the determinants of all adaptation options, thus limiting potential problems of correlation between the error terms. Correlation results between the error terms of different equations were significant indicating various adaptation strategies tend to be used by households in a complementary or substitute fashion. The results of the model showed that the likelihood of households to adapt improved varieties of crops, adjust planting date, use crop diversification and soil conservation practices were 58.73 %, 57.72 %, 35.61 % and 41.15 %, respectively. The results also showed that the joint probability of using all adaptation strategies was only 7.7 % and the joint probability of failure to use all of the adaptation strategies was 14.56 %.

The model results also confirmed that square of the age of the farmers, educational level of the household head, family size, maize production experience, size of Livestock units, frequency of extension contact, credit utilization, social responsibility, distance to the nearest market and distance between plot and home have significant impact on the choice of farmers' climate change adaptation strategies.

Thus, the results of the study provide information to policy makers and extension workers on how to improve farm level adaptation strategies and identify the determinants for adaptation strategies. These findings stress the need for appropriate policy formulation and implementation which enables farmers to reduce the impact of climate change as this is expected to have multiplier effects ranging from farm productivity growth to economic growth and poverty reduction at the macro level.

Acknowledgements

The author gratefully acknowledges Kumilachew Alamerie, Eden Andualem, anonymous referees and the editor of the journal for their insightful and valuable comments. Enumerators, sample respondents and extension staff of the ministry of agriculture are also highly appreciated for their efforts that led to the success of this study.

References

Acquah, H. D. (2011). Farmers' perception and adaptation to climate change: A willingness to pay analysis. *Journal of Sustainable Development in Africa*, 13 (5), 150–161.

Alemneh, D. (2011). Strengthening capacity for climate change adaptation in the agriculture sector in Ethiopia. Proceedings from National Workshop Held in Nazreth, Ethiopia 5–6 July 2010, Proceedings from National Workshop Held in Nazreth, Ethiopia 5–6 July 2010.

Apata, T. G., Samuel, K. D. & Adeola, A. O. (2009). Analysis of climate change perception and adaptation among arable food crop farmers in South Western Nigeria. Contributed paper prepared for presentation at the International Association of Agricultural Economists' 2009 conference, Beijing, China, August 16 (Vol. 22).

Aragie, E. A. (2013). Climate change, growth, and poverty in Ethiopia. The Robert S. Strauss Center for International Security and Law, University of Texas at Austin.

Arimi, K. (2014). Determinants of climate change adaptation strategies used by rice farmers in Southwestern, Nigeria. *Journal of Agriculture and Rural Development in the Tropics and Subtropics*, 115 (2), 91–99.

Bekele, W. & Drake, L. (2003). Soil and water conservation decision behavior of subsistence farmers in the Eastern Highlands of Ethiopia: a case study of the Hunde-Lafto area. *Ecological Economics*, 46 (3), 437–451.

Cairns, J. E., Hellin, J., Sonder, K., Araus, J. L., MacRobert, J. F., Thierfelder, C. & Prasanna, B. M. (2013). Adapting maize production to climate change in sub-Saharan Africa. *Food Security*, 5 (3), 345–360.

CIMMYT (2013). The drought tolerant maize for Africa project. DTMA Brief. URL http://dtma.cimmyt.org/index.php/about/background

Conway, D. & Schipper, E. L. F. (2011). Adaptation to climate change in Africa: Challenges and opportunities identified from Ethiopia. *Global Environmental Change*, 21 (1), 227–237.

Cooper, P. J. M., Dimes, J., Rao, K. P. C., Shapiro, B., Shiferaw, B. & Twomlow, S. (2008). Coping better with current climatic variability in the rain-fed farming systems of sub-Saharan Africa: An essential first step in adapting to future climate change? *Agriculture, Ecosystems & Environment*, 126 (1), 24–35.

CSA (2012a). Statistical report on area and crop production, Addis Ababa, Ethiopia. Central Statistical Agency (CSA).

CSA (2012b). Statistical report on population projected figures for the year 2012. Ethiopian Central Statistical Agency (CSA), Addis Ababa.

Daniel, K. (2008). Impacts of climate change on Ethiopia: a review of the literature. *In:* Green Forum (ed.), *Climate Change A Burning Issue for Ethiopia: Proceedings of the 2nd Green Forum Conference Held in Addis Ababa, 31 October–2 November 2007*. Green Forum, Addis Ababa.

Deressa, T., Hassan, R. M., Alemu, T., Yesuf, M. & Ringler, C. (2008). Analyzing the determinants of farmers' choice of adaptation methods and perceptions of climate change in the Nile Basin of Ethiopia. IFPRI Discussion Paper 00798, International Food Policy Research Institute (IFPRI), Washington D.C.

Deressa, T. T., Hassan, R. M. & Ringler, C. (2011). Perception of and adaptation to climate change by farmers in the Nile basin of Ethiopia. *The Journal of Agricultural Science*, 149 (1), 23–31.

Deressa, T. T., Hassan, R. M., Ringler, C., Alemu, T. & Yesuf, M. (2009). Determinants of farmers' choice of adaptation methods to climate change in the Nile Basin of Ethiopia. *Global Environmental Change*, 19 (2), 248–255.

Deschênes, O. & Greenstone, M. (2007). The economic impacts of climate change: evidence from agricultural output and random fluctuations in weather. *The American Economic Review*, 97 (1), 354–385.

Di Falco, S., Veronesi, M. & Yesuf, M. (2011). Does adaptation to climate change provide food security? A micro-perspective from Ethiopia. *American Journal of Agricultural Economics*, 93 (3), 829–846.

Emana, B., Gebremedhin, H. & Regassa, N. (2010). Impacts of improved seeds and agrochemicals on food security and environment in the rift valley of Ethiopia: implications for the application of an African green revolution. DCG Report No. 56, Drylands Coordination Group, Norway.

Ericksen, P. J., Thornton, P. K., Notenbaert, A., Cramer, L., Jones, P. & Herrero, M. (2011). Mapping hotspots of climate change and food insecurity in the global tropics. CCAFS Report No. 5, CGIAR Research Program on Climate Change.

Fosu-Mensah, B. Y., Vlek, P. L. & MacCarthy, D. S. (2012). Farmers' perception and adaptation to climate change: a case study of Sekyedumase district in Ghana. *Environment, Development and Sustainability*, 14 (4), 495–505.

Gebreegziabher, Z., Stage, J., Mekonnen, A. & Alemu, A. (2011). Climate change and the Ethiopian economy: A computable general equilibrium analysis. Environment for Development, Discussion Paper Series, EfD DP 11-09.

Greene, W. H. (2008). *Econometric Analysis. 7th Edn.*. Prentice Hall, New Jersey.

Hassan, R. & Nhemachena, C. (2008). Determinants of African farmers' strategies for adapting to climate change: Multinomial choice analysis. *African Journal of Agricultural and Resource Economics*, 2 (1), 83–104.

ICRA (2002). Food security among households in the different agro ecological zones in Arsi Negele Woreda, Ethiopia. International Center for Development Oriented Research in Agriculture (ICRA).

Jansen, H. C., Hengsdijk, H., Legesse, D., Ayenew, T., Hellegers, P. & Spliethoff, P. (2007). Land and water resources assessment in the Ethiopian Central Rift Valley: Project: Ecosystems for water, food and economic development in the Ethiopian Central Rift Valley. Alterra-rapport 1587, Alterra, Wageningen.

Kassie, M., Zikhali, P., Manjur, K. & Edwards, S. (2009). Adoption of sustainable agriculture practices: Evidence from a semi-arid region of Ethiopia. *Natural Resources Forum*, 33 (3), 189–198.

Legesse, B., Ayele, Y. & Bewket, W. (2013). Smallholder farmers' perceptions and adaptation to climate variability and climate change in Doba district, west Hararghe, Ethiopia. *Asian Journal of Empirical Research*, 3 (3), 251–265.

Legesse, D., Vallet-Coulomb, C. & Gasse, F. (2004). Analysis of the hydrological response of a tropical terminal lake, Lake Abiyata (Main Ethiopian Rift Valley) to changes in climate and human activities. *Hydrological Processes*, 18 (3), 487–504.

Lin, C. T. J., Jensen, K. L. & Yen, S. T. (2005). Awareness of foodborne pathogens among US consumers. *Food Quality and Preference*, 16 (5), 401–412.

Lobell, D. B., Bänziger, M., Magorokosho, C. & Vivek, B. (2011a). Nonlinear heat effects on African maize as evidenced by historical yield trials. *Nature Climate Change*, 1, 42–45.

Lobell, D. B., Schlenker, W. & Costa-Roberts, J. (2011b). Climate trends and global crop production since 1980. *Science*, 333 (6042), 616–620.

Maddison, D. J. (2007). The perception of and adaptation to climate change in Africa. World Bank Policy Research Working Paper 4308.

Makombe, G., Namara, R., Hagos, F., Awulachew, S. B., Ayana, M. & Bossio, D. (2011). A comparative analysis of the technical efficiency of rain-fed and smallholder irrigation in Ethiopia. IWMI Working Paper 143, International Water Management Institute, Colombo, Sri Lanka.

MoFED (2012). Growth and transformation Plan (2010/11–2014/15). Annual progress report for F.Y. 2011/12. Ministry of Finance and Economic Development (MoFED), Addis Ababa, Ethiopia.

Müller, C., Cramer, W., Hare, W. L. & Lotze-Campen, H. (2011). Climate change risks for African agriculture. *Proceedings of the National Academy of Sciences*, 108 (11), 4313–4315.

Newton, A. C., Johnson, S. N. & Gregory, P. J. (2011). Implications of climate change for diseases, crop yields and food security. *Euphytica*, 179 (1), 3–18.

Nhemachena, C. & Hassan, R. (2007). Micro-Level Analysis of Farmers' Adaptation to Climate Change in Southern Africa. IFPRI Discussion Paper No. 714, International Food Policy Research Institute, Washington, DC.

Okonya, J. S., Syndikus, K. & Kroschel, J. (2013). Farmers' perception of and coping strategies to climate change: evidence from six Agro-ecological zones of Uganda. *Journal of Agricultural Science*, 5 (8), 252–263.

Omoyo, N. N., Wakhungu, J. & Oteng'i, S. (2015). Effects of climate variability on maize yield in the arid and semi arid lands of lower eastern Kenya. *Agriculture & Food Security*, 4 (8).

Parry, M. L. (ed.) (2007). *Climate Change 2007: impacts, adaptation and vulnerability: contribution of Working Group II to the fourth assessment report of the Intergovernmental Panel on Climate Change (Vol. 4)*. Cambridge University Press.

Petheram, L., Zander, K. K., Campbell, B. M., High, C. & Stacey, N. (2010). 'Strange changes': Indigenous perspectives of climate change and adaptation in NE Arnhem Land (Australia). *Global Environmental Change*, 20 (4), 681–692.

Pryor, S. C., Scavia, D., Downer, C., Gaden, M., Iverson, L., Nordstorm, R., Patz, J. & Robertson, G. P. (2014). Midwest. *In:* Melillo, J. M., Terese (T.C.) Richmond & Yohe, G. W. (eds.), *Climate Change Impacts in the United States: The Third National Climate Assessment.* ch. 18, pp. 418–440, U.S. Global Change Research Program, Washington, D.C.

Quayum, M. A. & Ali, A. M. (2012). Adoption and diffusion of power tillers in Bangladesh. *Bangladesh Journal of Agricultural Research*, 37 (2), 307–325.

Rosenzweig, C., Elliott, J., Deryng, D., Ruane, A. C., Müller, C., Arneth, A., Boote, K. J., Folberth, C., Glotter, M., Khabarov, N., Neumann, K., Piontek, F., Pugh, T. A. M., Schmid, E., Stehfest, E., Yang, H. & Jones, J. W. (2014). Assessing agricultural risks of climate change in the 21st century in a global gridded crop model intercomparison. *Proceedings of the National Academy of Sciences*, 111 (9), 3268–3273.

Schlenker, W. & Lobell, D. B. (2010). Robust negative impacts of climate change on African agriculture. *Environmental Research Letters*, 5 (1), 1–8.

Shiferaw, B., Prasanna, B. M., Hellin, J. & Bänziger, M. (2011). Crops that feed the world 6. Past successes and future challenges to the role played by maize in global food security. *Food Security*, 3 (3), 307–327.

Skoufias, E., Rabassa, M. & Olivieri, S. (2011). The poverty impacts of climate change: a review of the evidence. World Bank Policy Research Working Paper No. 5622, The World Bank, Washington, DC.

Slingo, J. M., Challinor, A. J., Hoskins, B. J. & Wheeler, T. R. (2005). Introduction: food crops in a changing climate. *Philosophical Transactions of the Royal Society of London B: Biological Sciences*, 360 (1463), 1983–1989.

Sowunmi, F. A. (2010). Effect of climatic variability on maize production in Nigeria. *Research Journal of Environmental and Earth Sciences*, 2 (1), 19–30.

Storck, H., Emana, B., Adnew, B., Borowiccki, A. & Woldehawariat, S. (1991). *Farming systems and resource economics in the tropics: farming system and farm management practices of smallholders in the Hararghe Highland. Vol. II.* Wissenschaftsverlag Vauk, Kiel, Germany.

Tazeze, A., Haji, J. & Ketema, M. (2012). Climate change adaptation strategies of smallholder farmers: The case of Babilie district, East Harerghe zone of Oromia regional state of Ethiopia. *Journal of Economics and Sustainable Development*, 3 (14), 1–12.

Temesgen, D., Yehualashet, H. & Rajan, D. S. (2014). Climate change adaptations of smallholder farmers in South Eastern Ethiopia. *Journal of Agricultural Extension and Rural Development*, 6 (11), 354–366.

The World Bank Group (2010). Economics of Adaptation to Climate Change, Ethiopia Country Study. Washington, DC. URL http://climatechange.worldbank.org/content/Ethiopia

The World Bank Group (2011). Maize revolutions in Sub-Saharan Africa: Policy Research Working Paper. Agriculture and Rural Development Team, New York.

Tubiello, F. (2012). *Climate change adaptation and mitigation: Challenges and opportunities in the food sector*. Natural resources management and environment department, FAO, Rome.

Uddin, M. N., Bokelmann, W. & Entsminger, J. S. (2014). Factors Affecting Farmers' Adaptation Strategies to Environmental Degradation and Climate Change Effects: A Farm Level Study in Bangladesh. *Climate*, 2 (4), 223–241.

Zuluaga, V., Labarta, R. & Läderach, P. (2015). Climate Change Adaptation: The Case of the Coffee Sector in Nicaragua. International Center of Tropical Agriculture (CIAT), Selected Paper prepared for presentation at the Agricultural & Applied Economics Association and Western Agricultural Economics Association Annual Meeting, San Francisco, CA, July 26–28.

Appendix Table 1: *Conversion factors used to estimate tropical livestock unit (TLU) equivalents*

Animal Category	TLU
Calf	0.25
Donkey (young)	0.35
Weaned Calf	0.34
Camel	1.25
Heifer	0.75
Sheep and Goat (adult)	0.13
Cow and Ox	1
Sheep and Goat (young)	0.06
Horse	1.1
Chicken	0.013
Donkey (adult)	0.7

Source: Storck *et al.* (1991)

4

Alternate furrow irrigation of four fresh-market tomato cultivars under semi-arid condition of Ethiopia – Part I: Effect on fruit yield and quality

Ashinie Bogale [a,*], Wolfram Spreer [a,b],
Setegn Gebeyehu [c], Miguel Aguila [a], Joachim Müller [a]

[a]*Institute of Agricultural Engineering (440e), University of Hohenheim, Stuttgart, Germany*
[b]*Department of Highland Agriculture and Natural Resources, Faculty of Agriculture, Chiang Mai University, Thailand*
[c]*International Rice Research Institute, IRRI -WARDA Office, Dar Es Salaam, Tanzania*

Abstract

Scarcity of freshwater due to recurrent drought threatens the sustainable crop production in semi-arid regions of Ethiopia. Deficit irrigation is thought to be one of the promising strategies to increase water use efficiency (WUE) under scarce water resources. A study was carried out to investigate the effect of alternate furrow irrigation (AFI), deficit irrigation (DI) and full irrigation (FI) on marketable fruit yield, WUE and physio-chemical quality of four fresh-market tomato cultivars (*Fetan*, *Chali*, *Cochoro* and *ARP Tomato d2*) in 2013 and 2014. The results showed that marketable yield, numbers of fruits per plant and fruit size were not significantly affected by AFI and DI irrigations. WUE under AFI and DI increased by 36.7 % and 26.1 %, respectively with close to 30 % irrigation water savings achieved. A different response of cultivars to irrigation treatments was found for marketable yield, number of fruits and fruit size, WUE, total soluble solids (TSS) of the fruit juice, titratable acids (TA) and skin thickness. *Cochoro* and *Fetan* performed well under both deficit irrigation treatments exhibited by bigger fruit size which led to higher WUE. *ARP Tomato d2* showed good yields under well-watered conditions. *Chali* had consistently lower marketable fruit yield and WUE. TSS and TA tended to increase under deficit irrigation; however, the overall variations were more explained by irrigation treatments than by cultivars. It was shown that AFI is a suitable deficit irrigation practice to increase fresh yield, WUE and quality of tomato in areas with low water availability. However, AFI requires suitable cultivars in order to exploit its water saving potential.

Keywords: deficit irrigation, tomato (*Solanum lycopersicum* L.), tomato quality, water scarcity, water use efficiency

1 Introduction

Tomato (*Solanum lycopersicum* L.) is one of the most important vegetable crops worldwide and also among the important vegetable crops in Ethiopia with about 55,000 tons of fresh tomato produced on 7,000 ha annually (FAOSTAT, 2015). The demand for tomato has increased rapidly over the past years, as it has become the most profitable crop providing a higher income to small-scale farmers compared to other vegetable crops. However, the national average productivity is often low (7.9 t ha^{-1}), even below the African average (17.7 t ha^{-1}) - one reason for the fact that a substantial amount of irrigation water is required for tomato production. A better productivity is mandatory for sustainable increase in production. Moreover, increasing scarcity of freshwater along with forecasted increases in frequency and

---

[*] Corresponding author
Email: Ashinie.Gonfa@uni-hohenheim.de

severity of drought caused by climate change (Evans & Sadler, 2008; Patanè et al., 2011) and increasing competition from domestic and industrial uses (Strzepek & Boehlert, 2010) makes improving water use efficiency (WUE) in semi-arid and arid regions a primary concern.

Deficit irrigation (DI) is a strategy to increase on-farm water use efficiency (WUE) (Fereres & Soriano, 2007). There are, in principle, two DI techniques: regulated deficit irrigation (RDI) where a reduced amount of water is applied uniformly to the root-zone and partial root-zone drying (PRD), where the water is applied on a reduced area of the root-zone. The feasibility of both, RDI and PRD has been extensively studied in tomato with remarkable results in saving substantial amounts of irrigation water and increasing WUE (Zegbe et al., 2006; Patanè & Cosentino, 2010; Patanè et al., 2011). However, several other authors reported only marginal difference in yield response to PRD and well-watered greenhouse tomato (Campos et al., 2009; Yang et al., 2012). Also other benefits of PRD have been widely reported, namely promoting earlier crop maturity (Topcu et al., 2007), enhancing fruit quality in terms of taste and flavour (Haghighi et al., 2013), improving tomato plant resistance to disease (Xu et al., 2009) and reducing the incidence of blossom end rot development (Sun et al., 2013). On the other hand, also detrimental effects of PRD have been reported. Zegbe-Domínguez et al. (2003) and Casa & Rouphael (2014) found a significant reduction of processing tomato yields with PRD compared to a fully irrigated treatment. Similarly, Topcu et al. (2007) reported 20 % yield reduction as compared to full irrigation and saving about 50 % of irrigation water.

PRD was originally developed for micro irrigation systems. But meanwhile, it is also practiced as alternate furrow irrigation (AFI) in furrow irrigation studies. This irrigation technique is based on alternating wetting and drying of the opposite sides of the plant root system in subsequent irrigation events by watering one furrow and keeping dry the adjacent furrow until reversing in the next irrigation cycle. AFI has been proposed as a water saving technique with higher WUE without causing a significant yield reduction (Kang et al., 2000). As furrow irrigation is one of the most widely used surface irrigation technologies in Ethiopia, AFI is the method of choice for small or medium scale vegetable production in areas where irrigation water is scarce. Though drip irrigation has a higher water saving potential compared to furrow irrigation, AFI is inexpensive, easy to implement and also avoids the cost associated to investment and management of drip irrigation (Casa &

Rouphael, 2014). So far, AFI has been investigated in several cereal crops and grapes (Kang & Zhang, 2004; Du et al., 2013; Jia et al., 2014). Compared to conventional furrow irrigation, AFI saved 20–33 % irrigation water, shortened the time required for irrigation and substantially improved WUE. However, to our knowledge no experiments on the effect of AFI on field grown tomatoes have been reported until now.

Genotypic variations are not sufficiently addressed in most deficit irrigation studies and the studies conducted so far have generally focused on yield response of a single crop cultivar. A few studies compared response of maize and tomato genotypes under different deficit irrigations (Kaman et al., 2011; Patanè et al., 2014). Savic et al. (2011) found a variation in WUE and profit between two tomato cultivars under DI. A recent review by Chaves et al. (2010) has also highlighted that the efficiency of PRD or DI in modulating WUE depends on the varietal characteristics, soil type and prevailing weather conditions. The genotypic differences might be the result of differences in PRD-induced chemical signalling. Genotypes may also be different in the production of distinct fruit numbers and fruit sizes, which offers an opportunity to select water efficient genotypes depending on the kind of irrigation technique.

The use of different cultivars according to their level of tolerance to water stress employing DI strategies is a key for enhancing WUE in areas with growing water scarcity. Understanding responses of different cultivars to DI strategies is, therefore, necessary to optimize crop yield and quality of crops. In order to confirm whether alternate furrow irrigation is suitable for tomato cultivation and cultivars response differences, the study was conducted to investigate the agronomic response of four fresh tomato cultivars to moderate water deficit induced by AFI and DI under semi-arid condition of Ethiopia. The study is presented in two parts. Part I at hand addresses the agronomic response of tomato to deficit irrigation in terms of yield, WUE and quality. Part II presents the physiological response of tomato, which will allow additional insight to explain the agronomic response.

2 Materials and methods

2.1 Experimental site

The field experiment was carried out for two consecutive dry seasons in 2012/13 (thereafter 2013) and 2013/14 (thereafter 2014) from November to February and from December to May, respectively at Melkassa

Agricultural Research Center in the Central Rift Valley of Ethiopia (8°24′N, 39°21′E; 1,552 m asl). The climate of the area is semi-arid where rainfall is unpredictable in terms of onset, amount and distribution. Long-term (1977–2013) mean annual rainfall of the area is 829.5 mm, characterized by erratic inter-seasonal distribution, with a coefficient of variation of 28 % in August and 192 % in November. Weather data during the experimental periods is given in Table 1. The two growing seasons were different in rainfall amount and distribution. The rainfall amount was negligible in the growing season of 2013 whereas a total of 118 mm rainfall was recorded in March 2014 (Table 1). The average daily reference evapo-transpiration, ET_o (based on the Penman-Monteith FAO method), varied between 4.83 and 6.63 mm day^{-1} and total irrigation water was applied according to the irrigation treatments (Table 1).

The soil at the experimental field was a clay loam (sand 37 %, silt 42 %, and clay 21 %) moderately alkaline (pH 7.71) with low organic carbon (0.92–1.04 %), low to medium total soil nitrogen (0.058–0.080 %) and low available phosphorous (6.3–6.8 ppm) contents but with high extractable potassium (2.6–3.5 meq per 100 g soil). The volumetric soil moisture contents at field capacity (FC) and permanent wilting point (PWP) were 0.41 m^3 m^{-3} and 0.24 m^3 m^{-3}, respectively. The total available water (TAW) between FC and PWP and readily available water (RAW) for a to-

mato root extracting depth of 0.60 m were estimated to be 102.4 mm and 41.8 mm, respectively. The depletion factor (p) for tomato, the average fraction of the TWA that can be depleted from the root zone before water stress occurs, was assumed as 0.40 (Allen et al., 1998).

2.2　Plant material and growing conditions

Four fresh-market tomato cultivars, namely *Fetan*, *Chali*, *Cochoro* and *ARP Tomato d2*, were selected based on the similarity in phenology and growth habit from the rest of 27 tomato varieties that are officially recommended for commercial cultivation by Melkassa Agricultural Research Center. These table tomato cultivars are commonly grown in the central rift valley and other places in Ethiopia. The characteristics of the tomato cultivars used are given in Table 2.

Seedlings of these cultivars were raised in plastic trays containing peat moss for 24 days and transplanted to the experimental field on November 12, 2012 and December 27, 2013. Transplanting was performed manually at spacing of 0.4 m between plants and 0.9 m between rows. The plot size was 5.4 m × 4.0 m and each plot consisted of six rows and the middle four rows were used for data collection and final harvest. The distances between individual plots and between blocks were 1.5 and 4.0 m, respectively. A 1 m deep trench was constructed as a buffer-zone to prevent the lateral flow of irrigation water to the next experimen-

Table 1: *Mean monthly values of weather variables and the amount of irrigation water applied to full irrigation (FI), deficit irrigation (DI) and alternate furrow irrigation (AFI) at Melkassa Agricultural Research Center.*

Month	Average temperature (°C)		Relative humidity (%)	Cumulative daily wind speed (km day^{-1})	Sun shine (hr)	ET_o (mm day^{-1})	Total rainfall (mm)	Irrigation water applied (mm)		
	Max	Min						FI	DI	AFI
2012/13										
November	29.5	9.5	50.1	222.7	10.2	5.41	0.0	254.2	254.2	254.2
December	28.4	10.9	52.3	233.6	9.7	5.07	0.0	304.6	242.2	242.2
January	28.2	10.1	55.3	216.9	9.0	4.83	0.0	130.5	65.7	65.7
February	30.7	11.7	45.0	247.9	9.9	6.09	0.6	102.1	51.1	51.1
Total							0.6	791.4	613.2	613.2
2014										
January	29.1	11.6	45.4	249.4	9.7	5.58	0.0	211.0	211.0	211.0
February	30.4	15.6	50.0	247.1	8.8	5.76	8.3	180.6	90.3	90.3
March	30.0	15.6	51.7	247.9	8.2	5.85	118.6	165.9	83.0	83.0
April	31.9	16.1	45.3	266.1	9.0	6.63	4.1	73.2	36.6	36.6
Total							131.0	630.7	420.9	420.9

Table 2: *Characteristics of fresh-market tomato cultivars used for study.*

Cultivar	Growth habit	Fruit shape	Fruit size (g)	Maturity days	Yield (t ha^{-1})	Reaction to low water availability*
Fetan	Short, determinate	Cylindrical	110–120	75–80	45.4	NA
Chali	Short, determinate	Round	80–85	80–90	43.1	NA
Cochoro	Short, determinate	Round	70–76	80–110	46.3	NA
ARP Tomato d2	Short, determinate	Flat	80–100	80–90	48.6	NA

Source: Ministry of Agriculture and Rural Development, 2009–2013. Crop variety registers. Addis Ababa, Ethiopia.
* NA. Data not available

tal plots. Phosphorus at the rate of 92 kg P_2O_5 ha^{-1} was applied at transplanting using DAP fertiliser, which also contains 36 kg ha^{-1} of N. Additional nitrogen was applied using urea at the rate of 46 kg N ha^{-1} in two splits (23 kg N ha^{-1} at transplanting and the remaining 23 kg N ha^{-1} at flowering as side dressing). Disease and insect pest were controlled by spraying appropriate pesticides, equally to all experimental plots. Weeding and cultivations were done manually.

2.3 Experimental design and irrigation treatments

All plants were initially well watered as pre-irrigation for the first six irrigation events to ensure a good establishment of seedlings and subsequent plant growth. Deficit irrigation treatments were commenced when the plants developed their first truss: 36 days after transplanting (DAT) in 2013 and 39 DAT in 2014. Irrigation water was applied by furrow irrigation in a 3–5 days interval for the first four weeks after transplanting and every seven days thereafter. The field irrigation application efficiency was assumed to be 45 %. The different irrigation treatments were (1) full irrigation (FI): crop water requirements applied uniformly to all furrows, (2) deficit irrigation (DI): 50 % of crop water requirement applied uniformly to all furrows, and (3) alternate furrow irrigation (AFI): 50 % of crop water requirement applied to every other furrow and alternating the furrows at each irrigation event. The crop water requirement was calculated as the difference between measured volumetric soil water content (θAC) and soil water content at field capacity (θFC). The amount of water applied to each plot was measured via a three inch throat width Parshall flume installed at the inlet of the experimental field.

In 2013, the dynamics of soil-water content (θ, vol%) were monitored *in-situ* at three depths (0–20, 20–40, and 40–60 cm) using a TRIME-PICO profile TDR Probe (IMKO, Germany), twice a week right before irrigation and 24 h after irrigation. A total of 20 TECANAT access

tubes were installed down to a depth of 0.60 m along the side of the ridge equidistant between two plants. The average soil moisture content of FI treatment served as reference for the calculation of irrigation water needed for AFI and DI treatments. In the AFI treatment, two accesses tubes, one each on the left and right side, were placed to monitor soil moisture changes.

In 2014, 503DR neutron probe (CPN International, USA), previously calibrated for the experimental site, was used for monitoring soil moisture content. Eleven 0.80 m long access tubes were located per replicate, four in FI, three in DI and four in AFI plots. Soil moisture content was determined at depths of 0.15 m, 0.30 m, 0.45 m and 0.60 m. The top layer of 0.15 m was determined gravimetrically and then converted to volumetric water content by multiplying by bulk density of the soil (1.15 g cm^{-3}). The total of volumetric soil water content (m^3 m^{-3}) was summed up over the total rooting depth of 0.6 m.

The experimental lay out consisted in factorial combination of the four above mentioned commercial tomato cultivars and three irrigation treatments in a randomized block design with three replications.

2.4 Measurements

2.4.1 Agronomic data

Data on plant height and reproductive growth were collected from five tagged plants per plot. Plant height was measured at harvest. Mature and red ripe tomato fruits were manually harvested from the central four rows. Five harvests were carried out from 1st February to 8th March 2013 and from 1st April to 24th April 2014. Marketable fruit yield (tons ha^{-1}), numbers of fruit per plant and weight of 10 randomly selected fruits were recorded. Marketable and non-marketable yield were determined based on fruit size, presence of defects (malformed), disease and pest injuries. WUE was calculated as the ratio between marketable fruit yield and irrigation water applied. Total water use during the experimental period was the sum of irrigation water applied to each irrigation treatment in 2013 growing season. In 2014, effective rainfall was also taken into account.

2.4.2　Quality parameters

Ten ripe healthy fruits were randomly collected from each plot at the third harvesting time, weighed, washed and analysed for fruit quality traits. The following fruit biometric parameters were measured: fruit fresh weight, fruit longitudinal length, fruit width and skin thickness using a digital calliper (Harbor Freight Tools, USA). A fruit shape index was determined as the ratio between the fruit length and width. After measurements, juice was extracted by a juice extractor. The skin and solids were filtered out through muslin cloth and the juice content was expressed as ml juice per kg of fruit. The fruit juice extracts were analysed for pH, total soluble solids (TSS) and titratable acidity (TA). Juice pH was read with a Jenway 3520 pH meter (Bibby Scientific, UK) after standardization with buffer solutions of pH 4 and pH 7. TSS was determined using a TD-45 digital refractometer (Top Instrument, China). To determine TA, 10 ml of the extracted juice was thoroughly mixed with 50 ml of distilled water. Three drops of phenolphthalein as colorimetric indicator were added into each flask. The mixture was then titrated by adding 0.1 N NaOH. The volume of the sodium hydroxide, added to the solution, was multiplied by a correction factor of 0.064 to estimate TA as percentage of citric acid. TSS and TA were used to determine the sugar (TSS) and acid (TA) ratio as described by Beckles (2012). The determination was carried out in triplicate samples of each treatment.

2.5　Statistical analysis

Statistical analysis was performed through analysis of variance (ANOVA) using SAS statistical software of the SAS MIXED procedures (SAS Institute, 2004). Least significant difference (LSD) values at $P = 0.05$ were used to determine the significance of differences between treatment means.

3　Results

3.1　Dynamics of soil volumetric moisture contents

Changes in the volumetric soil water content (θ) of the irrigation treatments during the experimental periods are shown in Figure 1. Although the experiment was conducted during the dry season, there was a rainfall during 75–83 DAT in 2014 (Fig 1b), leading to an increase in soil moisture content of the deficit irrigation treatments.

In both years, different irrigation treatments showed distinct soil moisture content patterns in the top 0.60 m soil layers except for the rainy period in 2014. In FI, θ during the entire experimental period remained higher than in AFI and DI. The value of θ in DI was 6.9 % and 12.6 % lower than the FI in 2013 and 2014, respectively. However, the value of θ in AFI fluctuated depending on the one-week wetting and drying cycle, with the irrigated side closer to field capacity during the first and second irrigation cycle then steadily lowered as compared to FI. Except during 75–83 DAT in 2014, the θ of the dry side of AFI was also significantly lower than the values of the other two treatments. Even though both DI and AFI treatments received the same total volume of irrigation water, slightly greater reduction in θ (expressed as average wet side and dry side) was observed in AFI, about 18.9 % and 23.4 % lower than FI in 2013 and 2014, respectively.

3.2　Effects of irrigation techniques on fruit yield and water use efficiency

Significant differences were found among cultivars for marketable fruit yield under the different irrigation techniques (Fig 2a & b). In 2013, ARP Tomato d2 and Cochoro had the highest total fruit yield and marketable fruit yields under FI whereas Cochoro and Fetan were best performers under DI (Fig 2a). There was a different response of cultivars to deficit irrigation treatments. Relative to FI, ARP Tomato d2 encountered a marked reduction in marketable fruit yield under both DI and AFI (Fig 2a). The yield reductions in Chali were significant in DI and not in AFI. The reductions were 23.9 % and 6.4 % in Chali and 16.9 % and 13.9 % in ARP Tomato d2 under DI and AFI, respectively. The yield decrease was chiefly due to an increase of non-marketable fruits (Fig 3a). In 2014, Cochoro had the highest and Chali the lowest total and marketable fruit yields under all irrigation techniques (Fig 2b). On the other hand, Fetan grown under DI and AFI techniques, gave 32.2 % and 25.1 % more marketable fruit yield than under FI, respectively. Fetan had less fruits per plant but maintained bigger fruit size as compared to other cultivars (Table 3). The out performance of Cochoro over the other cultivars under deficit irrigation was mainly attributed to larger number of fruits per plant and medium-sized fruits (Table 3). Fetan and ARP Tomato d2 had less non-marketable fruits resulted from better fruit size distribution and highest relative fruit growth under low soil water availability (Fig 3b). Poor performance of Chali, as exhibited by its lower yielding potential and higher share of non-marketable fruit, was attributed to lower number of fruits per plant and smaller fruit size.

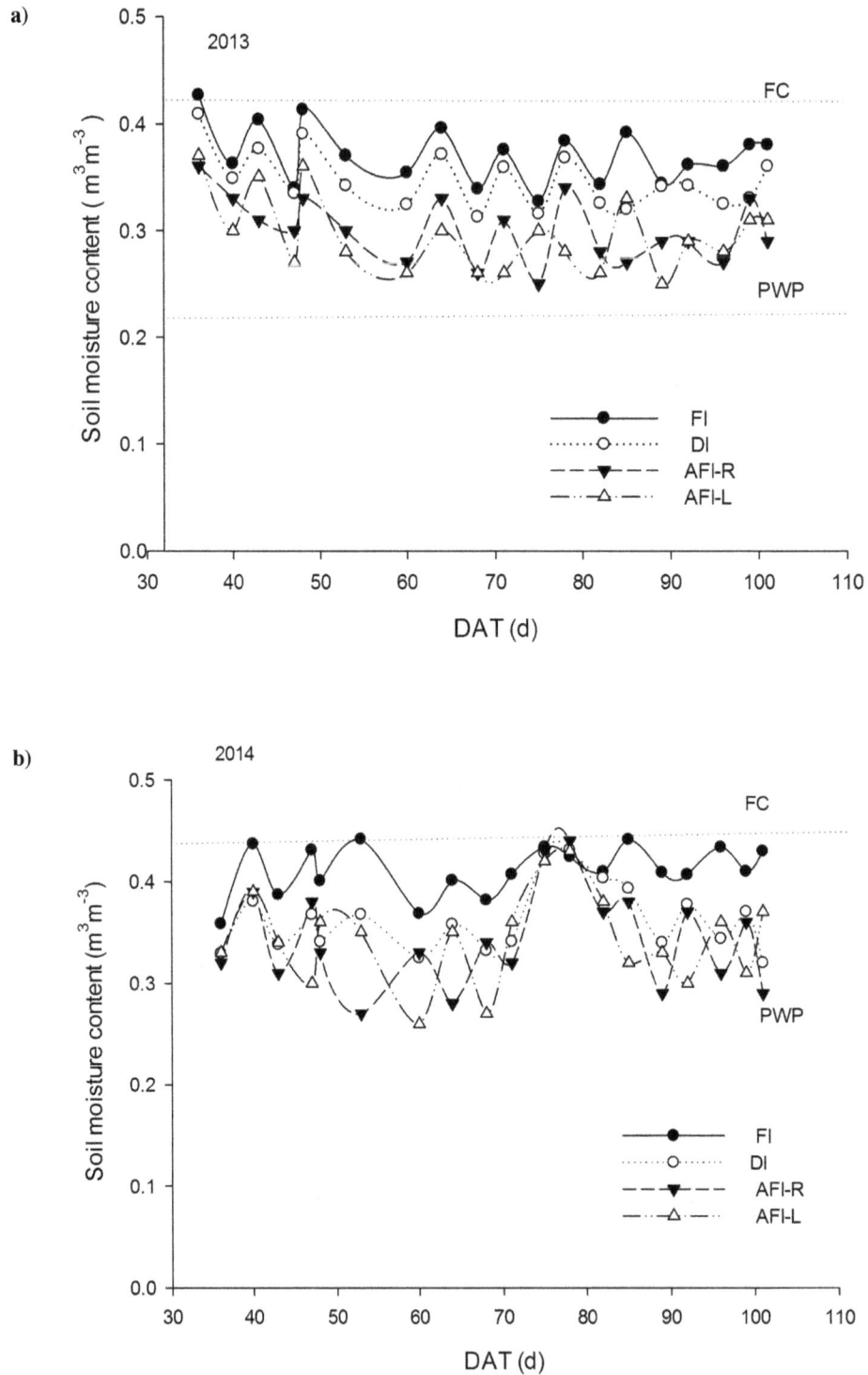

Fig. 1: *Soil moisture content vs. days after transplanting (DAT) under full irrigation (FI), deficit irriga-tion (DI) and right- and left-sides (AFI-right and AFI-left) of the plant root system of alternative furrow irrigation in 2013 (a) and 2014 (b) growing seasons at Melkassa Agricultural Research Center. FC=field capacity, PWP=permanent wilting point*

a)

b)

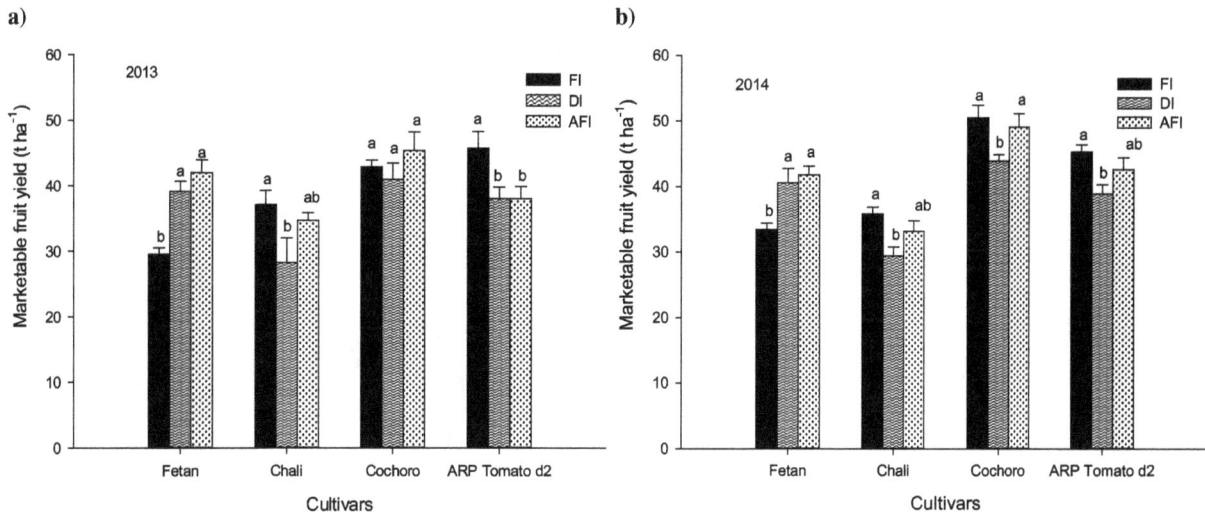

Fig. 2: *Marketable fruit yield of four tomato cultivars under full irrigation (FI), regulated deficit irrigation (DI) and alternate furrow irrigation (AFI) at Melkassa Agricultural Research Center in 2013 (a) and 2014 (b) growing seasons. Values represent means ± SD (n=3) and mean bars of the same cultivar followed by the same letter are not significantly different (P < 0.05).*

Table 3: *Number of fruits per plant and average fruit weight of four tomato cultivars under full irrigation (FI), regulated deficit irrigation (DI) and alternate furrow irrigation (AFI) at Melkassa Agricultural Research Center. Values represent means ± SD (n=3).*

Cultivar	Number of fruits per plant					Average fresh fruit weight (g)				
	FI	DI	AFI	Mean	LSD (5 %)	FI	DI	AFI	Mean	LSD (5 %)
2013										
Fetan	14.6±3.3	18.9±1.8	17.9±2.3	17.1	1.95	100.8±4.2	105.7±6.1	98.3±4.2	101.6	11.1
Chali	18.1±0.5	17.1±1.1	18.1±3.4	17.8	1.65	79.6±2.6	80.9±2.9	78.7±1.6	79.7	5.1
Cochoro	22.2±2.0	23.1±2.7	28.3±0.3	24.5	9.4	97.9±4.6	91.3±3.3	96.2±3.8	95.1	13.4
ARP Tomato d2	23.4±1.8	22.1±1.2	21.8±0.9	22.4	8.4	99.5±3.8	93.9±2.7	93.1±7.1	95.5	22.9
Mean	19.6	20.3	21.5			94.5	92.9	91.6		
LSD (5 %)	4.65	NS	7.58			14.66	9.32	6.29		
2014										
Fetan	15.8±3.7	18.1±0.8	21.8±0.8	18.6	5.6	92.3±5.3	92.2±1.7	88.1±1.0	90.9	5.4
Chali	18.6±2.8	17.9±0.8	17.5±1.5	18.0	3.7	73.9±4.3	65.3±0.6	75.2±0.5	71.5	5.7
Cochoro	32.3±1.7	24.6±1.0	27.6±1.2	28.2	7.5	95.2±1.6	102.3±4.9	95.3±4.7	97.6	5.5
ARP Tomato d2	19.4±1.1	20.9±0.7	22.9±1.1	21.1	2.2	86.4±5.0	82.9±4.6	94.9±6.1	88.1	10.1
Mean	21.5	20.4	22.5			86.9	85.7	88.4		
LSD (5 %)	3.08	2.84	2.62			9.46	7.92	8.86		

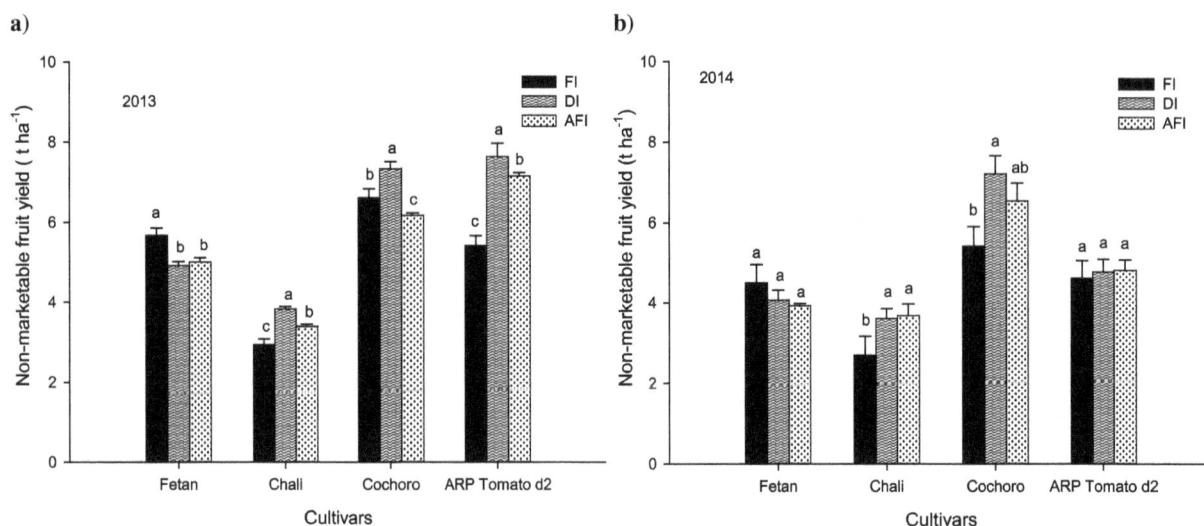

Fig. 3: *Non-marketable fruit yield of four tomato cultivars under full irrigation (FI), regulated deficit irrigation (DI) and alternate furrow irrigation (AFI) at Melkassa Agricultural Research Center in 2013 (a) and 2014 (b) growing seasons. Values represent means ± SD (n=3) and mean bars of the same cultivar followed by the same letter are not significantly different (P < 0.05).*

WUE considerably varied among irrigation techniques as well as cultivars (Fig 4a and b). It ranged from 3.73 to 7.40 kg m^{-3} and 4.70 to 9.80 kg m^{-3} in 2013 and 2014 growing seasons, respectively. Compared to FI, 27.9 % irrigation water was saved with the use of AFI and DI thereby improving WUE of the cultivars under the two irrigation techniques by 36.7 % and 26.1 %, respectively. Despite the same amount of irrigation water applied, AFI resulted in significantly higher WUE than DI (Fig 4a and b). WUE of the four cultivars was also significantly different with the same trends as for the productivity differences. In 2013, WUE of *Fetan* and *Cochoro* was significantly increased under DI and AFI compared to FI. However, *Chali* and *ARP Tomato d2* did not exhibit significant increment in WUE. On the other hand, in 2014, WUE of all cultivars was significantly higher under deficit irrigation treatments as compared to FI (Fig 4b). Pronounced increments were recorded in *Fetan* and *Cochoro*. Overall, the WUE of *Fetan* was increased by 77.8 % and 72.3 % under AFI and DI as compared to FI, respectively. *Cochoro* also exhibited an increase of 46.3 % and 26.1 % WUE compared to FI. Similar to the 2013 results, *Chali* and *ARP Tomato d2* had lower WUE, but the WUE of both cultivars was significantly higher under DI (15.3 % and 13.6 %, respectively) and AFI (23.5 % and 28.2 %, respectively) as compared to FI.

The irrigation treatments and cultivars had significant effect on vegetative growth as determined by plant height at harvest (Table 4). Significantly highest plant height was obtained in FI and reduction in growth was

evident under DI and AFI. However, cultivars differed in response to deficit irrigation and between seasons. In 2013, the vegetative growth of all cultivars was reduced under DI and AFI. In 2014, the vegetative growth of *Cochoro* and *ARP Tomato d2* was reduced significantly whereas it did not affected the vegetative growth of *Fetan* and *Chali*. Differences in fruit maturity periods were evident between years with fruits being ready for harvest after 82 and 91 days after transplanting (DAT) in 2013 and 2014, respectively. Nevertheless, deficit irrigation treatments had no significant effect on the maturity period for any of the cultivars in 2013 (Table 4) but it promoted earlier harvest in *Chali* and *ARP Tomato d2* in 2014. Over all, *Chali* was earlier mature while *Cochoro* was late by almost two weeks.

3.3 Effects of irrigation techniques on physio-chemical quality

Compared to the growing season 2013, the values of TSS and TA were lower in the 2014 (Table 5). TSS and TA were significantly higher under DI and AFI as compared to FI in both growing seasons. In 2013, TSS and TA were significantly higher in *Chali* and *Cochoro* under DI and AFI while the values of TSS and TA remained unchanged in *Fetan* and *ARP Tomato d2*. In 2014, TSS was significantly higher in all cultivars while TA increased only in *ARP Tomato d2*. However, years and irrigation treatments could explain 69.2 % and 11.5 % of the total variation of TSS in tomatoes fruits, respectively. Similarly, the total variation of TA and TSS was explained more by year variations and irriga-

a)

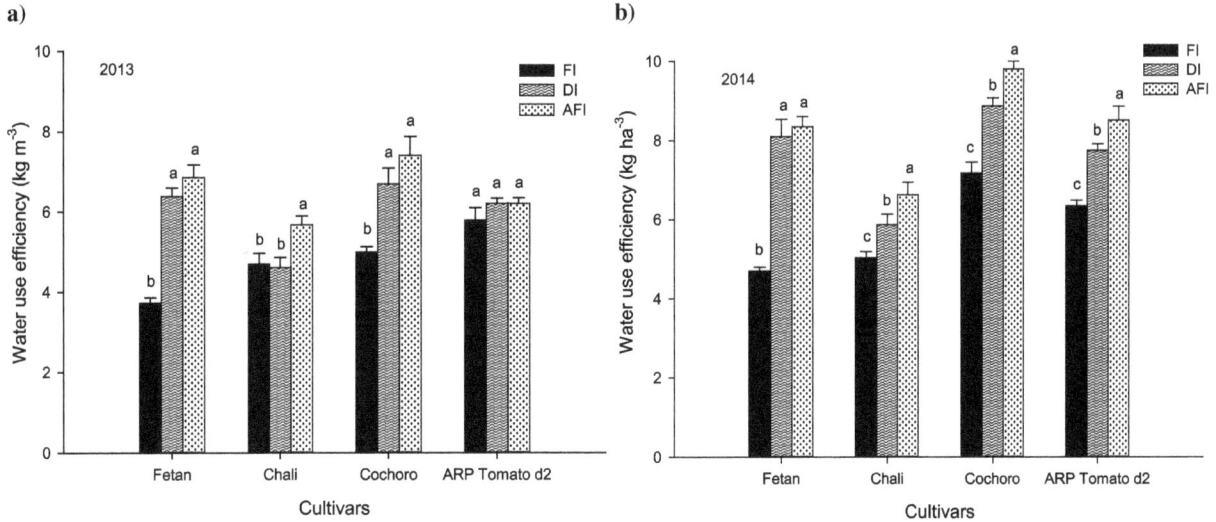

b)

Fig. 4: *Water use efficiency (kg m^{-3}) of four tomato cultivars under full irrigation (FI), regulated deficit irrigation (DI) and alternate furrow irrigation (AFI) at Melkassa Agricultural Research Center during 2013 (a) and 2014 (b). Values represent means ± SD (n=3) and mean bars of the same cultivar followed by the same letter are not significantly different (P < 0.05).*

Table 4: *Plant height and maturity period of four tomato cultivars under full irrigation (FI), regulated deficit irrigation (DI) and alternate furrow irrigation (AFI) at Melkassa Agricultural Research Center. Values represent means ± SD (n=3).*

Cultivar	Plant height (cm)					Maturity period (days)				
	FI	DI	AFI	Mean	LSD (5%)	FI	DI	AFI	Mean	LSD (5%)
2013										
Fetan	81.6±2.6	74.7±1.9	75.4±3.4	77.2	4.1	105.9±0.7	106.7±1.2	107.7±1.5	106.8	2.2
Chali	78.1±2.0	66.1±3.2	66.2±2.3	70.1	8.9	98.6±0.2	97.5±0.9	95.4±0.5	97.3	1.4
Cochoro	88.9±2.8	76.7±1.6	72.2±3.3	79.3	2.7	109.9±1.0	110.4±1.0	109.2±1.6	109.8	3.0
ARP Tomato d2	86.7±2.8	71.1±3.8	73.3±3.5	77.0	10.4	102.6±0.7	103.4±1.3	108.5±0.8	104.8	2.6
Mean	83.9	72.2	71.8			104.3	104.5	105.3		
LSD (5%)	8.8	5.1	7.9			1.34	1.25	2.7		
2014										
Fetan	74.3±0.7	68.3±2.2	68.6±1.3	70.4	17.8	99.5±3.5	102.0±5.6	98.1±4.0	99.9	4.3
Chali	70.4±1.0	65.6±1.4	67.7±1.6	67.9	6.1	97.5±6.8	91.5±1.3	93.8±4.9	94.3	5.8
Cochoro	81.1±1.9	69.4±1.0	67.4±2.0	72.6	9.7	109.2±3.5	105.2±5.0	108.0±1.7	107.5	10.1
ARP Tomato d2	87.3±1.1	67.3±0.8	71.2±1.8	75.3	9.5	105.1±2.9	97.7±3.4	99.5±3.5	100.8	3.9
Mean	78.3	67.7	68.7			102.8	96.6	99.9		
LSD (5%)	2.8	13.6	6.1			4.47	6.11	3.24		

Table 5: *Physio-chemical characteristics of four tomato cultivars under full irrigation (FI), regulated deficit irrigation (DI) and alternate furrow irrigation (AFI) at Melkassa Agricultural Research Center. Values represent means ± SD (n=3).*

	Cultivar	Quality parameter	FI	DI	AFI	LSD (5 %)
2013	Fetan	Fruit skin thickness (mm)	5.42 ±0.5	5.79 ±0.70	5.79 ±0.49	NS
		Juice content (ml/kg)	1049.0±15.2	1054.3±25.2	1046.3±15.2	NS
		Total solid soluble (Brix)	4.07±0.1	4.00±0.10	4.07±0.12	NS
		Titratable acid (%)	0.48±0.02	0.51±0.02	0.50±0.02	NS
		TSS:TA ratio	8.46±0.13	7.84±0.82	8.11±0.55	NS
	Chali	Fruit skin thickness (mm)	6.96±0.19	5.92±0.47	6.19±0.69	NS
		Juice content (ml/kg)	1049.2±13.5	1052.6±13.5	1053.1±13.5	NS
		Total solid soluble (Brix)	3.33±0.13	3.70±0.10	3.93±0.12	0.211
		Titratable acid (%)	0.55±0.05	0.59±0.03	0.66±0.03	0.091
		TSS:TA ratio	6.16±0.79	6.31±0.49	6.19±0.14	NS
	Cochoro	Fruit skin thickness (mm)	5.66±0.34	5.95±0.62	7.13±0.61	1.39
		Juice content (ml/kg)	1082.7±9.4	1036.0±9.4	1051.1±9.4	37.1
		Total solid soluble (Brix)	3.82±0.20	4.07±0.12	4.27±0.01	0.35
		Titratable acid (%)	0.52±0.01	0.54±0.04	0.59±0.01	0.06
		TSS:TA ratio	7.61±0.18	7.59±0.25	6.84±0.56	NS
	ARP Tomato d2	Fruit skin thickness (mm)	5.21±0.68	5.63±0.01	6.76±0.29	1.32
		Juice content (ml/kg)	1040.7±11.4	1059.1±11.4	1048.5±11.4	NS
		Total solid soluble (Brix)	3.86±0.13	3.93±0.12	4.00±0.20	NS
		Titratable acid (%)	0.53±0.01	0.50±0.01	0.58±0.01	0.075
		TSS:TA ratio	7.31±0.18	7.82±0.25	6.89±0.25	0.82
2014	Fetan	Fruit skin thickness (mm)	6.10±0.85	6.67±0.90	7.13±0.84	0.98
		Fruit diameter (mm)	56.8±0.78	54.3±0.77	56.0±0.75	NS
		Fruit length (mm)	61.0±0.73	55.1±0.73	55.8±0.70	NS
		Fruit shape index	0.94±0.02	0.99±0.02	1.01±0.02	NS
		Juice content (ml/kg)	1130.9±10.5	1111.4±10.5	1133.4±10.5	NS
		pH	4.49±0.05	4.46±0.06	4.49±0.04	NS
		Total solid soluble (Brix)	2.78±0.03	3.11±0.03	3.27±0.01	0.016
		Titratable acid (%)	0.48±0.03	0.49±0.02	0.51±0.04	NS
		TSS:TA ratio	5.88±0.19	6.38±0.19	6.46±0.19	NS
	Chali	Fruit skin thickness (mm)	6.90±0.78	6.37±0.84	6.03±0.84	NS
		Fruit diameter (mm)	57.9±0.77	58.1±0.77	59.6±0.75	NS
		Fruit length (mm)	50.8±0.73	50.8±0.73	50.9±0.70	NS
		Fruit shape index	1.14±0.02	1.14±0.02	1.17±0.02	NS
		Juice content (ml/kg)	1105.9±10.5	1095.8±10.5	1126.0±10.5	NS
		pH	4.52±0.04	4.55±0.09	4.54±0.09	NS
		Total solid soluble (Brix)	3.10±0.05	3.25±0.05	3.63±0.05	0.27
		Titratable acid (%)	0.54±0.01	0.52±0.02	0.53±0.01	NS
		TSS:TA ratio	5.80±0.24	6.22±0.14	6.70±0.66	0.71
	Cochoro	Fruit skin thickness (mm)	6.05±0.35	6.77±0.35	7.10±0.35	1.01
		Fruit diameter (mm)	61.6±0.77	63.5±0.77	61.4±0.75	NS
		Fruit length (mm)	54.7±0.73	51.2±0.73	52.7±0.70	NS
		Fruit shape index	1.13±0.02	1.24±0.02	1.17±0.02	NS
		Juice content (ml/kg)	1057.7±13.0	1064.9±13.0	1065.9±10.5	NS
		pH	4.56±0.11	4.53±0.010	4.51±0.07	NS
		Total solid soluble (Brix)	2.95±0.06	3.02±0.02	3.63±0.02	0.28
		Titratable acid (%)	0.41±0.01	0.45±0.01	0.44±0.01	NS
		TSS:TA ratio	7.22±0.23	6.88±0.23	8.22±0.19	NS
	ARP Tomato d2	Fruit skin thickness	6.00±0.73	5.29±0.73	6.70±0.53	NS
		Fruit diameter	62.2±0.77	57.6±0.77	59.6±0.75	NS
		Fruit length	58.4±0.73	56.4±0.73	57.4±0.70	NS
		Fruit shape index	1.07±0.02	1.02±0.02	1.04±0.02	NS
		Juice content (ml/kg)	1087.9±9.1	1102.0±9.1	1091.3±10.5	NS
		pH	4.55±0.06	4.44±0.09	4.48±0.09	NS
		Total solid soluble (Brix)	2.89±0.01	3.12±0.01	3.46±0.01	0.13
		Titratable acid (%)	0.46±0.02	0.50±0.02	0.53±0.02	0.037
		TSS:TA ratio	6.30±0.19	6.23±0.17	6.54±0.17	NS

tion treatments than variations due to cultivars. In general, tomato grown under AFI had the highest TSS and TA. A negative but non-significant correlation was noted between fruit yield and TSS. Nevertheless, fruit biometric parameters (fruit size, fruit length, fruit width, and fruit shape index), pH and sugar acid ratio were not significantly influenced by irrigation treatments. However, a significant reduction of juice content under both deficit irrigation treatments was observed in *Cochoro*. The fruit skin thickness was increased under both DI and AFI as compared to FI, but the response varied among cultivars and years. In 2013, the fruit skin thickness of *Cochoro* and *ARP Tomato d2* were increased while it was not affected in *Fetan* and *Chali*. In 2014, *Fetan* and *Cochoro* had significantly higher skin thickness under DI and AFI as compared to FI. However, other fruit biometric parameters (fruit length and fruit width) and pH were not affected by deficit irrigation treatments (Table 5). The pH of the juice ranged from 4.46 to 4.55 and exceeds the minimum permissible level (pH < 4.30) to define the product as 'good' according to the reference of analytical scales for processing tomatoes (Patanè & Cosentino, 2010).

4 Discussion

The results of this study show that the comparison of different cultivars with respect to their performance under different irrigation regimes is necessary to evaluate both water saving potential and yield expectation. This is especially true for crops with high breeding activities and frequent occurrence of new varieties such as tomato, where there is a high cultivar-to-cultivar variation in response to deficit irrigations as a function of stress tolerance levels (Kaman *et al.*, 2011; Patanè *et al.*, 2014). This underlines the necessity that optimisation of the plant available soil water should be done for a particular cultivar and not for a species in general, as suggested by Savic *et al.* (2011). Obviously, the cultivars specific fruit size distribution are crucial for the final marketable yield, as demonstrated by Xu *et al.* (2009) who found that fruit yield of large fruit-sized tomato cultivars can be increased under PRD but not the fruit yield of the cherry tomato cultivar.

Deficit irrigation during early reproductive stages often causes abnormal reproductive organs and results in failure of pollination and fruit abortion consequently reducing yield (Pulupol *et al.*, 1996; Zegbe *et al.*, 2006). However, cultivars employ different strategies to adapt to the water stress induced by deficit irrigation. *Cochoro* and *ARP Tomato d2* had considerably higher yields under full irrigation. *Cochoro* and *Fetan* performed well under both deficit irrigation treatments leading to higher levels of WUE. As the higher yield of *Cochoro* was due to both greater number of fruits per plant and bigger fruit size, it is concluded that this cultivar is well adapted to water stress. Similarly, the local cultivar *Fetan* had bigger fruit size but less fruit numbers and performed well under DI and AFI. Both cultivars are considered to be recommendable for potentially drought affected production systems, but with a generally higher yield potential of *Cochoro* than *Fetan*. Moderate fruit numbers and medium fruit size contributed to the higher yields in *ARP Tomato d2*, a cultivar which is best suited for the production under sufficient irrigation. The production potential of genotypes may play an important role in WUE variation under water deficit conditions (Grant *et al.*, 2010; Kaman *et al.*, 2011). Drought resistant crop cultivars have been reported to improve WUE, however, crops with higher drought resistance are often associated with lower crop yield (Blum, 2005). Nevertheless, it is noted that cultivars with higher marketable yields were generally associated with higher WUE. In this study, *Fetan* and *Cochoro* showed the highest marketable yield as well as higher WUE under DI and AFI. *Chali* consistently had lower marketable fruit yield and as consequence, with lowest WUE across the irrigation treatments. Therefore, under given limited water resource, cultivars with highest marketable fruit yield potential are superior in WUE, suggesting that cultivars with higher WUE thus combined both drought avoidance mechanisms and higher productivity. As low yielding cultivar, *Chali* has excellent handling properties and taste quality, e.g. a ticker fruit skin and higher TA; the variety is widely produced in Ethiopia and has good market potential. In this experiment it was shown that *Chali* performs best when well-irrigated. Consequently, *Chali* responded with better taste quality in terms of higher TSS and TA in the 2014 season, were unexpected rainfalls changed the water supply pattern.

In general it can be stated that the fruit yield increase observed under AFI was mainly attributed to the increase in number of fruits than to mean fruit size concurring with previous reports (Pék *et al.*, 2014). Besides saving substantial irrigation water, AFI improves WUE over DI and FI, what is also reported in other studies (Kirda *et al.*, 2007; Topcu *et al.*, 2007). Mean fruit size, fruit diameter and length are mainly determined by genotype but they are also affected by irrigation amount to some extent (Patanè & Cosentino, 2010; Liu *et al.*, 2013). The study showed that deficit irrigation treatments had promoted earlier harvest but the response was different between cultivars and growing season. Previous reports also showed that deficit irrigation has shortened (Zegbe-Domínguez *et al.*, 2003) or did not modify the maturity period (Casa & Rouphael, 2014). Overall variations in fruit quality parameters were more explained by irrigation treatments than by cultivar. TSS and TA, the biochemical fruit quality traits which contribute to the flavour of fresh tomatoes (Panthee *et al.*, 2013), increased significantly under deficit irrigation treatments, concurring several earlier reports (Zegbe-Domínguez *et al.*, 2003; Patanè & Cosentino, 2010). These increases are mainly due to reduced water content and concomitant increase in dry matter that leads to higher solute concentration of fruits (Pulupol *et al.*, 1996; Haghighi *et al.*, 2013). Water deficit initiated during flowering and fruit set reduces the number of reproductive organs but may increase quality due to increased availability of assimilates for the remaining fruits (Patanè & Cosentino, 2010). The response of TSS and TA toward deficit irrigation differs between cultivars. *Chali* was low yielding under deficit irrigation but had higher TSS and TA than other cultivars. Tomato fruits under AFI had also higher values of TSS and TA than those under DI and FI. Water deficit before or during repining has been reported to increase TSS and TA of tomato (Zegbe-Domínguez *et al.*, 2003; Sun *et al.*, 2014) or had no effects (Campos *et al.*, 2009). The responses of fruit quality parameters differ considerably depending on the differences in genotypes, stress intensity and phonological stages at which deficit irrigation was imposed (Ripoll *et al.*, 2016). TSS content was found to depend on cultivar but seasonal environmental variations during the fruit repining stages may also influence the contents.

5 Conclusion

The result of the present study confirmed that, in general, AFI has the potential to save close to 30 % of irrigation water relative to full irrigation, greatly improving WUE, some fruit quality aspects (TSS, TA) and postharvest handling properties (fruit skin thickness) without causing a detrimental effect on the fruit yield under the studied semi-arid climate of Ethiopia. It was further shown that cultivars respond differently to irrigation treatments. This demonstrates the need for cultivar specific irrigation management practices. *Cochoro* and *Fetan* performed well under AFI, while *Chali* and *ARP Tomato d2* performed relatively better under full irrigation. Cultivars with highest marketable fruit yield have better WUE suggesting that the selection of suitable cultivar for different irrigation methods is crucial for improving WUE in areas with water scarcity.

Acknowledgements

We greatly acknowledge the financial support by Dr. Hermann Eiselen PhD Grant of *fiat panis* Foundation for field research as well as the scholarship grant by German Academic Exchange Service (DAAD) and German Federal Ministry for Economic Cooperation and Development (BMZ). The authors would also like to express their gratitude to the Melkassa Agricultural Research Center for support in performing the field experiment.

References

Allen, R. G., Pereira, J. S., Raes, D. & Smith, M. (1998). *Crop evapotranspiration: Guidelines for computing crop water requirements. Irrigation and Drainage Paper 56*. Food and Agriculture Organization of the United Nations (FAO), Italy. Rome.

Beckles, D. M. (2012). Factors affecting the postharvest soluble solids and sugar content of tomato (*Solanum lycopersicum* L.) fruit. *Postharvest Biology and Technology*, 63, 129–140.

Blum, A. (2005). Drought resistance, water-use efficiency, and yield potentia–are they compatible, dissonant, or mutually exclusive? *Australian Journal of Agricultural Research*, 56, 1159–1168.

Campos, H., Trejo, C., Peña-Valdivia, C. B., Ramírez-Ayala, C. & Sánchez-García, P. (2009). Effect of partial rootzone drying on growth, gas exchange, and yield of tomato (*Solanum lycopersicum* L.). *Scientia Horticulturae*, 120, 493–499.

Casa, R. & Rouphael, Y. (2014). Effects of partial root-zone drying irrigation on yield, fruit quality, and water-use efficiency in processing tomato. *Journal of Horticultural Science and Biotechnology*, 89, 389–396.

Chaves, M. M., Zarrouk, O., Francisco, R., Costa, J. M., Santos, T., Regalado, A. P., Rodrigues, M. L. & Lopes, C. M. (2010). Grapevine under deficit irrigation: hints from physiological and molecular data. *Annals of Botany*, 105 (5), 661–676.

Du, T.-s., Kang, S.-z., Yan, B.-y. & Zhang, J.-h. (2013). Alternate Furrow Irrigation: A Practical Way to Improve Grape Quality and Water Use Efficiency in Arid Northwest China. *Journal of Integrative Agriculture*, 12 (3), 509–519.

Evans, G. R. & Sadler, E. J. (2008). Methods and technologies to improve efficiency of water use. *Water Resources Research*, 44, W00E04.

FAOSTAT (2015). Food and Agriculture Organization of the United Nations Statistcs Division. URL http://faostat3.fao.org

Fereres, E. & Soriano, M. (2007). Deficit irrigation for reducing agricultural water use. *Journal of Experimental Botany*, 58, 147–159.

Grant, O. M., Abigail, W. J., Michael, J. D., Celia, M. J. & David, W. S. (2010). Physiological and morphological diversity of cultivated strawberry (*Fragaria × ananassa*) in response to water deficit. *Environmental and Experimental Botany*, 68 (3), 264–272.

Haghighi, M., France, J., Behboudian, M. H. & Mills, T. M. (2013). Fruit quality responses of 'Petopride' processing tomato (*Lycopersicon esculentum* Mill.) to partial rootzone drying. *Journal of Horticultural Science and Biotechnology*, 88, 154–158.

Jia, D.-Y., Dai, X.-L., Men, H.-W. & He, M.-R. (2014). Assessment of winter wheat (*Triticum aestivum* L.) grown under alternate furrow irrigation in northern China: Grain yield and water use efficiency. *Canadian Journal of Plant Science*, 94, 349–359.

Kaman, H., Kirda, C. & Sesveren, S. (2011). Genotypic differences of maize in grain yield response to deficit irrigation. *Agricultural Water Management*, 98, 801–807.

Kang, S., Liang, Z., Pan, Y., Shi, P. & Zhang, J. (2000). Alternate furrow irrigation for maize production in arid area. *Agricultural Water Management*, 45, 267–274.

Kang, S. & Zhang, J. (2004). Controlled alternate partial root-zone irrigation: its physiological consequences and impact on water use efficiency. *Journal of Experimental Botany*, 55, 2437–2446.

Kirda, C., Topcu, S., Cetin, M., Dasgan, H. Y., Kaman, H., Topaloglu, F., Derici, M. R. & Ekici, B. (2007). Prospects of partial root zone irrigation for increasing irrigation water use efficiency of major crops in the Mediterranean region. *Annals of Applied Biology*, 150, 281–291.

Liu, H., Duan, A.-w., Li, F.-s., Sun, J.-s., Wang, Y.-c. & Sun, C.-t. (2013). Drip Irrigation Scheduling for Tomato Grown in Solar Greenhouse Based on Pan Evaporation in North China Plain. *Journal of Integrative Agriculture*, 12 (3), 520–531.

Panthee, D. R., Labate, J. A., McGrath, M. T., Breksa III, A. P. & Robertson, L. D. (2013). Genotype and environmental interaction for fruit quality traits in vintage tomato varieties. *Euphytica*, 193 (2), 169–182.

Patanè, C. & Cosentino, S. L. (2010). Effects of soil water deficit on yield and quality of processing tomato under a Mediterranean climate. *Agricultural Water Management*, 97, 131–138.

Patanè, C., La Rosa, S., Pellegrino, A., Sortino, O. & Saita, A. (2014). Water productivity and yield response factor in two cultivars of Processing tomato as affected by deficit irrigation under semi-arid climate conditions. *Acta Horticulturae*, 1038, 449–454.

Patanè, C., Tringali, S. & Sortino, O. (2011). Effects of deficit irrigation on biomass, yield, water productivity and fruit quality of processing tomato under semi-arid Mediterranean climate conditions. *Scientia Horticulturae*, 129, 590–596.

Pék, Z., Szuvandzsiev, P., Daood, H., Neményi, A. & Helyes, L. (2014). Effect of irrigation on yield parameters and antioxidant profiles of processing cherry tomato. *Central European Journal of Biology*, 9, 383–395.

Pulupol, L. U., Behboudian, M. H. & Fisher, K. J. (1996). Growth, yield and post harvest attributes of glass house tomatoes produced under deficit irrigation. *HortScience*, 31 (6), 926–929.

Ripoll, J., Urban, L., Brunel, B. & Bertin, N. (2016). Water deficit effects on tomato quality depend on fruit developmental stage and genotype. *Journal of Plant Physiology*, 190, 26–35.

Savic, S., Stikic, R., Zaric, V., Vucelic-Radovic, B., Jovanovic, Z., Marjanovic, M., Djordjevic, S. & Petkovic, D. (2011). Deficit irrigation technique for reducing water use of tomato under polytunnel conditions. *Journal of Centeral European Agriculture*, 12, 590–600.

Strzepek, K. & Boehlert, B. (2010). Competition for water for the food system. *Philosophical Transactions of the Royal Society B: Biological Sciences*, 365, 2927–2940.

Sun, Y., Feng, H. & Liu, F. (2013). Comparative effect of partial root-zone drying and deficit irrigation on incidence of blossom-end rot in tomato under varied calcium rates. *Journal of Experimental Botany*, 64, 2107–2116.

Sun, Y., Holm, P. E. & Liu, F. (2014). Alternate partial root-zone drying irrigation improves fruit quality in tomatoes. *Horticultural Science*, 41, 185–191.

Topcu, S., Kirda, C., Dasgan, Y., Kaman, H., Cetin, M., Yazici, A. & Bacon, M. A. (2007). Yield response and N-fertiliser recovery of tomato grown under deficit irrigation. *European Journal of Agronomy*, 26, 64–70.

Xu, H. L., Qin, F. F., Du, F. L., Xu, Q. C., Wang, R., Shaha, R. P., Zhoa, A. H. & Li, F. M. (2009). Applications of xerophytophysiology in plant production - Partial root drying improves tomato crop. *Journal of Food, Agriculture & Environment*, 7, 981–988.

Yang, L., Qu, H., Zhang, Y. & Li, F. (2012). Effects of partial root-zone irrigation on physiology, fruit yield and quality and water use efficiency of tomato under different calcium levels. *Agricultural Water Management*, 104, 89–94.

Zegbe, J. A., Behboudian, M. H. & Clothier, B. E. (2006). Yield and fruit quality in processing tomato under partial rootzone drying. *European Journal of Horticultural Science*, 71, 252–258.

Zegbe-Domínguez, J. A., Behboudian, M. H., Lang, A. & Clothier, B. E. (2003). Deficit irrigation and partial rootzone drying maintain fruit dry mass and enhance fruit quality in 'Petopride' processing tomato (*Lycopersicon esculentum*, Mill.). *Scientia Horticulturae*, 98, 505–510.

5

Alternate furrow irrigation of four fresh-market tomato cultivars under semi-arid condition of Ethiopia – Part II: Physiological response

Ashinie Bogale [a,*], Wolfram Spreer [a,b],
Setegn Gebeyehu [c], Miguel Aguila [a], Joachim Müller [a]

[a] *Institute of Agricultural Engineering (440e), University of Hohenheim, Stuttgart, Germany*
[b] *Department of Highland Agriculture and Natural Resources, Faculty of Agriculture, Chiang Mai University, Thailand*
[c] *International Rice Research Institute, IRRI -WARDA Office, Dar Es Salaam, Tanzania*

Abstract

Understanding the variation in physiological response to deficit irrigation together with better knowledge on physiological characteristics of different genotypes that contribute to drought adaptation mechanisms would be helpful in transferring different irrigation technologies to farmers. A field experiment was carried to investigate the physiological response of four tomato cultivars (*Fetan*, *Chali*, *Cochoro* and *ARP Tomato d2*) to moderate water deficit induced by alternate furrow irrigation (AFI) and deficit irrigation (DI) under semi-arid condition of Ethiopia during 2013 and 2014. The study also aimed at identifying physiological attributes to the fruit yield of tomato under different deficit irrigation techniques. A factorial combination of irrigation treatments and cultivar were arranged in a complete randomized design with three replicates. Results showed that stomatal conductance (g_s) was significantly reduced while photosynthetic performance measured as chlorophyll fluorescence (Fv'/Fm'), relative water content (RWC) and leaf ash content remained unaffected under deficit irrigations. Significant differences among cultivars were found for water use efficiency (WUE), g_s, chlorophyll content (Chl$_{SPAD}$), normal difference vegetation index (NDVI), leaf ash content and fruit growth rate. However, cultivar differences in WUE were more accounted for by the regulation of g_s, therefore, g_s could be useful for breeders for screening large numbers of genotypes with higher WUE under deficit irrigation condition. The study result also demonstrated that cultivar with traits that contribute to achieve higher yields under deficit irrigation strategies has the potential to increase WUE.

Keywords: alternate furrow irrigation, chlorophyll content, fruit growth rate, relative leaf water content, stomatal conductance, tomato

1 Introduction

Scarcity of freshwater and recurrent drought is one of the major bottlenecks that limit agricultural production in most arid and semi-arid regions of Ethiopia (Amede, 2015). Though Ethiopia has large water reserves that could be used for a wide range of irrigation development, statistical figures show that from estimated 5.3 Mio. ha irrigable land, less than 5 % is currently equipped for irrigation (Awulachew *et al.*, 2010).

It was shown that out of four of the most widespread tomato cultivars in Ethiopia, two, namely *Cochoro* and *Fetan*, performed well under alternate furrow irrigation (AFI), while *Chali* and *ARP Tomato d2* performed relatively better under full irrigation (FI). This data is an important baseline both for breeders to im-

* Corresponding author
Email: Ashinie.Gonfa@uni-hohenheim.de

prove drought resistance in tomatoes and also for producers to choose the optimal variety dependent on water availability. However, it is also of crucial importance to understand the underlying physiological functions of tomato subjected to AFI. In general, it is assumed that in AFI, as in as in partial root-zone drying irrigation (PRD), deliberately irrigates only part of the root-zone, while the remainder is allowed to dry and alternating subsequently these wet and dry zones in the next irrigation events, an increase of WUE comes through tight regulation of stomatal aperture as plants open their stomata for CO_2 uptake and at the same time lose water (Kang & Zhang, 2004). Consequently, biomass production may be reduced as gas exchange is restricted due to stomatal closure causing water savings. One of the options for the potential adaptation to such a situation is the use of genotypes with higher WUE.

Different response of cultivars to deficit irrigation treatments has been reported in tomatoes (Mahadeen et al., 2011; Patanè et al., 2014), which have been described in terms of total yield and WUE. The study carried out by Patanè et al. (2014) indicated that the cv. 'Season' exhibited nearly twice greater efficiency in the use of total water available compared to 'Solerosso' with an equal amount of water savings. Barrios-Masias & Jackson (2016) also reported that 'CXD255' had a 10 % greater WUE than 'AB2'. The differences among cultivars may be related to water economy through regulating their stomata aperture and maintaining their leaf water status (Riccardi et al., 2016). Understanding physiological responses to irrigation associated within different climate and soil conditions is helpful in transferring different irrigation technologies to farmers and optimize regional water management (Morison et al., 2008). Studies have shown that stomatal conductance was reduced while the photosynthetic rate was not greatly impaired under PRD treated plants (Campos et al., 2009; Yang et al., 2012). Substantial reduction in stomatal conductance coupled with little effect on photosynthesis, including photosystem efficiency (PSII), could lead to improved crop WUE (Dry & Loveys, 1998). Nevertheless, there is no evidence that a characteristic decrease of stomatal conductance under mild stress condition as deliberately induced in deficit irrigation is associated with a characteristic decrease in Fv/Fm values.

Water deficit also often results in premature induction of leaf senescence and consequently leads to inefficient conversion of resources and finally to yield losses (Ramírez et al., 2014). Leaf chlorophyll concentration and a variation in normalized difference vegetation index (NDVI) are used as integrative method to assess the canopy senescence rates (Rolando et al., 2015). Maintenance of green leaf (a low rate of leaf senescence) under low water availability is a desirable trait because it may reflect the photosynthetic activity and capacity for light harvesting during the fruit development and ripening period leading to larger fruit size.

The final fruit size of horticultural crops determines the quality, profitability and customers' acceptance (Wubs et al., 2012). Studies have shown that deficit irrigation strategies can save substantial irrigation water but caused small fruit size (Favati et al., 2009; Patanè & Cosentino, 2010). Water deficit during the linear fruit growth can often induce a reduction in total final yield due to smaller fruit sizes (Savić et al., 2008). Different genotypes may have different mechanisms to cope with water deficit induced by deliberate deficit irrigations; adoption of drought tolerant genotypes will sustain crop production under low water availability. Therefore, the study was conducted to identify cultivars to be used under water shortage conditions employing different deficit irrigation strategies. The study was also carried out to investigate physiological responses and variation in fruit growth patterns of different cultivars under deficit irrigation techniques. The study will facilitate an evaluation of water saving deficit irrigation strategies with neither compromising fruit yield nor fruit sizes. This second part of the study presents the physiological response of tomato, which allows additional insight to explain the agronomic responses presented in part I.

2 Materials and methods

2.1 Location and experimental setup

The experiment at Melkassa Agricultural Research Center was setup as a randomized complete block design comprising a factorial combination of four cultivars (Fetan, Chali, Cochoro and ARP Tomato d2) and three irrigation treatments with three replications. Irrigation treatments were (1) full irrigation, FI (crop water requirements applied uniformly to all furrows), (2) deficit irrigation, DI (50 % of crop water requirement applied uniformly to all furrows) and (3) alternate furrow irrigation, AFI (50 % of crop water requirement applied to every other furrow and alternating the furrows at each irrigation event). Total crop water requirement was estimated based on soil-water deficit as the difference between measured volumetric soil water content (θAC) and soil water content at field capacity (θFC) multiplied by plot area and rooting depth of 0.60 m. A detailed description of the location, varieties used and agronomic details can be found in part I of the study.

2.2 Plant growth parameters

Data on fruit growth was collected from three tagged plants per plot. Three fully opened flowers on the third truss were tagged for measuring the fruit diameter (FD) at a six days interval using a digital calliper (Harbor Freight Tools, USA) starting from 60 days after transplanting (DAT). Development of standardized fruit diameter (SFD) was calculated between 60 and 102 DAT as:

$$SFD(t) = \frac{FD(t) - FD60}{FD102 - FD60}$$

Fruit-growth duration (FGD) was taken as the time between anthesis and the final harvest. Absolute fruit growth rate (AGR) was also quantified by dividing the fruit diameter at harvest by FGD. Plant height was measured before commencement of deficit irrigation (39 DAT) and at harvest. Stem diameter (girth) of the main stem 5 cm above soil surface was measured using a calliper at ten day intervals starting at 60 DAT.

2.3 Relative leaf water content

Relative leaf water content (RWC) was measured at two occasions from the fully expanded leaves from each plot at 45 and 65 DAT in 2013 and 55 and 65 DAT in 2014. Immediately after cutting at the base of lamina, leaves were sealed within a plastic bag and transferred to the laboratory. Fresh weight (FW) was measured after excision and the full turgid weight (TW) after hydration of the leaves by placing them in a plastic flask containing 100 ml distilled water for 24 h at room temperature (about 21 °C). Dry weight (DW) was measured after oven drying at 70 °C for 72 hrs.

$$RWC = \frac{FW - DW}{TW - DW} * 100$$

2.4 Stomatal conductance (g_s) and chlorophyll fluorescence

Stomatal conductance (g_s) was measured on five fully expanded leaves per plot on the abaxial leaf surface with a steady state diffusion porometer SC-1 (Decagon, USA) on five occasions at ten days interval from 45 through 85 DAT in 2013 and on four occasions in 2014. The measurements were conducted from 12:00 to 14:00. Chlorophyll fluorescence emission kinetics was also measured on three occasions in 2013 and 2014

using a portable fluorometer FluorPen FP 100 (Photon Systems Instruments, Czech Republic) from four most recently fully expanded leaves per plot. Continuous fluorescence yield in non-actinic light adapted initial fluorescence (Ft) and Photosystem II quantum yield efficiency (Fv'/Fm') were used to assess the photosynthetic performance of the four tomato cultivars under different irrigation treatments.

2.5 Chlorophyll contents (Chl_{SPAD}) and normalized difference vegetation index (NDVI)

Chlorophyll content (Chl_{SPAD}) was measured with a portable chlorophyll meter SPAD-502 (Konica Minolta, Japan). The mean values of ten readings per plot were used for analysis. Measurement was done once at fruit development stage (65 DAT) in 2013 and six readings of Chl_{SPAD} were used to assess stay-greenness (senescence rate) at a five days interval starting from 90 through 110 DAT in 2014. Normalized difference vegetation index (NDVI) used to assess the maintenance of green foliage growth under deficit irrigation treatments was also undertaken using a commercial GreenSeeker® portable spectroradiometer (Trimble Navigation, USA) in 2014. Means of five readings of NDVI from the central four rows of each plot at distance of one meter above the canopy were used for analysis. NDVI and SPAD measurements were carried out on the same day.

2.6 Leaf ash content

Ash content, expressed on dry weight basis (%), was determined from leaves after complete combustion of the samples. Samples were collected from each plot, oven-dried for 48 h at 72 °C and grounded to a fine powder. Approximately 3.0 g of the samples were incinerated at 575 °C for 16 h in an electric muffle furnace.

2.7 Data analysis

Analysis of variance (ANOVA) was applied to examine the effects of cultivars and irrigation treatments on different physiological variables using SAS Proc MIXED procedure (SAS 8.02 Cary, NC, USA). Physiological attributes that were measured on two or more occasions during the growing season were also subjected to ANOVA with repeated measurement over time using SAS Proc GLM procedure. Means were compared by LSD ($P < 0.05$). Linear functions were employed and the slopes were used to compare the temporal trends of each physiological attribute among irrigation treatments. Each pair of slopes was compared using Student's t-test. Pearson correlation index was also employed to define the relative predictors of physiological attributes to marketable fruit yield and WUE.

3 Results

3.1 Fruit growth patterns

The dynamics of fruit diameter increase of four cultivars were similar under FI (Fig. 1). However, differences in relative growth pattern among cultivars under deficit irrigation treatments were visible during the time of rapid fruit development (66–84 DAT). Higher relative fruit growth was observed in *Fetan* and *ARP Tomato d2* whereas it was reduced in *Chali* and *Cochoro* under DI and AFI. On the other hand, irrigation treatments did not significantly affect FGD (Table 1). Nevertheless, cultivars differ in both FGD and AGR. The highest AGR recorded in *Fetan* followed by *ARP Tomato d2* whereas the lowest was recorded in *Chali* (Table 1). Thus *Fetan* and *ARP Tomato d2* maintained relatively higher fruit

size under deficit irrigation. *Chali* had shorter FGD and lower fruit size. *Cochoro* maintained a prolonged FGD and relatively higher AGR, resulting in larger fruit under DI and AFI.

3.2 Relative leaf water content (RWC)

RWC was not significantly affected by irrigation treatments, time factors and treatment by time interaction (Table 2 and 3), but cultivars articulated different responses of leaf tissue water status to different irrigation treatments (Table 2 and 3). Only *Chali* exposed lower leaf RWC while no distinct variation was observed among the other three cultivars. RWC was reduced in *Chali* under DI and AFI by 7.4 % and 3.7 % relative to FI, respectively.

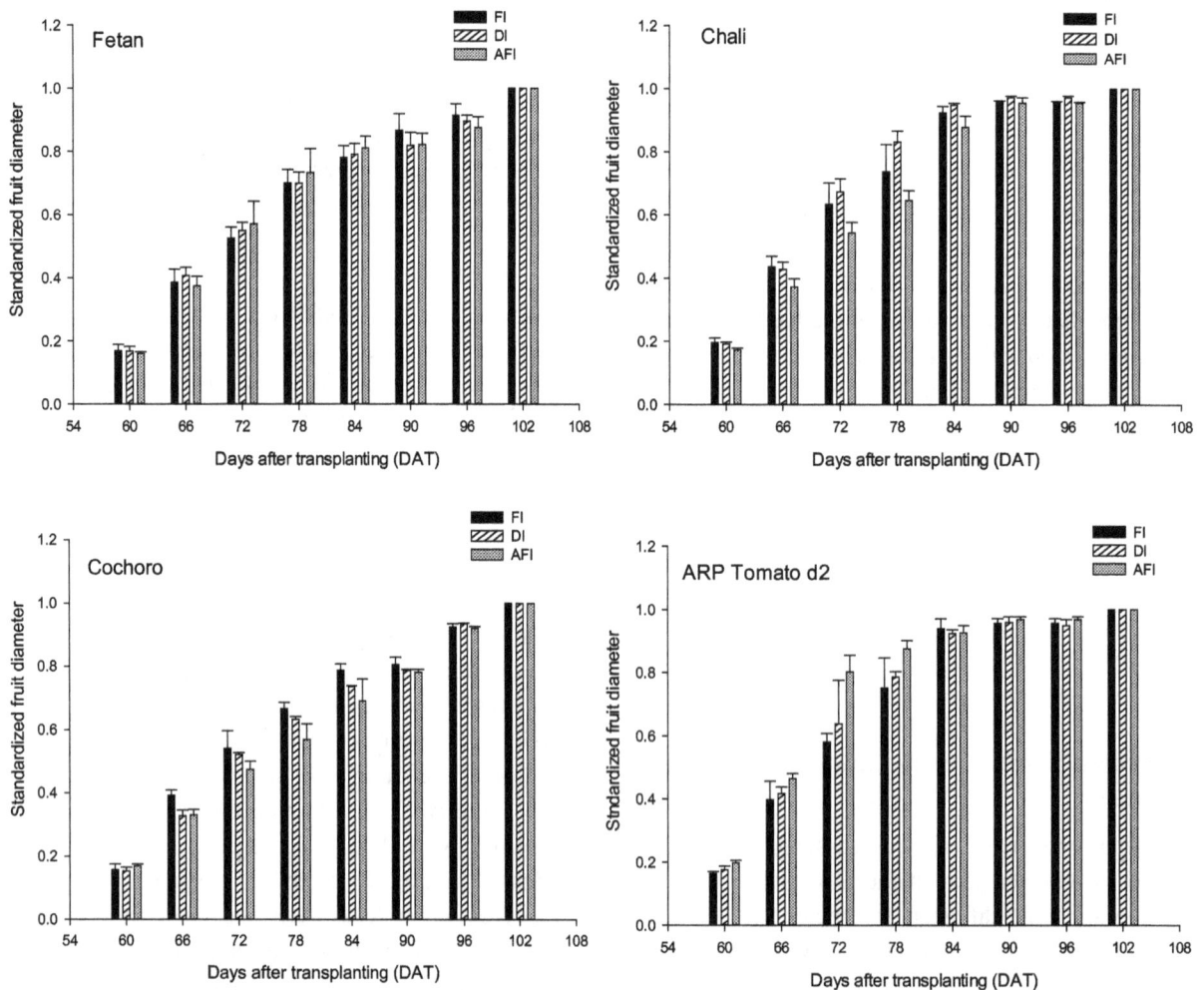

Fig. 1: *Standardized fruit diameter (SFD) between 60 and 102 days after transplanting (DAT) of four tomato cultivars under full irrigation (FI), deficit irrigation (DI) and alternate furrow irrigation (AFI) in 2014.*

Table 1: *Fruit growth duration (FGD) and absolute fruit growth rate (AGR) of four cultivars under full irrigation (FI), regulated deficit irrigation (DI) and alternate furrow irrigation (AFI) in 2014.*

Cultivar	Fruit growth duration (days after anthesis)				Absolute fruit growth rate (mm day^{-1})			
	FI	DI	AFI	Mean	FI	DI	AFI	Mean
Fetan	^{A}46^a	^{A}47^a	^{B}48^a	47	^{A}1.24^b	^{AB}1.14^b	^{A}1.45^a	1.28
Chali	^{B}39^a	^{B}38^a	^{C}38^a	39	^{A}1.14^a	^{B}1.04^a	^{C}1.17^a	1.12
Cochoro	^{A}50^a	^{A}51^a	^{A}50^a	50	^{A}1.18^a	^{A}1.19^a	^{C}1.21^a	1.19
ARP Tomato d2	^{AB}42^a	^{A}43^a	^{B}48^a	48	^{A}1.23^a	^{A}1.23^a	^{B}1.32^a	1.26

Means within column preceded by different capital letters or means within rows followed by different lower-case letters indicate significant differences among cultivars and irrigation treatments, respectively ($P < 0.05$).

Table 2: *Analysis of variance (ANOVA) showing F-value of relative leaf water content (RWC), stomatal conductance (g_s), chlorophyll content (Chl$_{SPAD}$), initial fluorescence (Ft) and quantum yield efficiency (Fv'/Fm') among four cultivars and three irrigation treatments in 2013.*

Physiological Parameter	Cultivar	Irrigation treatments (IT)	Time	IT*Time	Mean Value		
					FI	DI	AFI
RWC	Fetan	0.04 ^{NS}	1.66 ^{NS}	7.76 ***	^{A}0.89^a	^{A}0.90^a	^{A}0.89^a
	Chali	2.17 ^{NS}	1.57 ^{NS}	0.41 ^{NS}	^{B}0.75^a	^{B}0.78^a	^{B}0.80^a
	Cochoro	0.94 ^{NS}	2.94 ^{NS}	0.55 ^{NS}	^{A}0.88^a	^{A}0.87^a	^{B}0.82^a
	ARP Tomato d2	3.22 ^{NS}	0.16 ^{NS}	0.26 ^{NS}	^{A}0.89^a	^{A}0.87^a	^{A}0.93^a
g_s	Fetan	5.89 **	21.83 ***	1.06 ^{NS}	^{B}136.9^a	^{A}125.9^b	^{B}112.3^b
	Chali	23.4 ***	42.39 ***	2.85 *	^{A}164.0^a	^{A}127.9^b	^{A}125.9^b
	Cochoro	24.27 ***	51.61 ***	1.66 ^{NS}	^{B}136.6^a	^{B}114.8^b	^{B}109.6^b
	ARP Tomato d2	5.91 *	49.72 ***	1.88 ^{NS}	^{B}139.2^a	^{B}118.5^b	^{A}124.5^b
Chl$_{SPAD}$†	Fetan	5.13 *	–	–	^{A}44.1^b	^{A}42.9^b	^{AB}48.4^a
	Chali	8.40 **	–	–	^{A}42.3^b	^{A}44.5^b	^{AB}48.3^a
	Cochoro	104.86 ***	–	–	^{A}41.9^b	^{AB}41.5^b	^{A}50.0^a
	ARP Tomato d2	28.98 ***	–	–	^{B}38.1^b	^{B}39.1^b	^{B}46.0^a
Ft	Fetan	1.19 ^{NS}	21.99 ***	0.89 ^{NS}	^{A}247.5^a	^{B}258.0^a	^{B}259.8^a
	Chali	3.78 *	5.19 *	1.26 ^{NS}	^{A}261.7^b	^{A}286.4^{ab}	^{A}300.4^a
	Cochoro	3.75 *	2.86 ^{NS}	1.74 ^{NS}	^{A}259.8^b	^{A}277.5^a	^{B}269.7^{ab}
	ARP Tomato d2	1.21 ^{NS}	32.05 ***	4.09 **	^{A}259.8^a	^{B}256.7^a	^{B}274.2^a
Fv'/Fm'	Fetan	3.42 ^{NS}	16.60 ***	2.45 ^{NS}	^{A}0.57^a	^{A}0.56^a	^{A}0.54^a
	Chali	0.10 ^{NS}	4.19 *	0.80 ^{NS}	^{A}0.56^a	^{A}0.56^a	^{A}0.56^a
	Cochoro	2.66 ^{NS}	6.18 **	2.14 ^{NS}	^{A}0.57^a	^{A}0.56^a	^{A}0.54^a
	ARP Tomato d2	0.54 ^{NS}	8.20 ***	0.30 ^{NS}	^{A}0.56^a	^{A}0.55^a	^{A}0.56^a
Leaf ash content †	Fetan	1.36 ^{NS}	–	–	^{A}0.30^a	^{A}0.30^a	^{A}0.31^a
	Chali	1.26 ^{NS}	–	–	^{A}0.31^a	^{B}0.29^a	^{A}0.30^a
	Cochoro	9.76 **	–	–	^{A}0.30^b	^{A}0.32^a	^{B}0.28^c
	ARP Tomato d2	9.96 *	–	–	^{A}0.31^a	^{B}0.29^b	^{AB}0.29^b

(NS) non significant, *, **, *** significantly different at 5 %, 1 %, 0.1 % probability level, respectively. Means within column preceded by different capital letters or means within rows followed by different lower-case letters indicate significant differences among cultivars and irrigation treatments, respectively ($P < 0.05$). † sampled once during the experimental year.

Table 3: *Analysis of variance (ANOVA) showing F-value of relative leaf water content (RWC), stomatal conductance (g_s), chlorophyll content (Chl_{SPAD}), normalized difference vegetation index (NDVI), initial fluorescence (Ft) and quantum yield efficiency (Fv'/Fm') among four cultivars and three irrigation treatments in 2014.*

Physiological Parameter	Cultivar	Irrigation treatments (IT)	Time	IT*Time	Mean Value FI	DI	AFI
RWC	Fetan	0.04 ^{NS}	0.78 ^{NS}	0.17 ^{NS}	^{A} 0.89 ^{a}	^{A} 0.90 ^{a}	^{A} 0.90 ^{a}
	Chali	2.79 ^{NS}	0.02 ^{NS}	0.32 ^{NS}	^{B} 0.81 ^{a}	^{B} 0.75 ^{a}	^{B} 0.78 ^{a}
	Cochoro	0.08 ^{NS}	1.08 ^{NS}	1.05 ^{NS}	^{A} 0.89 ^{a}	^{A} 0.91 ^{a}	^{A} 0.92 ^{a}
	ARP Tomato d2	1.43 ^{NS}	2.30 ^{NS}	2.36 ^{NS}	^{AB} 0.86 ^{a}	^{A} 0.92 ^{a}	^{A} 0.92 ^{a}
g_s	Fetan	64.9 **	98.54 ***	8.79 ***	^{A} 196.1 ^{a}	^{B} 137.8 ^{b}	^{B} 115.3 ^{c}
	Chali	4251.5 ***	61.45 ***	2.92 ^{NS}	^{A} 199.2 ^{a}	^{A} 160.8 ^{b}	^{A} 134.7 ^{c}
	Cochoro	110.7 *	67.37 ***	5.25 ^{NS}	^{B} 158.5 ^{a}	^{C} 123.4 ^{b}	^{A} 130.6 ^{b}
	ARP Tomato d2	188.1 ***	135.66 ***	8.15 ***	^{A} 192.8 ^{a}	^{C} 126.9 ^{b}	^{A} 138.3 ^{b}
Chl_{SPAD}	Fetan	3.52 ^{NS}	57.4 ***	1.15 ^{NS}	^{A} 43.0 ^{a}	^{B} 42.3 ^{a}	^{B} 42.9 ^{a}
	Chali	17.1 ***	76.9 ***	1.13 ^{NS}	^{B} 33.5 ^{b}	^{C} 31.4 ^{b}	^{B} 39.7 ^{a}
	Cochoro	12.0 ***	64.3 ***	5.14 ***	^{A} 42.4 ^{b}	^{A} 47.7 ^{a}	^{A} 46.5 ^{a}
	ARP Tomato d2	7.15 ***	127.7 ***	2.37 *	^{A} 41.5 ^{a}	^{B} C35.6 ^{b}	^{B} 37.2 ^{b}
NDVI	Fetan	0.11 ^{NS}	0.24 ^{NS}	51.3 ***	^{B} 0.46 ^{a}	^{B} 0.45 ^{a}	^{A} 0.45 ^{a}
	Chali	1.79 ^{NS}	0.43 ^{NS}	82.1 ***	^{B} 0.43 ^{a}	^{C} 0.41 ^{a}	^{B} 0.40 ^{a}
	Cochoro	3.15 ^{NS}	2.63 ^{NS}	68.9 ***	^{A} 0.51 ^{a}	^{A} 0.53 ^{a}	^{A} 0.47 ^{a}
	ARP Tomato d2	2.92 ^{NS}	2.03 ^{NS}	128.9 ***	^{B} 0.48 ^{a}	^{BC} 0.43 ^{a}	^{A} 0.45 ^{a}
Ft	Fetan	0.80 ^{NS}	43.85 ***	0.80	^{C} 259.8 ^{a}	^{A} 264.1 ^{a}	^{B} 260.4 ^{a}
	Chali	2.07 ^{NS}	63.49 ***	10.80 ***	^{B} 270.8 ^{a}	^{A} 264.8 ^{a}	^{B} 262.4 ^{a}
	Cochoro	22.10 ***	25.19 ***	0.63 ^{NS}	^{A} 295.6 ^{a}	^{A} 271.9 ^{b}	^{A} 283.3 ^{ab}
	ARP Tomato d2	16.55 ***	82.05 ***	3.65 **	^{D} 247.1 ^{b}	^{A} 262.9 ^{a}	^{A} 275.6 ^{a}
Fv'/Fm'	Fetan	1.03 ^{NS}	50.96 ***	11.87 ***	^{A} 0.55 ^{a}	^{A} 0.57 ^{a}	^{B} 0.56 ^{a}
	Chali	0.10 ^{NS}	45.26 ***	3.73 **	^{A} 0.57 ^{a}	^{A} 0.55 ^{a}	^{B} 0.55 ^{a}
	Cochoro	1.15 ^{NS}	6.78 **	3.91 ^{NS}	^{A} 0.57 ^{a}	^{A} 0.57 ^{a}	^{A} 0.58 ^{a}
	ARP Tomato d2	0.04 ^{NS}	38.04 ***	4.29 ***	^{A} 0.56 ^{a}	^{A} 0.56 ^{a}	^{B} 0.56 ^{a}
Leaf ash content †	Fetan	14.81 **	–	–	^{A} 0.21 ^{b}	^{A} 0.27 ^{a}	^{A} 0.23 ^{b}
	Chali	1.29 ^{NS}	–	–	^{A} 0.24 ^{a}	^{A} 0.26 ^{a}	^{A} 0.22 ^{a}
	Cochoro	1.69 ^{NS}	–	–	^{A} 0.20 ^{a}	^{B} 0.18 ^{a}	^{B} 0.18 ^{a}
	ARP Tomato d2	3.81 *	–	–	^{A} 0.22 ^{ab}	^{B} 0.20 ^{b}	^{A} 0.25 ^{a}
Stem diameter	Fetan	0.24 ^{NS}	27.88 ***	0.73 ^{NS}	^{A} 12.8 ^{a}	^{A} 12.9 ^{a}	^{A} 13.5 ^{a}
	Chali	0.25 ^{NS}	140.52 ***	2.91 **	^{A} 12.2 ^{a}	^{A} 13.2 ^{a}	^{A} 12.4 ^{a}
	Cochoro	1.92 ^{NS}	28.51 ***	0.13 ^{NS}	^{A} 13.0 ^{a}	^{A} 12.1 ^{a}	^{A} 13.8 ^{a}
	ARP Tomato d2	1.84 ^{NS}	49.92 ***	0.43 ^{NS}	^{A} 12.3 ^{a}	^{A} 13.5 ^{a}	^{A} 13.2 ^{a}

NS is non-significant, *, **, *** significantly different at 5 %, 1 %, 0.1 % probability level, respectively. Means within columns preceded by different capital or means within rows followed by different lower-case letters indicate significant differences among cultivars and irrigation treatments, respectively ($P < 0.05$). † ash content sampled once.

3.3　Stomatal conductance (g_s)

Marked temporal variations were observed for g_s (mmol m^{-2} s^{-1}) among the different irrigation treatments and cultivars in both years (Fig. 2). g_s progressively declined over the course of sampling date (Fig. 2) with maximum and minimum values observed at 45 and 75 DAT, respectively. g_s was significantly reduced by deficit irrigation treatments relative to FI, and the trends of reduction were similar for both DI and AFI (Table 2 and 3). In some cases, however, slightly lower g_s was observed under AFI (Fig. 2a and b), resulting in reduction of 16.7 % and 20.4 % g_s compared to FI under DI and AFI, respectively (Table 2). In 2014, on average, FI plants had g_s values of 186.7 mmol m^{-2} s^{-1} while the corresponding g_s values for DI and AFI plants were 26.5 % and 30.5 % lower (Table 3). Statistically significant differences were found among cultivars with respect to g_s. *Chali* had the highest values of g_s under DI (Table 2 and 3) and *Cochoro* had the lowest. *Fetan* also showed significantly lower values of g_s under AFI. Generally, cultivars with higher g_s under deficit irrigation treatments had lower WUE compared to other cultivars.

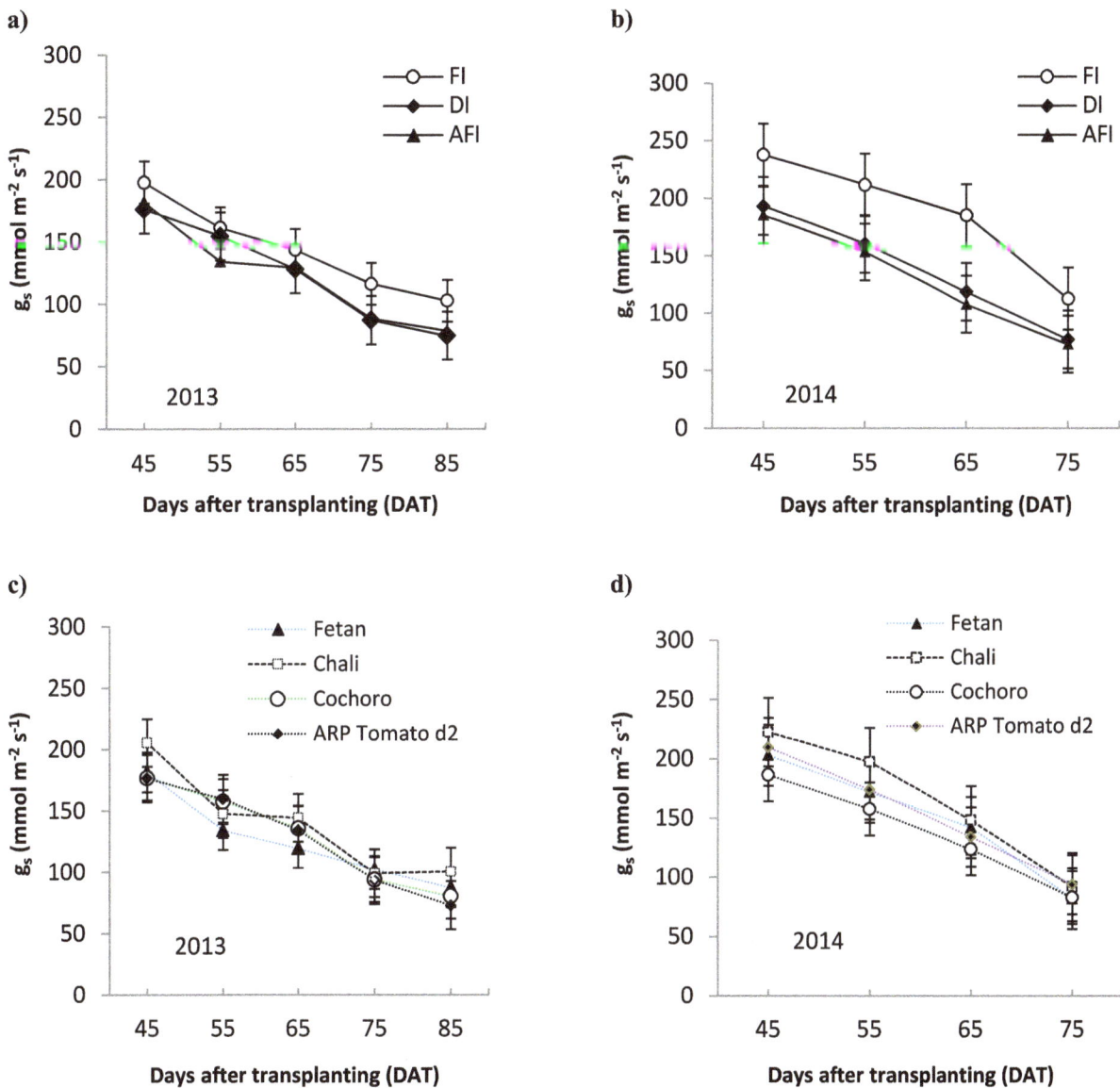

Fig. 2: *Stomatal conductances (g_s) vs. days after transplanting (DAT) of four tomato cultivars (c and d) grown under full irrigation (FI), deficit irrigation (DI), and alternate furrow irrigation (AFI) (a and b) during 2013 and 2014.*

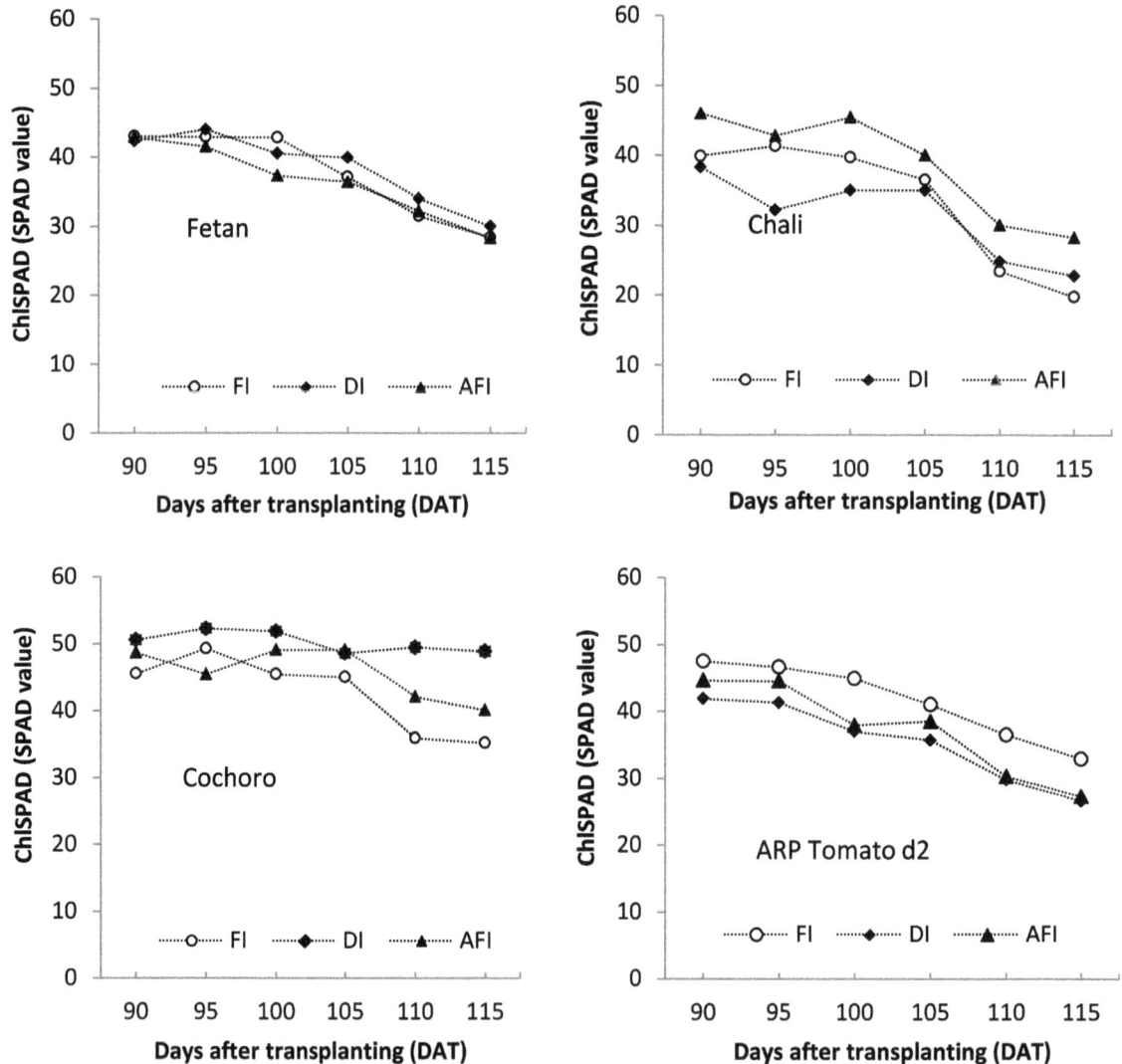

Fig. 3: *Chlorophyll content (Chl$_{SPAD}$) vs. days after transplanting (DAT) of four cultivars under full irrigation (FI), deficit irrigation (DI) and alternate furrow irrigation (AFI) in 2014.*

3.4   Chlorophyll contents (Chl$_{SPAD}$) and NDVI

In 2013, significant differences were found for Chl$_{SPAD}$ between the irrigation treatments ($P < 0.001$) and cultivars ($P < 0.01$) (Table 2). However, irrigation by cultivar interaction was not significant ($P > 0.09$). Plants under AFI maintained significantly higher Chl$_{SPAD}$ values whereas the values of FI and DI did not show any distinct differences. Significant differences in Chl$_{SPAD}$ were found among cultivars under deficit irrigation treatments. *ARP Tomato d2* displayed lower Chl$_{SPAD}$ whereas *Fetan*, *Chali* and *Cochoro* exhibited higher values.

In 2014, the Chl$_{SPAD}$ values progressively declined over time. This trend was observed for all cultivars and irrigation treatments (Fig. 3). Different response of cultivars to irrigation treatments was exhibited by Chl$_{SPAD}$

(Table 3). Except *Fetan*, the Chl$_{SPAD}$ of the other cultivars was significantly affected by irrigation treatments. Chl$_{SPAD}$ values of *Cochoro* and *Chali* were higher under DI and AFI relative to FI, while the values of *ARP Tomato d2* remained higher under FI and lower in deficit irrigation treatments. The comparison of the magnitude of the slope of the linear function of Chl$_{SPAD}$, indicated that significant differences were observed among the irrigation treatments (Table 4). However, no significant difference was detected between the slopes for AFI and FI.

Deficit irrigation treatments had no significant effect on leaf greenness when assessed by NDVI (Table 2), but the slope of NDVI was significantly higher for AFI than for FI and DI (Table 4). But NDVI varied between cultivars (Table 3). *ARP Tomato d2* displayed the most pronounced drop in the green canopy vegetation at late ripening stages while *Cochoro* maintained a relatively stable and higher NDVI (albeit at reduced level) during 95 to 115 DAT (Fig. 4). *Chali* consistently exhibited a lower NDVI.

Table 4: *Linear functions of chlorophyll contents (Chl$_{SPAD}$) and NDVI vs days after transplanting (DAT) for full irrigation (FI), deficit irrigation (DI) and alternate furrow irrigation (AFI) in 2014.*

Parameter	Irrigation treatment	Equation	R^2
Chl$_{SPAD}$	FI	$Y = -0.669x + 172.2$	0.879
	DI	$Y = -0.477x + 87.4$	0.963
	AFI	$Y = -0.565x + 97.5$	0.948
NDVI	FI	$Y = -0.008x + 1.294$	0.922
	DI	$Y = -0.008x + 1.326$	0.892
	AFI	$Y = -0.009x + 1.420$	0.882

3.5 Chlorophyll fluorescence parameters

A temporal trend of light adapted initial fluorescence (Ft) over time under different irrigation treatments in 2013 and 2014 is presented in Fig. 5. The value was consistently higher under both deficit irrigation treatments (Fig. 5a and b) and tended to decline over time (45 to 65 DAT). Irrigation treatment and cultivar had significant effects on Ft in both years. In 2013, no significant differences in Ft were found among cultivars under FI. Ft values of all cultivars were significantly higher under DI and AFI relative to FI (with exception of *ARP Tomato d2* under DI) (Table 2). In 2014, significant differences in Ft were found among cultivars only under FI and AFI (Table 3). Among the cultivars, *Cochoro* exhibited significantly the highest Ft values whereas *ARP Tomato d2* displayed the lowest under FI. The Ft values in *Cochoro* and *ARP Tomato d2* were higher under AFI and did not change in *Fetan* and *Chali* (Table 3).

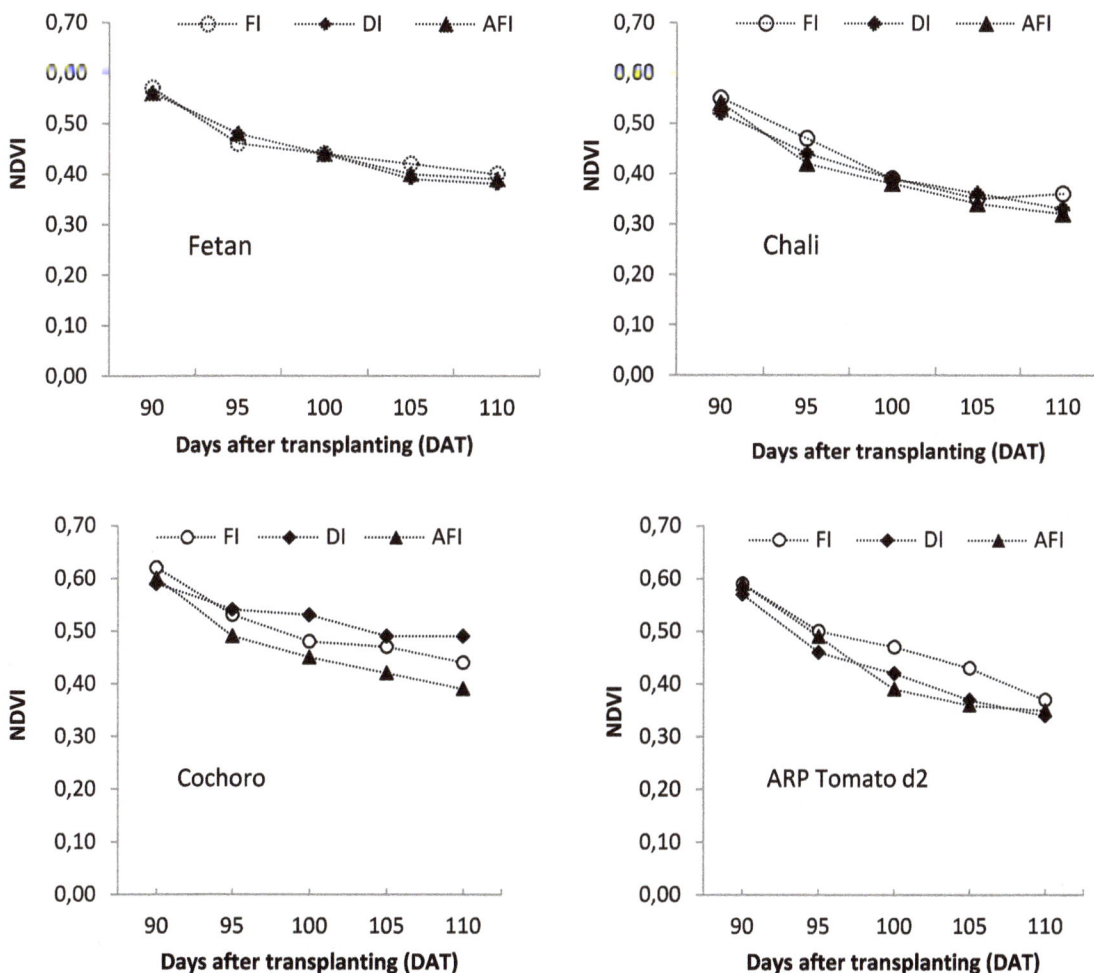

Fig. 4: *Normalized difference vegetation index (NDVI) vs. days after transplanting (DAT) of four cultivars under full irrigation (FI), deficit irrigation (DI) and alternate furrow irrigation (AFI) in 2014.*

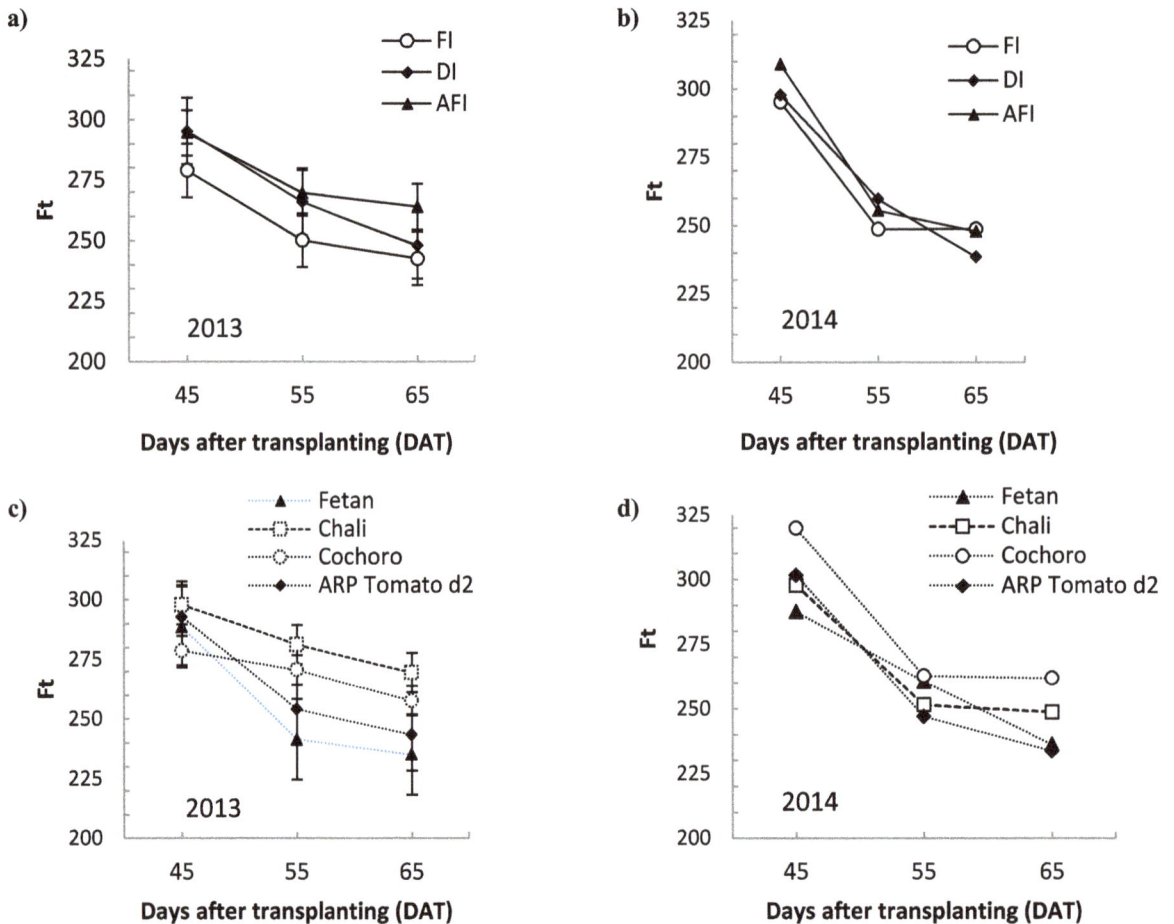

Fig. 5: *Light adapted initial chlorophyll fluorescence (Ft) vs. days after transplanting (DAT) in tomato growing under full irrigation (FI), deficit irrigation (DI) and alternate furrow irrigation (AFI) (a and b) and cultivars (c and d) in 2013 and 2014.*

Fv'/Fm' was not significantly influenced by irrigation and cultivars in 2013 (Table 2). However, significant cultivar differences were observed under AFI in 2014. *Cochoro* had the highest Fv'/Fm' value (0.58) (Table 2).

3.6 Stem diameter and ash content

Stem diameter was not significantly influenced by deficit irrigation treatments or cultivars. There were significant differences ($P < 0.05$) in leaf ash content among cultivars as well as irrigation treatments. However, cultivars variation in leaf ash content was only found under DI and AFI (Table 2 and 3). Overall, *Fetan* maintained the highest leaf ash content (27 %) while *Cochoro* showed the lowest (19.0 %) under DI.

3.7 Relationship among physiological attributes, fruit yield and WUE

There was an inverse relationship between WUE and g_s (Fig. 6). Further, a positive and significant correlation was found between average NDVI and final marketable fruit yield but the correlation with WUE was

non-significant (Table 5). Positive and significant associations between NDVI and Chl_{SPAD} were noted and to each other at all sampling dates.

No correlation was found between ash content and fruit yield or any other physiological trait in 2013 (Table 5), but leaf ash content was significantly negatively correlated with marketable fruit yield and fruit weight in 2014 (Table 6). The correlation with WUE was weak negative. RWC exhibited positive and significant correlation with marketable fruit yield, WUE, fruit number per plant and average fruit weight (Table 5 and 6). Similarly, significant and positive correlations were found between light adapted initial fluorescence (Ft) and maximum quantum yield of PSII (Fv'/Fm') with a number of physiological and agronomic attributes, the latter were presented in part I. Fv'/Fm' was significantly positively correlated with marketable fruit yield, WUE, fruit numbers per plant, Chl_{SPAD} index, NDVI and RWC and inverse correlation with g_s. The positive Fv/Fm correlation with SPAD and NDVI indicated that photosynthetic performance is related to leaf greenness.

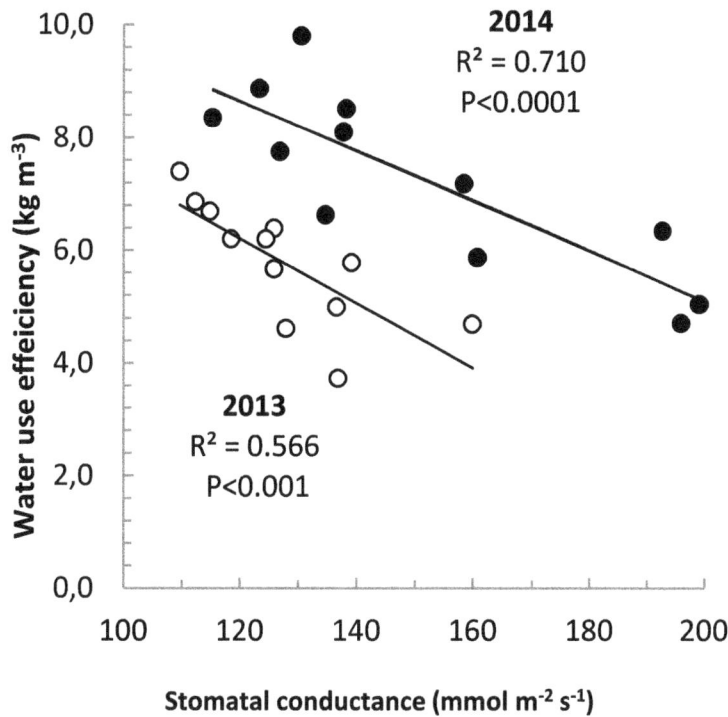

Fig. 6: *Relationship between water use efficiency (WUE) and stomatal conductance (g_s) of four cultivars and three irrigation treatments (full irrigation, regulated deficit irrigation and alternate furrow irrigation).*

Table 5: *Correlation coefficients between physiological attributes and fruit yield and their agronomical components of four tomato cultivars in 2013.*

	MFY	WUE	FN	Avwt
WUE	0.810 ***			
FN	0.606 ***	0.668 ***		
Avwt	0.218 NS	0.081 NS	−0.055 NS	
Chl$_{SPAD}$	0.007 NS	0.261 NS	−0.125 NS	−0.240 NS
RWC	0.362 *	0.370 *	0.079 NS	0.455 **
g_s	−0.007 NS	−0.356 **	−0.148 NS	−0.049 NS
Ash	0.165 NS	0.030 NS	−0.125 NS	−0.109 NS
Ft	0.172 NS	0.373 *	0.316 *	−0.342 *
Fv'/Fm'	0.042 NS	0.230 NS	0.093 NS	0.133 NS

MFY: marketable fruit yield); WUE: irrigation water use efficiency; FN: numbers of fruit per plant; Avwt: average fruit weight; Chl$_{SPAD}$: SPAD chlorophyll content; g_s: stomatal conductance; Ash: ash content %; RWC: leaf relative water content; Ft: light adapted initial fluorescence; Fv'/Fm': PSII quantum yield efficiency of light adapted leaf;
NS: non significant, *, **, ***: significantly different at 5 %, 1 %, 0.1 % probability level, respectively.

Table 6: *Correlation coefficients between physiological attributes and fruit yields and their agronomical components of four tomato cultivars in 2014.*

	MFY	WUE	FN	Avwt
WUE	0.751 ***			
FN	0.500 **	0.490 **		
Avwt	0.213 NS	0.106 NS	−0.551 ***	
Chl$_{SPAD}$	0.540 ***	0.425 **	−0.072 NS	0.522 ***
NDVI	0.526 ***	0.278 NS	0.100 NS	0.437 **
g_s	−0.156 NS	−0.730 ***	−0.235 NS	−0.096 NS
Ash	−0.317 *	−0.227 NS	0.073 NS	−0.359 *
SDM	0.061 NS	0.189 NS	0.284 NS	−0.257 NS
RWC	0.398 **	0.474 ***	0.437 **	0.117 NS
Ft	0.363 *	0.400 *	0.395 *	−0.009 NS
Fv'/Fm'	0.465 **	0.492 **	0.414 **	−0.064 NS

MFY: marketable fruit yield; WUE: irrigation water use efficiency; FN: numbers of fruit per plant; Avwt: average fruit weight; Chl$_{SPAD}$: SPAD chlorophyll content; NDVI: normalized difference vegetation index; g_s: stomatal conductance; Ash (ash content, %); SDM: stem diameter; RWC: relative water content; Ft: light adapted initial fluorescence; Fv'/Fm': PSII quantum yield efficiency of light adapted leaf;
NS: non significant, *, **, ***: significantly different at 5 %, 1 %, 0.1 % probability level, respectively.

4 Discussion

Whenever water deficit occurs in plants, the water balance is disturbed and leaf RWC and water potential usually decreases. Indeed, significant difference was found among cultivars in RWC under deficit irrigation. *Fetan* and *ARP Tomato d2*, *Cochoro* consistently displayed the highest RWC and lowest g_s. The reduction in stomatal conductance under both deficit irrigated treatments was not related to plant water status as there was no significant change in leaf RWC between irrigation techniques. It may be assumed that root borne abscisic acid (ABA) triggered the stomatal closure; a well-documented effect to cause substantial water savings in PRD (Sun *et al.*, 2013; Wang *et al.*, 2010).

Though substantial variations exhibited between cultivars, all cultivars responded to water deficit irrigation by lowering g_s (Fig. 2c and d). Since g_s controls both photosynthetic assimilation and transpiration in plants, it was found as the main factor affecting WUE under moderate drought stress (Mei *et al.*, 2013). In this study, *Cochoro* gave the highest fruit yield and WUE along with the lowest g_s under deficit irrigation. This indicates that the cultivar combines higher productivity, WUE and stress avoidance mechanism. In contrast, *Chali* exhibited a lower fruit yield and WUE but with the highest g_s showing little adaptability to water stress. While g_s was considerably reduced under both deficit irrigation treatments (Fig. 2a and b), the photosynthetic capacity of the four cultivars, determined as Fv'/Fm', was not reduced under AFI and DI (Table 2 and 3). Several studies have also shown that g_s was substantially reduced while the photosynthetic rate was not greatly affected under deficit irrigation particularly in PRD treated plants (Campos *et al.*, 2009; Yang *et al.*, 2012). Mild drought associated with stomatal closure will often not result in a reduction of a photosynthetic capacity (Baker & Rosenqvist, 2004). Therefore, the apparent reduction in g_s of different cultivars is often greater than a respective change in photosynthetic rate under deficit irrigations resulting in differential WUE responses of the cultivars.

Under water deficit, *Fetan* and *ARP Tomato d2* had the highest relative fruit growth (Fig. 1). Remarkably, those cultivars maintain a higher growth rate under water deficit during the time of rapid fruit growth. Besides, *Fetan* was observed to react by pronounced stomatal closure and maintained higher yields under deficit irrigation. *Cochoro* had medium fruit growth rate but had longer fruit growth duration (Table 1), associated with the highest yield potential, while *Chali* had the lowest fruit growth rate and shortest duration, and produced smaller fruit under deficit irrigation. Differences in fi-

nal fruit size among cultivars are effect of differences in fruit growth rates and fruit growth duration, concurring with earlier findings (Okello *et al.*, 2015). This was illustrated by spectral reflectance measurements of the leaf greenness of individual plants and at canopy level (Fig. 3 and 4). Cultivars which had higher Chl_{SPAD} and NDVI values throughout the fruit development and ripening stages had higher fruit yields due to prolonged leaf metabolic activity. It can be seen from this study that the highest yielding cultivar, *Cochoro* maintained the highest Chl_{SPAD} and NDVI whereas the low yielding cultivar, *Chali* had the lowest under DI and AFI, concurring with other previous studies, who reported that genotypes which delayed senescence rate often out yielded genotypes without this trait (Lu *et al.*, 2011; Lopes & Reynolds, 2012; Rolando *et al.*, 2015). On the other hand, *ARP Tomato d2* displayed a higher level of Chl_{SPAD} and NDVI under FI, and this cultivar performed best in areas with better water availability. As Chl_{SPAD} and NDVI are significantly correlated with average fruit weight (Table 6), differences in N mobilisation might partly explain genotype differences in delayed senescence (Shahnazari *et al.*, 2008; Gregersen *et al.*, 2013). Besides, significant positive correlations between photosynthetic performance measured by Fv'/Fm', Chl_{SPAD} and NDVI indicate that plants can maintain source activity by delaying leaf senescence rate under deficit irrigation.

Defining the relationship between physiological functions and agronomic performance of different genotypes can be used to identify physiological traits that can assist as indirect selection criteria of genotypes with good adaptability to water deficit. Significant positive associations found between NDVI and Chl_{SPAD} indicate that both physiological tools are good indicators for assessing the greenness or delayed senescence rate of different genotypes, and may be used either conjunctively or separately for phenotyping physiological traits. Besides, the heritability of NDVI was reported to be higher than that of grain yield in maize (Lu *et al.*, 2011), indicating its usefulness in the selection of secondary traits under water-limited environments. Leaf ash content (Misra *et al.*, 2010; Glenn, 2014) and stem diameter (Bhattarai & Midmore, 2007) were previously suggested as surrogates of WUE under water deficit, but significant correlations were not found between those traits and WUE in our study. Previous studies also showed that significant relationship between ash content and WUE was found only in drier environments and such relationships tend to disappear under moderate to non-water stress conditions (Monneveux *et al.*, 2005; Cabrera-Bosquet *et al.*, 2009). The result imply that the relationship between ash and yield depends more on stress intensity than stress modality and, consequently, ash content and stem diameter could not be reliable putative physiological reference for WUE under mild-water stress of AFI and DI.

5 Conclusion

The present study evidences that the differential responses of four tomato cultivars in terms of total fruit yield and WUE were ascribed to cultivars differences in physiological attributes. However, the overall differences in WUE were more accounted for by the variations in stomatal conductance (g_s). Therefore, g_s could be useful for breeders for screening large numbers of genotypes with higher WUE under deficit irrigation condition. Nevertheless, employing deficit irrigation strategies needs suitable cultivars that possess traits achieving higher yields and WUE. Due to combing higher productivity, WUE and stress adaptation mechanisms, the cultivar *Cochoro* can be used both, by producers and breeders under low water availability.

Acknowledgements

We greatly acknowledge the financial support by Dr. Hermann Eiselen PhD Grant of fiat panis Foundation for field research as well as the scholarship grant by German Academic Exchange Service (DAAD) and German Federal Ministry for Economic Cooperation and Development (BMZ). The authors would also like to express their gratitude to the Melkassa Agricultural Research Center for support in performing the field experiment.

References

Amede, T. (2015). Technical and institutional attributes constraining the performance of small-scale irrigation in Ethiopia. *Water Resources and Rural Development*, 6, 78–91.

Awulachew, S., Erkossa, T. & Namara, R. (2010). Irrigation potential in Ethiopia: Constraints and opportunities for enhancing the system. Unpublished Report to the Bill and Melinda Gates Foundation.

Baker, N. R. & Rosenqvist, E. (2004). Applications of chlorophyll fluorescence can improve crop production strategies: an examination of future possibilities. *Journal of Experimental Botany*, 55 (403), 1607–1621.

Barrios-Masias, F. H. & Jackson, L. E. (2016). Increasing the effective use of water in processing tomatoes through alternate furrow irrigation without a yield decrease. *Agricultural Water Management*, 107–117.

Bhattarai, S. P. & Midmore, D. J. (2007). Carbon Isotope Discrimination and Other Surrogates of Water Use Efficiency for Tomato Under Various Soil Moistures. *International Journal of Vegetable Science*, 13, 19–40.

Cabrera-Bosquet, L., Sánchez, C. & Araus, J. L. (2009). How yield relates to ash content, $\Delta^{13}C$ and $\Delta^{18}O$ in maize grown under different water regimes. *Annals of Botany*, 1004, 1207–1216.

Campos, H., Trejo, C., Peña-Valdivia, C. B., Ramírez-Ayala, C. & Sánchez-García, P. (2009). Effect of partial rootzone drying on growth, gas exchange, and yield of tomato (*Solanum lycopersicum* L.). *Scientia Horticulturae*, 120, 493–499.

Dry, P. & Loveys, B. (1998). Factors influencing grapevine vigour and the potential for control with partial rootzone drying. *Australian Journal of Grape and Wine Research*, 4, 140–148.

Favati, F., Lovelli, S., Galgano, F., Miccolis, V., Di Tommaso, T. & Candido, V. (2009). Processing tomato quality as affected by irrigation scheduling. *Scientia Horticulturae*, 122, 562–571.

Glenn, D. M. (2014). An analysis of ash and isotopic carbon discrimination ($\Delta^{13}C$) methods to evaluate water use efficiency in apple. *Scientia Horticulturae*, 171, 32–36.

Gregersen, P. L., Culetic, A., Boschian, L. & Krupinska, K. (2013). Plant senescence and crop productivity. *Plant Molecular Biology*, 82 (6), 603–622.

Kang, S. & Zhang, J. (2004). Controlled alternate partial root-zone irrigation: its physiological consequences and impact on water use efficiency. *Journal of Experimental Botany*, 55, 2437–2446.

Lopes, M. S. & Reynolds, M. P. (2012). Stay-green in spring wheat can be determined by spectral reflectance measurements (normalized difference vegetation index) independently from phenology. *Journal of Experimental Botany*, 63, 3789–3798.

Lu, Y., Hao, Z., Xie, C., Crossa, J., Araus, J.-L., Gao, S., Vivek, B. S., Magorokosho, C., Mugo, S., Makumbi, D., Taba, S., Pan, G., Li, X., Rong, T., Zhang, S. & Xu, Y. (2011). Large-scale screening for maize drought resistance using multiple selection criteria evaluated under water-stressed and well-watered environments. *Field Crops Research*, 124, 37–45.

Mahadeen, A. Y., Mohawesh, O. E., Al-Absi, K. & Al-Shareef, W. (2011). Effect of irrigation regimes on water use efficiency and tomato yield (*Lycopersicon esculentum* Mill.) grown in an arid environment. *Archives of Agronomy and Soil Science*, 57, 105–114.

Mei, X.-r., Zhong, X.-l., Vincent, V. & Liu, X.-y. (2013). Improving Water Use Efficiency of Wheat Crop Varieties in the North China Plain: Review and Analysis. *Journal of Integrative Agriculture*, 12 (7), 1243–1250.

Misra, S. C., Shinde, S., Geerts, S., Rao, V. S. & Monneveux, P. (2010). Can carbon isotope discrimination and ash content predict grain yield and water use efficiency in wheat? *Agricultural Water Management*, 97, 57–65.

Monneveux, P., Reynolds, M. P., Trethowan, R., González-Santoyo, H., Peña, R. J. & Zapata, F. (2005). Relationship between grain yield and carbon isotope discrimination in bread wheat under four water regimes. *European Journal of Agronomy*, 22, 231–242.

Morison, J. I., Baker, N. R., Mullineaux, P. M. & Davies, W. J. (2008). Improving water use in crop production. *Philosophical Transactions of the Royal Society of London B: Biological Sciences*, 363, 639–658.

Okello, R. C., de Visser, P. H., Heuvelink, E., Lammers, M., de Maagd, R. A., Struik, P. C. & Marcelis, L. F. (2015). A multilevel analysis of fruit growth of two tomato cultivars in response to fruit temperature. *Physiologia plantarum*, 153, 403–418.

Patanè, C. & Cosentino, S. L. (2010). Effects of soil water deficit on yield and quality of processing tomato under a Mediterranean climate. *Agricultural Water Management*, 97, 131–138.

Patanè, C., La Rosa, S., Pellegrino, A., Sortino, O. & Saita, A. (2014). Water productivity and yield response factor in two cultivars of Processing tomato as affected by deficit irrigation under semi-arid climate conditions. *Acta Horticulturae*, 1038, 449–454.

Ramírez, D. A., Yactayo, W., Gutiérrez, R., Mares, V., De Mendiburu, F., Posadas, A. & Quiroz, R. (2014). Chlorophyll concentration in leaves is an indicator of potato tuber yield in water-shortage conditions. *Sci-*

entia Horticulturae, 168, 202–209.

Riccardi, M., Pulvento, C., Patanè, C., Albrizio, R. & Barbieri, G. (2016). Drought stress response in long-storage tomatoes: Physiological and biochemical traits. *Scientia Horticulturae*, 200, 25–35.

Rolando, J. L., Ramírez, D. A., Yactayo, W., Monneveux, P. & Quiroz, R. (2015). Leaf greenness as a drought tolerance related trait in potato (*Solanum tuberosum* L.). *Environmental and Experimental Botany*, 110, 27–35.

Savić, S., Stikić, R., Radović, B. V., Bogičević, B., Jovanović, Z. & Šukalović, V. H.-T. (2008). Comparative effects of regulated deficit irrigation (RDI) and partial root-zone drying (PRD) on growth and cell wall peroxidase activity in tomato fruits. *Scientia Horticulturae*, 117, 15–20.

Shahnazari, A., Ahmadi, S. H., Laerke, P. E., Liu, F., Plauborg, F., Jacobsen, S.-E., Jensen, C. R. & Andersen, M. N. (2008). Nitrogen dynamics in the soil-plant system under deficit and partial root-zone drying irrigation strategies in potatoes. *European Journal of Agronomy*, 28, 65–73.

Sun, Y., Feng, H. & Liu, F. (2013). Comparative effect of partial root-zone drying and deficit irrigation on incidence of blossom-end rot in tomato under varied calcium rates. *Journal of Experimental Botany*, 64, 2107–2116.

Wang, Y., Liu, F., Andersen, M. N. & Jensen, C. R. (2010). Improved plant nitrogen nutrition contributes to higher water use efficiency in tomatoes under alternate partial root-zone irrigation. *Functional Plant Biology*, 37, 175–182.

Wubs, A. M., Ma, Y. T., Heuvelink, E., Hemerik, L. & Marcelis, L. F. (2012). Model Selection for Nondestructive Quantification of Fruit Growth in Pepper. *Journal of the American Society for Horticultural Science*, 137, 71–79.

Yang, L., Qu, H., Zhang, Y. & Li, F. (2012). Effects of partial root-zone irrigation on physiology, fruit yield and quality and water use efficiency of tomato under different calcium levels. *Agricultural Water Management*, 104, 89–94.

6

Meat yield and quality of Tanzania Shorthorn Zebu cattle finished on molasses/maize grain with agro-processing by-products in 90 days feedlot period

Lovince Asimwe [a,*], Abiliza Kimambo [a], Germana Laswai [a], Louis Mtenga [a], Martin Weisbjerg [b], Jorgen Madsen [c], John Safari [d]

[a] *Department of Animal Science and Production, Sokoine University of Agriculture, Morogoro, Tanzania*
[b] *Department of Animal Science, AU Foulum, Aarhus University, Tjele, Denmark*
[c] *Department of Larger Animal Sciences, University of Copenhagen, Frederiksberg, Denmark*
[d] *Institute of Rural Development Planning, Dodoma, Tanzania*

Abstract

This study was conducted to evaluate the effects of feeding molasses or maize grain with agro-processing by-products on yield and quality of meat from Tanzania shorthorn zebu (TSZ) cattle. Forty five steers aged 2.5 to 3.0 years with 200 ± 5.4 kg body weight were allocated into five dietary treatments namely hominy feed with molasses (HFMO), rice polishing with molasses (RPMO), hominy feed with maize meal (HFMM), rice polishing with maize meal (RPMM) and maize meal with molasses (MMMO). *Ad libitum* amount of each dietary treatment and hay were offered to nine steers for 90 days. Cooking loss (CL) and Warner Bratzler shear force (WBSF) values were determined on *M. longissimus thoracis et lumborum* aged for 3, 6, 9 and 12 days. Steers fed on HFMO diet had higher ($P < 0.05$) nutrient intake (86.39 MJ/d energy; 867 g/d CP), weight gain (919 g/d) and half carcass weight (75.8 kg) than those fed other diets. Meat of steers from all diets was tender with average WBSF values of 47.9 N cm^{-2}. The CL (22.0 ± 0.61 %) and WBSF (53.4 ± 0.70 N cm^{-2}) were highest in meat aged for 3 days followed by 6, 9 and 12 days. WBSF values for meat aged for 9 and 12 days from steers fed HFMO and RPMM diets were similar and lower than those on other dietary treatments × aging periods. Overall, molasses and hominy feed can be used to replace maize meal in feedlot finishing diets to spare its use in animal feeds.

Keywords: agro-processing by-products, feedlot, retail cuts, steers

1 Introduction

Meat quality is a complex parameter with varying properties and is influenced by several intrinsic and extrinsic factors. The intrinsic factors include breed, sex, and slaughter weight, whereas pre- and post-slaughter carcass handling, type and feeding level are some of the extrinsic factors (Mushi *et al.*, 2009; Safari, 2010). In Tanzania, meat is mainly produced from Tanzania shorthorn zebu (TSZ) cattle which are extensively managed on range lands. In most of these range lands, pastures are in short supply with low nutritive values especially during the dry seasons resulting in low animal growth rates, and late attainment of slaughter weights. Studies have reported loss in weight leading to poor body condition which affects meat yield (du Plessis & Hoffman, 2004) and quality (Muchenje *et al.*, 2008). The constraint of poor nutrition in beef production could

* Corresponding author
Department of Animal Science and Production, Sokoine University of Agriculture, P.O. Box 3004, Morogoro, Tanzania
Email: lovinceasimwe@yahoo.com

be minimised by finishing animals in feedlot using diets based on agro-processing by-products. Rice polishing and hominy feed are agro-processing by-products that can substitute part of maize grain in concentrate feed for beef cattle (Matz, 1991). A study by Sani *et al.* (2014) on Bunaji bulls fed rice offal as energy source showed rice offal to yield high quality carcass with lean meat (less amount of fat) compared to maize offal. Rice polishing and hominy feed have been found to give consistent results when combined with molasses and urea (Lopez & Preston, 1977; Larson *et al.*, 1993). A study by Hunter (2012) on beef cattle found that inclusion of sugar cane molasses in feedlot diets has the potential of producing meat with good eating quality. Earlier study in Tanzania by Mwilawa (2012) on beef fattening using maize meal and molasses has shown that TSZ cattle responds well to fattening by increasing carcass quantity from 90 kg for sole grazing group to 154 kg for maize meal and molasses fed group. Similarly, the quality of meat was improved from 60 N cm^{-2} in the sole grazing group to 45 N cm^{-2} for maize meal and molasses fed group. However, the major limitation of using maize meal as cattle diet in Tanzania is the role it plays as the main food source for humans. Thus, it is of interest to find out alternatives to maize meal as feedlot diet which is less competitive to humans' food. The available by-product feeds that could be used for finishing cattle under feedlot at national level were estimated to be (tones DM) maize bran (5.6×10^5), rice-polishing (8.9×10^4), cotton seedcake (6.2×10^4), wheat-bran (1.4×10^4), sunflower seedcake (2.1×10^4), wheat-pollard (4.2×10^3) and molasses (4.5×10^4) (Nandonde, 2008). We hypothesize that molasses with other agro-processing by-products can substitute maize meal in feedlot diets without reducing meat yield and quality. Therefore, the objective of this study was to assess the yield and quality of meat produced from TSZ cattle finished on molasses or maize meal with agro-processing by-products under feedlot conditions.

2 Materials and methods

2.1 Animals and diets

A total of 45 TSZ steers aged 2.5–3.0 years with initial body weight 200 ± 5.4 kg (mean ± SEM) were allocated in a completely randomised design to feedlot experiment performed at Kongwa ranch, located in Dodoma region, central Tanzania. These animals were grazing on pastures available at the ranch area for back grounding before starting the experiment. Animals were also assessed for health and nutritional status prior to the experiment to minimise variation errors. Five concentrate diets were formulated from different proportions of the raw materials of the agro-processing by products to contain hominy feed with molasses (HFMO), rice polishing with molasses (RPMO), hominy feed with maize meal (HFMM), rice polishing with maize meal (RPMM) and a control maize meal with molasses (MMMO). The proportions of the ingredients composition shown in Table 1 were determined by trial and error method using nutritive values of ingredients from feedstuff tables for ruminants developed by Doto *et al.* (2004). Each formulated diet was fed together with hay *ad libitum* to 9 steers for a period of 90 days. All animals were provided with free access of fresh drinking water throughout the experiment and were housed in individual pens. During the experimental period, all animals were weighed after every two weeks and the amount of feed eaten was weighed daily. The amount of consumed compounded diet was 6.20 kg DM/day for MMMO, 6.91 kg DM/day for HFMO, 7.05 kg DM/day for RPMO, 5.29 kg DM/day for HFMM and 6.18 kg DM/day for RPMM diet while the consumed hay was 1.20 kg DM/day for MMMO, 1.15 kg DM/day for HFMO, 1.24 kg DM/day for RPMO, 1.31 kg DM/day for HFMM and 1.32 kg DM/day for RPMM diet. The metabolisable energy (ME) content of hay was 4.8 MJ/kg DM while the crude protein (CP) content was 33 g/kg DM. The ME contents of compounded feeds were estimated using the equation by MAFF (1975), that was ME (MJ/kg DM) = 0.012 CP + 0.031 EE + 0.005 CF + 0.014 NFE, whereas that of hay was determined by the McDonald *et al.* (2002) equation: ME (MJ/kg DM) = 0.016 DOMD, where DOMD = g digestible organic matter per kg dry matter.

2.2 Slaughter procedures

After 90 days of feeding, all animals were weighed for three days consecutively to obtain the final body weight and transported by truck to Dodoma Abattoir (82 km from Kongwa ranch). At the abattoir, the animals were deprived of feed for 24 hours, but had access to fresh drinking water. Slaughtering and dressing of the animals followed the abattoir procedures as described by Bourguet *et al.* (2011). The animals were stunned using electrical stunner, slaughtered and suspended on an overhead rail system for bleeding, de-hiding and evisceration. The head was removed at the atlanto-occipital joint, the forefeet and hind feet were removed at the carpal–metacarpal and tarsal–metatarsal joints, respectively. The dressed carcasses were halved into two us-

Table 1: *Ingredient composition (kg/100 kg), Crude protein (g/kg DM) and Metabolisable energy (MJ/kg DM) levels of the dietary treatments*

Ingredients	Compounded diets				
	MMMO	*HFMO*	*RPMO*	*HFMM*	*RPMM*
Ingredient composition (kg/100 kg as fed)					
Maize meal	38	–	–	38	38
Hominy feed	–	40	–	50	–
Rice polishing	–	–	41	–	51
Molasses	47	47	47	–	–
Cotton seed cake	13	11	10	10	09
Mineral mix	1	1	1	1	1
Urea	0.5	0.5	0.5	0.5	0.5
Salt	0.5	0.5	0.5	0.5	0.5
Nutritive values					
Crude protein (g/kg DM)	122	120	97	145	111
Metabolisable energy (MJ/kg DM)	11.0	11.7	10.1	12.6	10.2

MMMO: maize meal with molasses; HFMO: hominy feed with molasses; RPMO: rice polishing with molasses; HFMM homing feed with maize meal; RPMM rice polishing with maize meal.

ing an electrical saw, weighed within 45 minutes post-mortem to obtain the hot carcass weight. Temperature and pH measurements were recorded after 45 minutes and 6 hours post-mortem at room temperature then carcasses were transferred to a cold room set at 0°C where further measurements after 24 hours and 48 hours were recorded.

2.3 Fabrication of retail cuts

Twenty five (25) right half carcasses (five carcasses from each treatment (5*5)) were randomly selected from the total of 45 TSZ steers for retail cuts fabrication. After 48 hours in a chilling room set at 0°C, the carcasses were weighed to obtain cold carcass weight and then quartered between 10th and 11th ribs into fore and hind quarters. After recording the weights, the hind quarters were fabricated into nine bone-in retail cuts as described by FAO (1991), the cuts were hind shin, topside, silverside, fillet, rump, loin strip, thin flank, wing ribs and thick flank. All visible fat were trimmed and similar procedure was carried out with the fore quarters, which were fabricated into eight bone-in retail cuts as brisket, prime ribs, hump, flat ribs, chuck, shoulder blade, neck and fore shin. The weight of each individual cut was recorded. The retail cuts from the fore and

hind quarters were grouped as prime and nonprime cuts based on guidelines issued by VETA (2006). The prime cuts comprised the following cuts: shoulder blade, loin strip, prime ribs, rump, wing ribs, topside, silverside, thick flank and fillet. The nonprime cuts included fore and hind shin, hump, neck, flat ribs, chuck, brisket, and thin flank.

2.4 Meat quality measurements

2.4.1 Temperature and pH measurements

Temperature and pH of the carcasses were taken at 45 min, 6, 24 and 48 hours post-mortem at the same point on the 10th rib of left side carcass in the *M. longissimus thoracis et lumborum* (LL). The temperature was measured by inserting a digital meat thermometer (FUNKO-TION Digital stegetermometer, HA 250K, Japan), while the pH was measured by inserting an electrode (Mettler Toledo) of a portable pH-meter (Knickportamess 911, Germany) in the same muscle. The temperature and pH readings at 45 min and 6 hours post-mortem were taken at room temperature while temperature and pH readings at 24 and 48 hours post-mortem were taken when the carcasses were in the chilling room.

2.4.2   Determination of cooking loss (CL) and Warner-Bratzler shear force (WBSF)

At the abattoir the LL muscle from 7th to 13th rib of the left side carcass was removed 48 hours post-mortem and prepared for cooking loss (CL) and Warner-Bratzler shear force (WBSF) determination. The LL muscle was cut into 4 pieces measuring approximately 9 cm long which were labelled and placed in the chilling room set at 0°C for ageing for 3, 6, 9 and 12 days. After each ageing time the samples were taken from the chilling room, vacuum packed in polythene bags and frozen at −20°C in a deep freezer. Thereafter the samples were transported to the laboratory at the Department of Animal Science and Production of Sokoine University of Agriculture, where CL and WBSF values were determined. The LL muscle samples were thawed at 4°C overnight, removed from the polythene bags, wiped up with paper towel trimmed down to reduce the size to approx mean weight ($W1$) of 298 ± 36 g and then re-sealed using a vacuum pack machine. The samples were heated at 75°C for 1 hour in a thermostatically controlled water bath. The heated samples were left to cool under running tap water for 2 hours, and then transferred to a refrigerator set at 4°C and stored overnight. The samples were removed from the polythene bags, dried with paper towel and re-weighed ($W2$). CL was calculated as $((W1 - W2)/W1) * 100$. Muscle sample for WBSF assessment were prepared from the cooked samples by cutting seven cubes measuring $1 \times 1 \times 1$ cm, 5 cm long in fibre direction. Warner Bratzler shear blade attached to Zwick/Roell (Z2.5, Germany) instrument was used to determine the force ($N\,cm^{-2}$) required for shearing through a muscle cube at a right angle to the muscle fibre direction. The Zwick was set with 1 kN load cell with a crosshead speed of $100\,mm\,min^{-1}$.

2.5   Statistical analysis

In data analysis, individual animal was considered as experimental unit in all the variables analysed and initial weight was used as covariate. Contrast statements were used to compare molasses vs maize meal and hominy feed vs rice polishing. The GLM procedure of SAS (version 9.3, 2002) was used to analyse the data on yield, with dietary treatment as the main effect. The differences were considered significant at $P < 0.05$ and least squares means were separated by LSD. For meat quality parameters (CL, WBSF, pH and temperature) dietary treatments, time and their interactions were regarded as fixed effects. The MIXED procedure of SAS (Version 9.3, 2002) was used with repeated statement where compound symmetry was used as covariance structure. Differences were considered significant at $P < 0.05$ and least squares means were separated by Turkey-Kramer protection.

Table 2: *Performance of Tanzania Shorthorn Zebu steers fed molasses or maize meal with rice or maize by-products*

Parameters	Diets					SEM	P-values	P-contrasts	
	MMMO	HFMO	RPMO	HFMM	RPMM			MO-MM	HF-RP
Initial live weight (kg)	198	203	199	198	203	5.37	0.9253	0.8937	0.9263
Final live weight(kg)	259^c	283^a	264^{bc}	271^b	258^c	4.12	0.0005	0.0368	0.0003
Average daily gain (g)	658^c	919^a	709^{bc}	791^{ab}	639^c	0.05	0.0005	0.0368	0.0003
ME intake (MJ /day)	73.89^b	86.39^a	77.18^b	72.83^b	69.38^b	2.76	0.0013	0.0004	0.0264
CP intake (g/day)	795^b	867^a	725^b	809^{ab}	737^b	29.6	0.0095	0.421	0.0007
Hot half carcass weight (kg)	69.2^b	75.8^a	68.0^b	69.9^b	65.8^b	1.70	0.0052	0.0339	0.0021
Cold half carcass weight (kg)	67.4^b	74.4^a	66.2^b	68.1^b	64.1^b	1.70	0.0062	0.0314	0.0019
Chilling shrinkage (%)	2.63	1.93	2.61	2.49	2.55	0.45	0.7872	0.6030	0.4136

^{abc} Least squares means with a common superscript in the same raw are not significantly different ($P > 0.05$); SEM, Standard error of the mean; MMMO, maize meal with molasses; HFMO, hominy feed with molasses; RPMO, rice polishing with molasses; HFMM, hominy feed with maize meal, and RPMM, rice polishing with maize meal; MO-MM, molasses versus maize meal; HF-RP, hominy feed versus rice polishing.

3 Results

3.1 Carcass yield

Steers from different dietary treatments differed ($P <$ 0.05) in CP and ME intake, growth rate, final live weight and carcass yield. Steers fed on HFMO diet had higher ($P < 0.05$) values than their counterparts from other diets (Table 2). There was a positive correlation between growth rate and final live weight ($r = 0.99$) and between growth rate and hot/cold carcass weight ($r = 0.89$). The tendency of dietary treatments to give high values observed on carcass weight was similar to that observed in trimmed fat weight as proportion of half carcass. It was observed that the amount of trimmed fat increased with increase in growth rates. At the highest growth rate (919 g/day) for HFMO steers the amount of trimmed fat exceeded that of the lowest growth rate (639 g/day) for RPMM steers by 65 %. When expressed as proportion of half carcass weight, hind and fore quarter weight, saleable cuts, prime and non-prime cuts did not differ ($P > 0.05$) between dietary treatments despite the differences in half carcass weight (Table 3). Dietary treatments had no effects ($P > 0.05$) on yield of retail cuts as proportion of side carcass weight (Table 4). When comparisons were made between the diets, it was observed that molasses and hominy feed based diets had higher performance than maize meal and rice polishing based diets in all the parameters measured.

3.2 Meat quality attributes

3.2.1 Temperature and pH

The dietary treatments had no effects on the rate of carcass temperature decline post-mortem (Table 5). The decline in temperature was faster in the first 6–24 hours after which very little change was observed. No difference ($P > 0.05$) was observed between dietary treatments on post-mortem muscle pH decline, but there was a sharp decline in pH values in the first six hours followed by a gradual decrease till ultimate pH (pHu) was attained at 24 hours post-mortem (Table 5).

3.2.2 Cooking loss (CL) and Warner-Bratzler shear force (WBSF)

The CL and WBSF values of *M. longissimus thoracis et lumborum* (LL) were not ($P > 0.05$) influenced by dietary treatments, but were affected ($P < 0.05$) by ageing time (Table 6). Higher ($P < 0.05$) CL and WBSF values were observed on muscle aged for 3 days compared to muscles aged for 6, 9 and 12 days. The lowest values for both CL and WBSF were observed on meat aged for 9 and 12 days. An interaction ($P < 0.05$) between dietary treatments and muscle ageing time was observed for WBSF values. The meat from steers fed HFMO and RPMM diets and aged for 9 and 12 days had the lowest WBSF values compared to those from other dietary treatments × aging times (Table 7).

Table 3: *Carcass yield from Tanzania Shorthorn Zebu steers fed molasses or maize meal with rice or maize by-products*

Parameters	Diets					SEM	P-values	P-contrasts	
	MMMO	HFMO	RPMO	HFMM	RPMM			MO-MM	HF-RP
Carcass yield (% side carcass weight):									
Fore quarter	47.5	48.3	47.8	47.8	46.9	0.87	0.8499	0.4523	0.4015
Hind quarter	49.9	50.0	49.6	49.7	50.6	0.75	0.9418	0.5689	0.6220
Saleable cuts	87.1	85.7	88.5	88.0	87.7	1.20	0.5619	0.5718	0.2897
Prime cuts	46.7	46.4	45.2	46.5	47.6	1.13	0.8286	0.3008	0.9482
Non-prime cuts	40.4	39.3	43.3	41.4	40.1	1.34	0.4659	0.6978	0.3215
Aitch bone	3.08	3.29	3.47	3.12	3.47	0.28	0.7790	0.7927	0.3532
Fat trimmings	4.33 [b]	6.36 [a]	4.18 [b]	4.09 [b]	3.84 [b]	0.59	0.0394	0.0487	0.0488
Waste trimmings	2.50	2.42	1.36	1.99	2.56	0.43	0.4882	0.4112	0.5755
Uncountable loss	0.39	0.29	0.26	0.39	0.12	0.30	0.9689	0.9507	0.6208

[a b] Least squares means with a common superscript in the same raw are not significantly different ($P > 0.05$); SEM, Standard error of the mean; MMMO, maize meal with molasses; HFMO, hominy feed with molasses; RPMO, rice polishing with molasses; HFMM, hominy feed with maize meal, and RPMM, rice polishing with maize meal; MO-MM, molasses versus maize meal; HF-RP, hominy feed versus rice polishing.

Table 4: *Yield of retail cuts from Tanzania Shorthorn Zebu steers fed molasses or maize meal with rice or maize by-products*

Parameters	Diets					SEM	P-values	P-contrasts	
	MMMO	HFMO	RPMO	HFMM	RPMM			MO-MM	HF-RP
Prime cuts (% side carcass weight):									
Shoulder blade	9.60	9.50	8.60	9.60	9.20	0.90	0.9261	0.6869	0.5016
Prime ribs	5.33	5.56	5.93	5.86	5.97	0.40	0.6814	0.6720	0.5082
Wing ribs	2.02	2.44	1.95	1.95	2.05	0.36	0.8007	0.5638	0.5406
Loin strip	7.99	7.99	7.40	8.11	8.90	0.49	0.5435	0.1328	0.8366
Rump	4.02	3.99	3.70	3.58	4.85	0.34	0.2593	0.2934	0.1465
Topside	6.13	6.53	6.24	6.38	6.30	0.23	0.6763	0.8206	0.3606
Silverside	6.36	5.62	6.56	5.99	5.51	0.31	0.1322	0.2536	0.4025
Thick flank	4.25	3.90	3.72	4.28	3.88	0.33	0.5729	0.3919	0.3209
Fillet	0.90	0.94	1.07	0.83	0.89	0.13	0.7032	0.2561	0.3995
Non-prime cuts (% of side carcass weight)									
Fore shine	5.85	5.15	6.78	6.15	5.76	0.93	0.7981	0.9930	0.4458
Hump	2.17	1.79	1.72	2.31	2.30	0.21	0.2103	0.0252	0.8562
Neck	6.58	6.41	6.72	6.80	6.27	0.42	0.9454	0.9434	0.7904
Chuck	4.92	5.17	5.06	5.12	5.59	0.49	0.8979	0.6166	0.6873
Flat ribs	4.25	4.05	4.15	4.21	4.01	0.34	0.9881	0.9710	0.8958
Brisket	7.87	8.20	7.69	7.35	7.09	0.43	0.2896	0.0840	0.3100
Thin flank	3.95	4.50	6.38	4.84	4.51	0.71	0.1901	0.2566	0.2204
Hind shine	4.80	4.01	4.83	4.65	4.54	0.24	0.0920	0.4534	0.1032

SEM, Standard error of the mean; MMMO, maize meal with molasses; HFMO, hominy feed with molasses; RPMO, rice polishing with molasses; HFMM, hominy feed with maize meal, and RPMM, rice polishing with maize meal; MO-MM, molasses versus maize meal; HF-RP, hominy feed versus rice polishing.

Table 5: *Effect of time change on temperature decline and pH change post-mortem for five dietary treatments*

Dietary treatments	Variables							
	45 min		6 (hrs)		24 (hrs)		48 (hrs)	
	Temp	pH	Temp	pH	Temp	pH	Temp	pH
MMMO	35.7	6.46	26.2	5.88	1.4	5.62	1.1	5.52
HFMO	37.0	6.36	27.0	5.87	1.5	5.67	1.1	5.56
RPMO	36.0	6.41	26.1	5.84	1.2	5.71	1.0	5.52
HFMM	36.6	6.43	26.6	5.95	1.4	5.69	1.2	5.51
RPMM	35.8	6.30	26.3	5.86	1.0	5.63	0.8	5.50
SE	0.37	0.04	0.37	0.04	0.37	0.04	0.37	0.04
P-value	0.7973	0.4313	0.7973	0.4313	0.7973	0.4313	0.7973	0.4313

SE, Standard error of the mean; MMMO, maize meal with molasses; HFMO, hominy feed with molasses; RPMO, rice polishing with maize meal; HFMM, hominy feed with maize meal and RPMM, rice polishing with maize meal; MO-MM, molasses versus maize meal; HF-RP, hominy feed versus rice polishing.

Table 6: *Cooking losses and Warner-Bratzler shear force values for Tanzania Shorthorn Zebu steers fed molasses or maize meal with rice or maize by-products*

	Cooking loss (%)	Shear force ($N\,cm^{-2}$)
Dietary treatments (T)		
MMMO	21.5	49.0
HFMO	21.0	46.4
RPMO	20.3	50.8
HFMM	19.6	49.4
RPMM	19.7	44.0
SEM	0.82	3.72
P-value	0.42	0.72
Ageing (A, days)		
3	22.0[b]	53.4[c]
6	20.3[a]	48.7[b]
9	20.1[a]	45.1[a]
12	19.3[a]	44.3[a]
SEM	0.61	1.70
P-value	0.01	<0.0001
T Y A	0.18	<0.0001
P-contrasts (MO-MM)	0.24	0.61
P-contrasts (HF-RP)	0.75	0.89

[abc] Least squares means with a common superscript in the same raw are not significantly different ($P > 0.05$); SEM, Standard error of the mean; MMMO, maize meal with molasses; HFMO, hominy feed with molasses; RPMO, rice polishing with molasses; HFMM, hominy feed with maize meal and RPMM, rice polishing with maize meal; MO-MM, molasses versus maize meal; HF-RP, hominy feed versus rice polishing.

Table 7: *Effect of ageing time on tenderness for the five dietary treatments ($N\,cm^{-2}$)*

Ageing time	Dietary treatments					
	MMMO	HFMO	RPMO	HFMM	RPMM	Mean
3 days	53.4[a]±3.5	56.3[a]±1.2	54.1[a]±1.2	53.9[a]±1.2	50.1[a]±1.2	53.5±0.6
6 days	50.9[a]±3.5	47.4[b]±1.2	50.3[a]±1.2	50.3[ab]±1.2	43.0[bc]±1.2	48.4±0.6
9 days	50.0[a]±3.5	40.6[c]±1.2	50.0[b]±1.2	44.5[a]±1.2	40.8[c]±1.2	45.2±0.6
12 days	44.3[b]±3.5	40.5[c]±1.2	48.8[b]±1.2	44.5[b]±1.2	40.6[b]±1.2	44.3±0.6
Mean	49.0±3.4	46.4±0.7	50.8±0.7	49.4±0.7	44.0±0.7	

P-values for main effects: Dietary treatments, 0.72; Ageing time, 0.61;[abc] Least squares means with a common superscript in the same column are not significantly different ($P > 0.05$); MMMO, maize meal with molasses; HFMO, hominy feed with molasses; RPMO, rice polishing with molasses; HFMM, hominy feed with maize meal and RPMM, rice polishing with maize meal.

4 Discussion

4.1 Carcass yield

The observed higher final body weight and carcass weight for steers fed HFMO diet compared to those fed other diets could be associated with high intake levels of both energy and protein nutrients. Likewise, steers fed HFMO diet had the highest level of fat deposition (highest amount of trimmed fat). Growth rate and fat deposition are directly related with the level of energy and protein intake as they increase muscle and fat mass (Safari, 2010; Khalid *et al.*, 2012). These findings coincide with observations by Pazdiora *et al.* (2013) who found the degree of fat cover to increase with the body weight of animals. Part of high final weights observed could be due to compensatory growth, which is commonly seen for old animals of this age given *ad libitum* access to feed following a period of restricted feeding, since these animals were grazing on poor range land pasture before they were taken in feedlot. The observed increased amount of trimmed fat was found to decrease the amount of saleable cuts. This implies economic losses to beef producers, since excess fat deposits are not part of usable carcass under certain market conditions (Kitts, 2011). It is however, important to note that preference for fat content varies with place and culture. For instance, Kamugisha (2014) reported that consumers in Arusha, Tanzania showed high preference to meat with high fat content which suggests the need for taking into account preferences of targeted consumers in feedlot finishing.

The observed chilling shrinkage of 2.4 % in the present study was comparable to the range of 2.4–2.7 % observed by Khalafalla *et al.* (2011) and Fadol & Babiker (2010) in Sudan Baggra bulls, and is slightly above the standard cold shrinkage of 2.0 % in 24 hours chilling (Pascoal *et al.*, 2010). The absence of dietary effects on yield of different retail cuts including primal and non-primal cuts suggests that differences in levels of dietary energy and protein in the present study were not large enough to elucidate differences in yield of retail cuts. Similar findings have been observed in several other studies involving beef cattle (Fadol & Babiker, 2010; de Souza Duarte *et al.*, 2011; Turki *et al.*, 2011). On the other hand, studies have shown that variation in terms of breed (Sharaf Eldin *et al.*, 2013), sex (Lazzaroni & Biagini, 2008) and age (Pazdiora *et al.*, 2013) of animals has significant effects on the distribution of cuts.

4.2 Meat quality attributes

4.2.1 Temperature and pH

The rate and extent of decline of muscle pH and temperature during the immediate post-mortem periods seriously influence meat quality development, mainly tenderness (Safari, 2010; Frylinck *et al.*, 2013), because temperature changes can initiate cold or heat shortening. A rapid decline in temperature during early post-mortem when muscle pH is still high can cause cold shortening (Frylinck *et al.*, 2013), which will lead to tough meat. In the present study, pH reading reached 6 while the temperature was still high (>26.5°C) which shows unfavourable conditions for cold shortening to occur. Lack of dietary effect on pH values for concentrate fed cattle observed in the present study is in agreement with findings from other studies (Mapiye *et al.*, 2010; Lage *et al.*, 2012). The findings indicate that there was sufficient glycogen content in the muscles of animals in these different dietary treatments. The average pH of 5.67 for carcasses in the present study is within the quality range of 5.5 to 5.8 which is considered normal and optimal according to shelf life and eating properties (Silva *et al.*, 1999; Mach *et al.*, 2008).

4.2.2 Cooking loss and Warner-Bratzler shear force

Cooking loss in muscles depend on ultimate pH, cooking conditions (Mushi *et al.*, 2009), and intramuscular fat content (Safari, 2010). Although intramuscular fat content was not measured in the present study, it can be argued that the magnitude of pH variation was not large enough to elicit differences in CL values. The values observed for CL are comparable to the range of 16.8 to 24.6 % reported by Mwilawa (2012) on TSZ steers fed 100 % concentrate, but are slightly lower than the 22.5 to 25.2 % reported by Mapiye *et al.* (2010) on Nguni steers. This deviation may be attributed to differences in feeding systems, breeds and ageing time used between studies. The observed WBSF values are within the range of 41.8 to 50.9 N cm^{-2} reported on carcases of TSZ steers fed concentrate diet with 125 g CP and 12 MJ ME per kg DM (Mwilawa, 2012). Lack of dietary effects on CL and WBSF concurs with previous studies (Lage *et al.*, 2012; Neto *et al.*, 2012).

Meat samples from carcasses from cattle in the five treatments in the current study are considered tender as WBSF values are less than 50 N cm^{-2} (Devitt *et al.*, 2002). The lowest values for CL and WBSF observed on meat aged for longer days (9 and 12 days) may be associated with enzymatic reactions that disintegrate the myofibrillar proteins with increasing ageing time. The influence of ageing time on CL and WBSF has also been observed in previous studies (Florek *et al.*, 2009; Filipčík *et al.*, 2009).

In general, post-mortem storage resulted in decreased CL and WBSF values. Observations from this study have shown that dietary treatment and ageing time are two factors that independently affect meat quality characteristics of steers but jointly influenced the WBSF values which decreased most for HFMO and RPMM diets on 9th and 12th days of ageing. Ageing increases tenderisation by degrading and weakening structural integrity of myofibrillar proteins brought by the calpain proteolytic enzyme system especially μ-calpain (Safari, 2010; Kemp & Parr, 2012). On the other hand, energy concentration consumed by the animal influences glycogen stores which decrease ultimate muscle pH, and thus influencing meat tenderness. The influence of consumed energy on glycogen reserve has also been reported by Koger *et al.* (2010). In addition, high energy diets influences tenderness by increasing intramuscular fat that gives rise to the dilution of muscle structure (Wood *et al.*, 1999; Cardeno *et al.*, 2006). It can be concluded that molasses with hominy feed can substitute maize meal in feedlot diets without reducing meat yield and quality of TSZ cattle finished in feedlots in 90 days feed lot period.

Acknowledgements

Authors are grateful to the financial support provided by the SUA-IGMAFU project and positive collaboration from National Ranching Company (NARCO)-Tanzania, Tanzania Meat Company (TMC) and Vocational Education Training Authority (VETA)-Tanzania.

References

Bourguet, C., Deiss, V., Tannugi, C. C. & Terlouw, E. M. C. (2011). Behavioural and physiological reactions of cattle in a commercial abattoir: Relationships with organizational aspects of the abattoir and animal characteristics. *Meat Science*, 88, 158–168.

Cardeno, A., Vieira, C., Serranol, E. & Manteconl, A. R. (2006). Carcass and meat quality in Brown fattened young bulls: effect of rearing method and slaughter weight. *Czech Journal of Animal Science*, 51, 143–150.

Devitt, C. J. B., Wilton, J. W., Mandell, I. B., Fernandes, T. L. & Miller, S. P. (2002). Genetic evaluation of tenderness of the longissimus in multi-breed populations of beef cattle and the implications of selection. 7th World congress on genetics applied to livestock production, August 19–23, 2002, Montpellier, France.

Doto, S. P., Kimambo, A. E., Mgheni, D. M., Mtenga, L. A., Laswai, G. H., Kurwijila, L. R., Pereka, A. E., Kombe, R. A., Weisbjerg, M. R., Hvelplund, T., Madsen, J. & Petersen, P. H. (2004). *Tanzania feedstuff table for ruminants*. Sokoine University of Agriculture, Morogoro, Tanzania. 71pp.

Fadol, S. R. & Babiker, S. A. (2010). Effect of feedlot regimen on performance and carcass characteristics of Sudan Baggara Zebu cattle. *Journal of Livestock Research for Rural Development*, 22, Article #27. URL http://www.lrrd.org/lrrd22/2/fado22027.htm (last accessed: 10.05.2012).

FAO (1991). Guideline for slaughtering, meat cutting and further processing. FAO Animal Production and Health Paper 91, Food and Agriculture Organization of United Nation, Rome, Italy. URL http://www.fao.org/docrep/004/t0279e/T0279E00.htm (last accessed: 09.2011).

Filipčík, R., Šubrt, J. & Bjelka, M. (2009). The factors influencing beef quality in bulls, heifers and steers. *Slovak Journal of Animal Science*, 42, 54–61.

Florek, M., Litwińczuk, Z. & Skałecki, P. (2009). Influence of slaughter season of calves and ageing time on meat quality. *Polish Journal of Food and Nutrition Science*, 59, 309–314.

Frylinck, L., Strydom, P. E., Webb, E. C. & du Toit, E. (2013). Effect of South African beef production systems on post-mortem muscle energy status and meat quality. *Meat Science*, 93, 827–837.

Hunter, R. A. (2012). High-molasses diets for intensive feeding of cattle. *Animal Production Science*, 52, 787–794.

Kamugisha, P. P. (2014). *Quality beef supply chain efficiency and consumption of quality beef in Arusha and Dar es salaam cities, Tanzania*. Ph.D. thesis, Sokoine University of Agriculture Morogoro, Tanzania.

Kemp, C. M. & Parr, T. (2012). Review advances in apoptotic mediated proteolysis in meat tenderization. *Meat Science*, 92, 252–259.

Khalafalla, I. E. E., Atta, M., Eltahir, I. E. & Mohammed, A. M. (2011). Effect of body weight on slaughtering performance and carcass measurements of Sudan Baggara bulls. *Journal of Livestock Research for Rural Development*, 23, Article #47. Retrieved May 13, 2012, URL http://www.lrrd.org/lrrd23/3/khal23047.htm (last accessed: 13.05.2012).

Khalid, M. F., Sarwar, M., Rehman, A. U., Shahzad, M. A. & Mukhtar, N. (2012). Effect of dietary protein sources on lamb's performance: A Review. *Iranian Journal of Applied Animal Science*, 2, 111–120.

Kitts, S. E. (2011). *Effects of adipogenic compounds on growth performance and fat deposition in finishing beef steers*. Ph.D. thesis, University of Kentucky, Lexington, United States.

Koger, T. J., Wulf, D. M., Weaver, A. D., Wright, C. L., Tjardes, K. E., Mateo, K. S., Engle, T. E., Maddock, R. J. & Smart, A. J. (2010). Influence of feeding various quantities of wet and dry distillers grains to finishing steers on carcass characteristics, meat quality, retail-case life of ground beef, and fatty acid profile of longissimus muscle. *Journal of Animal Science*, 88, 3399–3408.

Lage, J. F., Paulino, P. V. R., Valadares Filho, S. C., Souza, E. J. O., Duarte, M. S., Benedeti, P. D. B., Souza, N. K. P. & Cox, R. B. (2012). Influence of genetic type and level of concentrate in the finishing diet on carcass and meat quality traits in beef heifers. *Meat Science*, 90 (3), 770–774.

Larson, E. M., Stock, R. A., Klopfenstein, T. J., Sindt, M. H. & Shain, D. H. (1993). Energy value of hominy feed for finishing ruminants. *Journal of Animal Science*, 71, 1092–1099.

Lazzaroni, C. & Biagini, D. (2008). Effect of pre- and post-pubertal castration on Piemontese male cattle. II: Carcass measures and meat yield. *Meat Science*, 80, 442–448.

Lopez, J. & Preston, T. R. (1977). Rice polishing as supplement in Sugar cane diets for fattening cattle: effects of different combinations with blood meal. *Tropical Animal Production*, 2, 143–147.

Mach, N., Bach, A., Velarde, A. & Devant, M. (2008). Association between animal, transportation, slaughterhouse practices, and meat pH in beef. *Meat Science*, 78, 232–238.

MAFF (1975). *Energy allowances and feeding systems for ruminants*. Technical Bulletin No. 33, Ministry of Agriculture, Fisheries and Food (MAFF), London, UK. Pp. 62–67

Mapiye, C., Chimonyo, M., Dzama, K., Muchenje, V. & Strydom, P. E. (2010). Meat quality of Nguni steers supplemented with Acacia karroo leaf-meal. *Meat Science*, 84, 621–627.

Matz, S. A. (1991). *The Chemistry and Technology of Cereals as Food and Feed. 2nd edition*. Pen. Tech International Inc., USA.

McDonald, P., Edwards, R. A., Greenhalgh, J. F. D. & Morgan, C. A. (2002). *Animal Nutrition*. 6th edition. Longman Science and Technology, Essex, England.

Muchenje, V., Dzama, K., Chimonyo, M., Raats, J. G. & Strydom, P. E. (2008). Meat quality of Nguni, Bonsmara and Aberdeen Angus steers raised on natural pasture in the Eastern Cape, South Africa. *Meat Science*, 79, 20–28.

Mushi, D. E., Safari, J., Mtenga, L. A., Kifaro, G. C. & Eik, L. O. (2009). Effects of concentrate levels on fattening performance, carcass and meat quality attributes of Small East African × Norwegian crossbred goats fed low quality grass hay. *Livestock Science*, 124, 148–155.

Mwilawa, A. J. (2012). *Effects of breed and diet on performance, carcass characteristics and meat quality of beef cattle*. Ph.D. thesis, Sokoine University of Agriculture, Morogoro, Tanzania.

Nandonde, S. W. (2008). *Assessment of the availability of major resources for production of quality beef in Tanzania*. Master's thesis, Sokoine University of Agriculture Morogoro, Tanzania. 111 pp.

Neto, O. R. M., Ladeira, M. M., Chizzotti, M. L., Jorge, A. M., Mendes de Oliveira, D., de Calvalho J. R., R. & do Sacramento Ribeiro, J. (2012). Performance, carcass traits, meat quality and economic analysis of feedlot of young bulls fed oilseeds with and without supplementation of vitamin E. *Revista Brasileira de Zootecnia*, 41 (7), 1756–1763.

Pascoal, L. L., Lobato, J. F. P., Restle, J., Vaz, F. N., Vaz, R. Z. & de Menezes, L. F. G. (2010). Beef cuts yield of steer carcasses graded according to conformation and weight. *Revista Brasileira de Zootecnia*, 39 (6), 1363–1371.

Pazdiora, R. D., de Resende, F. D., de Faria, M. H., Siqueira, G. R., de Souza Almeida, G. B., Sampaio, R. L., Pacheco, P. S. & Prietto, M. S. R. (2013). Animal performance and carcass characteristics of Nellore young bulls fed coated or uncoated urea slaughtered at different weights. *Revista Brasileira de Zootecnia*, 42, 273–283.

du Plessis, I. & Hoffman, L. C. (2004). Effect of chronological age of beef steers of different maturity types on their growth and carcass characteristics when finished on natural pastures in the arid sub-tropics of South Africa. *South African Journal of Animal Science*, 34, 1–12.

Safari, J. G. (2010). *Strategies for improving productivity of small ruminants in Tanzania*. Ph.D. thesis, University of Life Sciences, Norway.

Sani, R. T., Lamidi, O. S. & Jokhtan, G. E. (2014). Feed intake and carcass characteristics of Bunaji bulls fed raw or parboiled rice offal as energy source. *Journal of Biology, Agriculture and Healthcare*, 4, 53–59.

SAS (2002). *Statistical Analysis System. User's guide, version 9.3*. SAS Institute, (INC. Cary. NC. USA).

Sharaf Eldin, I. M. I., Babiker, S. A., Elkkhidir, O. A. & El-Bukhary, H. A. A. (2013). Characteristics of beef from intensively fed western Baggara cattle: carcass yield and composition. *Iraqi Journal of Veterinary Science*, 27, 39–43.

Silva, J. A., Patarata, L. & Martins, C. (1999). Influence of ultimate pH on bovine meat tenderness during ageing. *Meat Science*, 52, 453–459.

de Souza Duarte, M., Paulino, P. V. R., de Campos Valadares Filho, S., Paulino, M. F., Detmann, E.,

Zervoudakis, J. T., dos Santos Monnerat, J. P. I., da Silva Viana, G., Silva, L. H. P. & Serão, N. V. L. (2011). Performance and meat quality traits of beef heifer fed with two levels of concentrate and ruminally undegradable protein. *Tropical Animal Health and Production*, 43 (4), 877–886.

Turki, I. Y., Elkadier, O. A., Elamin, M., El. Zuber, D. & Hassabo, A. A. (2011). Effect of Guar meals and oil seed cakes on carcass characteristics and meat quality attributes of beef cattle. *ACT-Biotechnology Research Communications*, 1 (2), 66–75.

VETA (2006). Curriculum guide for meat processing. Vocational Education Training Authority (VETA), Tanzania. , pp 424.

Wood, J. D., Enser, M., Fisher, A. V., Nute, G. R., Richardson, R. I. & Sheard, P. R. (1999). Animal nutrition and metabolism group symposium on 'Improving meat production for future needs' Manipulating meat quality and composition. *Proceedings of the Nutrition Society*, 58, 363−370.

7

Effects of charcoal-enriched goat manure on soil fertility parameters and growth of pearl millet (*Pennisetum glaucum* L.) in a sandy soil from northern Oman

Melanie Willich [a,**], Anne Kathrin Schiborra [b],
Laura Quaranta [b], Andreas Buerkert [a,*]

[a] *Organic Plant Production and Agroecosystems Research in the Tropics and Subtropics, Universität Kassel, Witzenhausen, Germany*
[b] *Animal Husbandry in the Tropics and Subtropics, Universität Kassel, Witzenhausen and Georg-August-Universität Göttingen, Göttingen, Germany*

Abstract

The effect of charcoal feeding on manure quality and its subsequent application to enhance soil productivity has received little attention. The objectives of the present study therefore were to investigate the effects of (i) charcoal feeding on manure composition, and (ii) charcoal-enriched manure application on soil fertility parameters and growth of millet (*Pennisetum glaucum* L.). To this end, two experiments were conducted: First, a goat feeding trial where goats were fed increasing levels of activated charcoal (AC; 0, 3, 5, 7, and 9 % of total ration); second, a greenhouse pot experiment using the manure from the feeding trial as an amendment for a sandy soil from northern Oman. We measured manure C, N, P, and K concentrations, soil fertility parameters and microbial biomass indices, as well as plant yield and nutrient concentrations. Manure C concentration increased significantly ($P < 0.001$) from 45.2 % (0 % AC) to 60.2 % (9 % AC) with increasing dietary AC, whereas manure N, P, and K concentrations decreased ($P < 0.001$) from 0 % AC (N: 2.5 %, P: 1.5 %, K: 0.8 %) to 9 % AC (N: 1.7 %, P: 0.8 %, K: 0.4 %). Soil organic carbon, pH, and microbial biomass N showed a response to AC-enriched manure. Yield of millet decreased slightly with AC enrichment, whereas K uptake was improved with increasing AC. We conclude that AC effects on manure quality and soil productivity depend on dosage of manure and AC, properties of AC, trial duration, and soil type.

Keywords: activated charcoal, goat manure, microbial biomass C, SOC, subtropical soils

1 Introduction

Under the arid subtropical conditions of the Batinah region in northern Oman, regular additions of organic soil amendments such as manure and compost, and careful irrigation management are determining soil productivity and sustainability of cropping systems. It is well known that microbial activity, and consequently soil organic matter (SOM) turnover, is strongly affected by wet-dry cycles (Lundquist *et al.*, 1999a,b). High microbial turnover is often reflected in rapid decomposition of organic matter (OM) followed by a breakdown of soil fertility (Ghoshal & Singh, 1995; Zech *et al.*, 1997; Wichern *et al.*, 2004). One possibility to counteract soil organic matter decay under year-round conditions of high mineralisation is the exploitation of the *terra preta* concept (Glaser *et al.*, 2001, 2002), adding charred organic material to the soil. Key features of biochar (BC) amended soils are higher levels of

---

* Corresponding author
Email: tropcrops@uni-kassel.de

** Current affiliation: College of Agriculture, Forestry and Natural Resource Management, University of Hawai'i at Hilo, Hilo, HI, USA

SOM, enhanced nutrient retention capacity, and higher moisture-holding capacity than in the surrounding soils (Glaser et al., 2001; Lehmann & Rondon, 2006; Liang et al., 2006). However, in order to substantially affect the aforementioned physico-chemical soil parameters, large quantities of charred material are needed, particularly in view of possible losses through wind and water erosion. Also, there is still little knowledge about the nutrient release dynamics from BC used as a manure-amendment which is subsequently applied to soils. In many cases it is unclear whether BC is benefiting plants by providing nutrients or inhibiting plant growth by sequestering them (Mukherjee & Zimmerman, 2013). In contrast to BC, activated carbon or charcoal (AC) is a homogeneous, technically refined type of BC with effects on soil properties similar to those of BC (Lehmann & Rondon, 2006; Braendli et al., 2008). AC is produced from coal, peat, bamboo, coconut shells or other organic materials by incomplete combustion followed by steam activation (Braendli et al., 2008). It is used as a strong sorbent for a wide range of organic compounds in many different applications such as gas and water purification, medicine, sewage treatment, and air filters (Norit Americas Inc., 2006). It is well documented that BC or AC can be used effectively as a gastrointestinal absorbent for treating forage-induced intoxications such as mycotoxins (Buck & Bratich, 1986; Huwig et al., 2001). Positive effects on feed intake and nutrient utilisation were also reported for animals feeding on low quality forages containing compounds such as alkaloids, phenols, and terpenes (Banner et al., 2000; Poage et al., 2000; Rogosic et al., 2006). However, so far BC studies primarily addressed effects of BC addition to soils (Glaser et al., 2002; Lehmann & Rondon, 2006), whereas little is known about the effect of AC as feed additive on manure quality, and how ingested, manure-bound AC affects soil properties, C sequestration, and plant growth. Recent results by Ingold et al. (2011) suggest that charcoal-enriched manure has a greater recalcitrant capacity than AC mixed with faeces outside the animal. Moreover, manure-bound AC seems better suited for no-till applications as it is less likely to disintegrate and erode from soil if bound to a carrier.

The objectives of this study consisting of a goat feeding trial followed by a greenhouse pot experiment were (i) to determine if AC fed daily to goats had negative effects on animal performance and how ingested AC affects the nutrient composition of the manure, and (ii) to measure AC-manure effects on soil microbial biomass, nutrient and water retention capacities, and growth of a test crop under controlled conditions.

2 Materials and methods

2.1 Soil collection and characterisation

Soil (0–40 cm) was collected from a private experimental farm (24°20′ N, 56°46′ E) located on the northeastern Batinah coast of the Sultanate of Oman (Siegfried et al., 2013). The soil had been classified as a mixed hyperthermic typic Torrifluvent (US Soil Taxonomy, Al-Farsi, 2001) derived from recent fluviatile wadi deposits with a gravel-rich subsoil and partly coverage by aeolian sand veneers. The soil had a coarse texture (82 % sand, 16 % silt, 2 % clay), a pH_{H_2O} of 8.5, a bulk density of 1.44 $g\,cm^{-3}$, a total C content of 12.7 $g\,kg^{-1}$, an organic C content (C_{org}) of 4.9 $g\,kg^{-1}$, a total N content of 0.6 $g\,kg^{-1}$, a $CaCO_3$ content of 6.5 %, a C : N ratio of 21.2, an Olsen P of 0.04 $g\,kg^{-1}$, and $K_{(CAL)}$ of 0.26 $g\,kg^{-1}$. Before the experiment started, the soil was air dried, sieved (< 2 mm) and shipped to Germany.

2.2 Experimental design

2.2.1 Goat feeding and manure collection

Manure was derived from a feeding trial with four male boer goats (Capra aegagrus hircus L.; 22.8 + 3.9 kg) conducted at the Department of Animal Sciences, Georg-August-Universität Göttingen. In this trial, each goat was offered increasing levels of AC (0, 3, 5, 7, 9 % of total diet, dry matter basis) together with concentrate feed over five consecutive 14-day-periods (one AC level per 14-day-period, in increasing order) and effects on feeding behaviour and goats' health and faecal excretion patterns were observed (Quaranta et al., 2013). To this end, the goats were kept in individual cages and fed twice per day with a mixture of 50 % hay (Lolium perenne L.) and 50 % concentrate, while hay was offered after concentrate was completely taken up. The concentrate was composed of 35 % barley, 35 % wheat, 15 % rapeseed extraction meal, and 15 % sugar beet molasse chips. AC powder was mixed with the ground ingredients at 0 (control), 6, 10, 14, and 18 % level (corresponding to 0, 3, 5, 7, and 9 % of total diet, dry matter basis) and pressed into pellets. AC powder was manufactured from coconut shells followed by steam activation (AquaSorb® CP1, Jacobi Carbons Service GmbH, Premnitz, Germany) and contained 92.1 % C, 0.1 % N, 0.03 % P, and 0.8 % K. It had a pH of 9.1, a particle size of 44 μm, a surface area of 1050 $m^2\,g^{-1}$, and a total pore volume of 0.62 $cm^3\,g^{-1}$ (Jacobi Carbons Service GmbH, Premnitz, Germany). Feed quantity was administered at 1.5 times energy maintenance requirements and adjusted every two weeks according to goats' weight gain.

Feeding behaviour (consumption rate, refusals) and manure characteristics (colour, odour, consistency) were observed during the first five days of each period. At days 8 to 10 of each 14-day-period, goats were equipped with faecal collection bags, which were attached by harnesses. The complete amount of manure in the collection bags was sampled twice daily before each feeding, weighted and stored immediately at $-20\,°C$.

2.2.2 Greenhouse trial

A greenhouse trial over 12 weeks with six replicates was carried out using the manure produced in the feeding trial: Treatments were (i) manure with 0 % AC in goat feed (control), (ii) manure with 3 % AC in goat feed, (iii) manure with 5 % AC in goat feed, (iv) manure with 7 % AC in goat feed, and (v) manure with 9 % AC in goat feed. In the following, these treatments will be referred to as AC 0 (or control), AC 3, AC 5, AC 7, and AC 9, respectively. For all treatments, 4.2 kg dry soil was filled into respective PVC pots (17.5 cm diameter, 22 cm height).The experiment was fertilised equivalent to 160 kg N ha^{-1} and thus received between 6.4 t (AC 0) and 9.4 t (AC 9) of manure per hectare depending on the AC treatment. Per pot, this amounted to 15.3 g (AC 0), 19.5 g (AC 3), 20.6 g (AC 5), 21.0 g (AC 7), and 22.1 g (AC 9) manure. Prior to the treatment application, soil and manure were analysed separately for microbial and chemical properties. From this analysis, the initial values for soil and all manure treatments were determined. Two days before sowing, the water content was adjusted to 40 % water holding capacity (WHC), which was gravimetrically controlled and adjusted every third to fourth day throughout the experiment, so that the water content remained at > 35 % of WHC. One day before sowing, the manure treatments were buried at 5 cm soil depth. In each pot, ten seeds of millet (*Pennisetum glaucum* L.) were sown at 2 cm depth, and thinned to four seedlings nine days later. Millet was chosen as indicator plant, because it grows well under greenhouse conditions and is fertility responsive. Climatic conditions were regulated to $28 \pm 1\,°C$ during the day and $16 \pm 1\,°C$ during the night with a 12/12 h day/night light regime and a light summation per day of at least 120 klx. The pots were re-randomized every fourth day throughout the duration of the trial.

2.3 Analytical methods

2.3.1 Manure nutrient contents

Upon thawing, manure samples were oven dried to constant weight at $60\,°C$ and ground with a ball mill, then homogenized, and combined to form a composite sample for each AC level. Two subsamples per treatment were analysed for C and N contents using a Vario MAX elemental analyser (Elementar GmbH, Hanau, Germany). The concentrations of P and K were determined photometrically (P; Hitachi U-2000, Hitachi Co Ltd., Tokyo, Japan) and by flame photometry (K; Instrument Laboratory 543, Bedford, USA) in coloured ash solution (32 % HCl) after burning the manure in a muffle furnace ($550\,°C$, 24 h; Murphy & Riley, 1962). Organic matter was calculated as difference between dry weight ($105\,°C$, 24 h) and ash content of a sample.

2.3.2 Soil organic carbon and total N

Before any soil samples were taken, plant roots were thoroughly removed. For capacity reasons, four out of the six replicates were randomly assigned from each treatment for soil analyses. Total C and N of soil was determined by gas chromatography after dry combustion to CO_2 and N_2 using a Vario Max CN analyser (Elementar GmbH, Hanau, Germany). Carbonate (CO_3^{2-}) was measured gas-volumetrically after addition of 1:2 diluted 32 % HCl. Soil organic carbon was calculated as difference between total C and CO_3-C.

2.3.3 Microbial biomass indices

Microbial biomass C (Vance *et al.*, 1987) and microbial biomass N (Brookes *et al.*, 1985) were measured on fresh soil by chloroform fumigation extraction with 0.5 M K_2SO_4 and subsequent analysis of organic C and total N using a CN analyser (Multi N/C 2100S, Analytik Jena AG, Jena, Germany). Microbial biomass C was calculated as E_C/k_{EC}, where E_C is (organic C extracted from fumigated soils) – (organic C extracted from non-fumigated soils), and k_{EC} is 0.45 (Wu *et al.*, 1990). Microbial biomass N was calculated as E_N/k_{EN}, where E_N is (total N extracted from fumigated soils) – (total N extracted from non-fumigated soils), and k_{EN} is 0.54 (Brookes *et al.*, 1985).

2.3.4 Soil physico-chemical parameters

Soil pH (H_2O) was measured using a glass electrode at a 1:2.5 soil-to-water ratio. Water holding capacity (WHC) of the soil was calculated as gravimetrical difference following complete saturation with deionized water and subsequent drying ($105\,°C$, 24 h). For the determination of cation exchange capacity (CEC), 2.5 g soil were saturated with 30 ml 0.1 M $BaCl_2$ buffered (pH 7) solution and shaken for 2 h. Then, samples were centrifuged (Centrikon T-124, Kontron Instruments, Milan, Italy) at 9000 rpm for 15 minutes, and filtered through a black band filter. K and Na were measured according to Murphy & Riley (1962) with a flame photometer (Flame Photometer 543, Instrumentation Laboratory, Bedford, MA, USA). Concentrations of Ca, Mg, and Al were determined by atomic absorption spectrometry (AAS 906AA, GBC Scientific Equipment, Melbourne, Australia). CEC was calculated as the sum of the exchangeable cations (K, Na, Ca, Mg, Al) and expressed in $cmol\,kg^{-1}$.

2.3.5 Plant nutrient uptake and biomass yield

At day 29, 36, 43, 58, 65, and 81 after sowing, shoot height was determined after stretching the longest leaf. Upon harvest, plants were cut above soil surface, dried (60 °C, 24 h), and ground with a ball mill. While a subsample was analysed for total C and N with a Vario MAX elemental analyser (Elementar, Hanau, Germany), the remainder of the samples were further dried to constant dry weight (105 °C, 24 h). Subsequently, 1.5 g dry matter was ashed in a muffle furnace (550 °C, 24 h) and dissolved in concentrated HCl for colorimetric analysis of P and K using the ascorbic acid method described by Murphy & Riley (1962). Individual plants per pot were treated as subsamples, pooled and values reported as averages per plant.

2.4 Statistical analysis

Significance of treatment effects was tested by analysis of variance (ANOVA) and post-hoc test statistics (Tukey HSD). Arithmetic means were compared at $P < 0.05$ using contrasts. All statistical analyses and graphs were performed using the statistical packages Statistica 7.0 (StatSoft GmbH, Hamburg, Germany), SPSS 17.0 (SPSS Inc., Chicago, IL, USA) and Sigma Plot (Systat Software Inc., San José, CA, USA).

3 Results

3.1 Manure quality and faecal scores

Based on feeding behaviour observations, intake of concentrate was not affected by AC concentration (Quaranta et al., 2013). Percentage of manure C increased significantly with the amount of AC fed to goats while N and P concentrations steadily decreased (Table 1).

The increase in C and decline in N resulted in a widened CN ratio with rising AC levels in the goat feed. Manure K concentration was significantly lower in AC-manure compared with the control. AC addition to the feed also affected manure consistency, colour and odour (Table 2).

Table 2: *Mode faecal scores of goat manure from the feeding trial with AC (n = 4).*

	AC in goat feed (%)				
	0	3	5	7	9
consistency	3	3	4	4	4
colour	2	3	3	3	4
odour	2	1	1	1	1

AC: activated charcoal fed to goats in % of total ration.
Consistency: 1 ≅ Diarrhoea, 2 ≅ Soft, 3 ≅ Normal, 4 ≅ Hard
Colour: 1 ≅ Light, 2 ≅ Normal, 3 ≅ Dark, 4 ≅ Very dark/black
Odour: 1 ≅ Mild, 2 ≅ Normal, 3 ≅ Strong

Table 1: *Mean carbon and nutrient contents of goat manure from the feeding trial with AC. Values in parentheses represent ± one standard error of the mean (n = 2). Significance of treatment effects at P < 0.05 based on contrast tests.*

AC in goat feed (%)	C (%)	N (%)	P (%)	K (%)	CN ratio
0	45.2 (0.1)	2.5 (0.1)	1.5 (0.0)	0.8 (0.0)	18 (0.3)
3	51.4 (0.0)	2.0 (0.0)	1.2 (0.0)	0.5 (0.1)	26 (0.3)
5	56.7 (0.1)	1.9 (0.0)	1.1 (0.1)	0.5 (0.0)	30 (0.4)
7	57.9 (0.2)	1.8 (0.0)	1.0 (0.1)	0.7 (0.0)	32 (0.0)
9	60.2 (0.5)	1.7 (0.0)	0.8 (0.0)	0.4 (0.0)	35 (0.1)
Contrasts					
Manure −AC vs. +AC	< 0.001	< 0.001	< 0.001	< 0.001	< 0.001
High AC vs. low AC	< 0.001	0.006	0.002	0.175	< 0.001
CV (%)	10	15	23	28	22

CV: mean coefficient of variation between replicates of one column. AC: activated charcoal fed to goats in % of total ration. High AC = 9 %, low AC = 3 %.

3.2 Soil physico-chemical parameters

Irrespective of AC treatment, manure addition to the soil did not affect SOC concentrations (Table 3).

By the end of the trial, however, the SOC concentration in soils treated with AC-enriched manure was significantly higher than in soils which received manure without dietary AC. In contrast, soil total nitrogen (TN) concentrations remained constant with only marginal increases for the two highest AC treatments. Consequently, the soil CN ratio tended to increase with increasing AC fed. Water holding capacity tended to increase with AC-enriched manure. The increase of the average WHC from the initial soil to the soils amended with the two lower AC treatments (3 and 5) was 3 %, and statistically not significant. However, in the soils of the two higher AC treatments (7 and 9) WHC increased by 9 and 25 % as compared with the initial soils' WHC; and by 6 and 21 % as compared with the soils that received un-amended manure. CEC dropped by 2 % in soils treated with AC 9 manure, and by 6 % in soils treated with AC 7 manure as compared with the control. However, compared with the initial soils' CEC, average CEC of the AC 9 treatment was still slightly higher (9.7 and 10 $cmol\,kg^{-1}$, respectively). There was no difference in pH level between the initial soil and the soil that received un-amended manure (both pH 8.5), but pH was significantly lower in soils that received AC-enriched manure.

3.3 Soil microbial biomass

AC-enriched manure applications did not affect microbial biomass C (Table 4). Compared with the initial value, mean microbial biomass C concentrations were nevertheless higher in all manure amended soils, except for soils that received AC 5 manure, where microbial biomass C was slightly lower than in the initial soil (4 %) and considerably lower than in all other fertilised soils (26 %). Microbial biomass N was significantly higher in all manure treated soils than before manure application. With increasing manure AC levels, microbial biomass N decreased by 36 % from 28 $\mu g\,g^{-1}$ (AC 3) to 18 $\mu g\,g^{-1}$ (AC 9). The ratio of microbial biomass C to SOC indicated an increase from initially 2 % to 3 % in the soils of the AC 5 and AC 7 treatments, respectively, and declined in the soils of the AC 9 treatment again to about 2 %. The contribution of K_2SO_4-extractable N to total N decreased during the 12 weeks of the experiment from 4.6 % to 2.3 % and was lowest in the two higher AC treatments (AC 7 and 9).

Table 3: *Mean concentrations of soil organic carbon (SOC) and total nitrogen (TN); CN ratio, water holding capacity (WHC) and cation exchange capacity (CEC) of soil samples. Values in parentheses represent ± one standard error of the mean (initial concentrations before treatment application: n = 4; treatment replicates: n = 4). Significance of treatment effects at P < 0.05 based on contrast tests.*

	SOC ($mg\,g^{-1}$ soil)	TN ($mg\,g^{-1}$ soil)	CN ratio	WHC (%)	CEC ($cmol\,kg^{-1}$ soil)	pH
Initial concentrations	4.9 (0.6)	0.6 (0.0)	21 (0.7)	32 (2)	9.7 (0.3)	8.5 (0.0)
Manure (% AC fed)						
0	3.7 (0.8)	0.5 (0.0)	22 (0.8)	33 (3)	10.2 (0.2)	8.5 (0.0)
3	4.3 (0.3)	0.5 (0.0)	22 (0.4)	33 (3)	10.2 (0.1)	8.4 (0.0)
5	4.6 (0.8)	0.5 (0.0)	23 (0.2)	33 (2)	10.2 (0.1)	8.3 (0.0)
7	5.9 (0.8)	0.6 (0.0)	23 (0.1)	35 (2)	9.6 (0.2)	8.4 (0.0)
9	6.2 (0.5)	0.6 (0.0)	23 (0.3)	40 (1)	10.0 (0.2)	8.3 (0.1)
Contrasts						
Initial *vs.* treatment	NS	NS	NS	NS	0.073	0.001
Manure −AC *vs.* +AC	0.040	0.062	NS	NS	NS	0.004
High AC *vs.* low AC	0.050	0.092	NS	0.044	NS	NS
CV (%)	31	12	4	14	4	1

CV: mean coefficient of variation between replicates of one column. AC: activated charcoal fed to goats in % of total ration. High AC = 9 %, low AC = 3 %. Treatment: all manures. NS: not significant.

Table 4: *Mean concentrations of K_2SO_4-extractable organic carbon and nitrogen and microbial biomass carbon and nitrogen of soil samples. Values in parentheses represent ±one standard error of the mean (initial concentrations before treatment application: n = 4; treatment replicates: n = 4). Significance of treatment effects at P<0.05 based on contrast tests.*

	K_2SO_4-C ($\mu g\,g^{-1}$ soil)	K_2SO_4-N (% TN)	Microbial biomass C ($\mu g\,g^{-1}$ soil)	Microbial biomass C (% SOC)	Microbial biomass N ($\mu g\,g^{-1}$ soil)
Initial concentrations	104 (7)	4.6 (0.0)	111 (6)	2.3 (0.3)	6 (0.2)
*Manure (% AC fed)**					
3	74 (6)	n.d.	158 (32)	n.d.	28 (4)
5	89 (5)	2.8 (0.6)	107 (19)	3.0 (0.8)	20 (6)
7	99 (5)	2.1 (0.2)	148 (37)	3.2 (1.3)	20 (4)
9	88 (2)	2.1 (0.3)	125 (14)	1.9 (0.1)	18 (1)
Contrasts					
Initial *vs.* treatment	0.020	0.002	NS	NS	0.024
High AC *vs.* low AC	0.056	n.d.	NS	n.d.	NS
CV (%)	12	35	39	64	41

* Values for 0 % AC are missing. n.d.: not determined. CV: mean coefficient of variation between replicates of one column. AC: activated charcoal fed to goats in % of total ration. High AC = 9 %, low AC = 3 %. Treatment: all manures. NS: not significant.

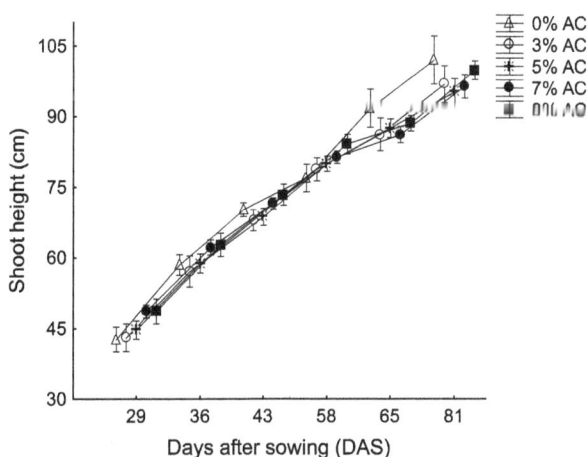

Fig. 1: *Effects of AC-amended manures on millet shoot growth. Data points show means (n = 6) ± one standard error at P < 0.05 measured 29, 36, 43, 58, 65, and 81 days after sowing. AC: activated charcoal fed to goats in % of total ration.*

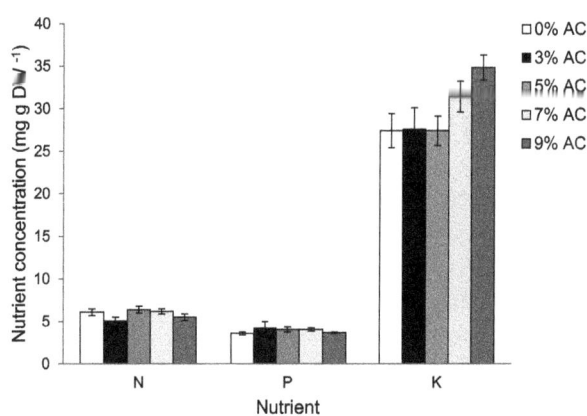

Fig. 2: *Effects of AC-amended manures on nitrogen (N), phosphorus (P), and potassium (K) concentration of millet shoot dry matter at harvest. Data show means (n = 6) and ± one standard error. AC: activated charcoal fed to goats in % of total ration.*

3.4 Plant growth and nutrient concentrations

No significant effects of manure AC enrichment on millet growth were found after twelve weeks (Fig. 1).

Nevertheless, towards the end of the experiment plants from the AC-enriched manure treatments showed a slightly smaller growth compared with plants amended with control manure. Plant aboveground biomass yield was on average 9.4 g DM per plant and remained unaffected by AC-manure. Nitrogen concentrations in plant material ranged between 5.1 (AC 3) and 6.4 mg g^{-1} DM (AC 5) and did not significantly differ between manure treatments (Fig. 2). Plant P concentration was with 4.3 ± 0.7 mg P g^{-1} DM highest for the AC 3 manure treatment and lowest in plants fertilised with AC 0 manure (3.6 ± 0.2 mg g^{-1} DM) and AC 9 enrichment (3.7 ± 0.1 mg g^{-1} DM), but these differences were not statistically significant. Plant K concentrations tended to increase with AC concentration: control (27 ± 2.0 mg g^{-1} DM), AC 3 (28 ± 2.5 mg g^{-1} DM), AC 5 (27 ± 1.7 mg g^{-1} DM), AC 7 (31 ± 1.8 mg g-1 DM), and AC 9 manure treatment (35 ± 1.5 mg g^{-1} DM), however, they only significantly differed between the control and AC 9 ($P = 0.027$).

4 Discussion

4.1 Effects of AC feeding on manure nutrient contents and faecal scores

Feeding AC is widely used in veterinary medicine to treat animals suffering from diarrhoea and different types of intoxication (Buck & Bratich, 1986). The use of AC in the daily diet of ruminants that feed on plants containing secondary compounds like phenolics or terpenes, which can reduce feed intake and decrease nutrient utilisation, was examined in several studies (Banner et al., 2000; Villalba et al., 2002). Even at higher AC levels than those of our study, effects of AC supplementation on feed intake (Rogosic et al., 2006, 2008) and nutrient utilisation (Murdiati et al., 1991; Van et al., 2006) were negligible or positive and there is no experimental evidence of negative effects on animal health such as constipation (Villalba et al., 2002). In our study AC intake did not affect feed ingestion by goats. This is in accordance with other studies that did not find differences in feed intake due to AC feeding (Van et al., 2006; Alkindi et al., 2013). As expected, with increasing AC rate faeces colour changed to dark, odour was reduced, and the consistency turned to hard, without any sign that the goats' digestion or health were negatively affected.

Our data confirm that daily feeding of goats with significant amounts of AC is physiologically feasible. We assume that AC is inert in the animals' organism, meaning that it is not digested or absorbed by the animal. This assumption is reflected by the results showing a linear increase in faeces C concentration with increasing AC supplementation. The decline in faeces N and P concentrations with increasing AC supplementation reflect a dilution effect due to higher amounts of excreted AC.

4.2 Effects of AC-enriched manure on soil fertility parameters and microbial biomass indices

AC-enriched manure increased SOC by up to 68 %, while the soil TN pool remained unaffected. The AC effect on SOC was particularly pronounced, because the SOC in the soil treated with control manure was lower than in the initial soil. Our applications ranged between $0.4\,t\,AC\,ha^{-1}$ (AC 3) and $1.4\,t\,AC\,ha^{-1}$ (AC 9), assuming that the difference in manure C concentration between the control manure and the AC-enriched manures derived from the undigestible AC. Sradnick et al. (2013) conducted a field trial on the same soil. They applied $15\,t$ dry goat manure ha^{-1} (control) and compared this with (i) goat manure and 2.5 % AC (AquaSorb® CP1) as feed additive, which corresponded to $1.65\,t\,ha^{-1}$,

and (ii) goat manure mixed with AC (AquaSorb® CP1; $1.7\,t\,ha^{-1}$). After the first seven months season, they reported an increase in SOC from 7.6 (initial) to 9.1 (control), 10.6 (i), and $10.5\,mg\,g^{-1}$ (ii); and an increase in TN from 0.4 (initial) to 0.6 (control), 0.7 (i), and $0.6\,mg\,g^{-1}$ (ii). The positive relationship between AC-enriched manure and soil TN concentrations found by these authors may be caused by the larger input of both manure and AC, as compared with our application rates. AC-manure effects on the soil CN ratio were surprisingly small, most likely because the application doses were too low. It is often discussed that the widening of CN ratios in biochar amended soils can lead to immobilisation of N (Lehmann et al., 2003; Ding et al., 2010; Nelson et al., 2011; Kloss et al., 2014). The results of this study show that TN did not differ significantly between treatments, hence we assume that differences in the immobilisation of N played a negligible role in overall nutrient availability. The WHC of AC-manure amended soils increased by up to 25 % (AC 9) compared with the manure amended or initial soil. The sensitivity of our soil towards manure-induced changes in WHC supports data of Rawls et al. (2003), who found the highest response of water retention to changes in soil organic matter on sandy soils with low SOC contents. AC-manure application did not increase soil pH, but even decreased it by up to 0.4 units. This is in contrast to other findings that show clear pH increases following BC additions (Kloss et al., 2014). However, most studies on biochar effects on soil quality were conducted on highly weathered acid soils (Glaser et al., 2002; Lehmann et al., 2003; Steiner et al., 2007), where the well-known proton buffering capacity of BC was effective. This is consistent with findings of Kloss et al. (2014), where the application of biochar increased the pH of an acid Planosol from 5.3 to 6.9 after seven months, whereas biochar addition at the same rate to an alkalic Chernozem only led to a minor pH increase from 7.4 to 7.6. Similarly, Liu et al. (2012) found on a German Cambisol of pH 6 a significant increase in pH and CEC as response to the addition of compost, but not of BC. Our results are in accordance with data by Ingold et al. (2015), who recorded a slight pH decrease of 0.1 units after AC-enriched manure addition to the same soil under field conditions. Similarly, the CEC in their study declined from $10.2\,cmol\,kg^{-1}$ to $9.4\,cmol\,kg^{-1}$ after manure addition, and to $9.5\,cmol\,kg^{-1}$ after the addition of AC-enriched manure (Ingold, unpublished data). These results are in contrast to numerous reports showing considerable CEC increases after BC amendment, an effect attributed to surface oxidation of BC particles as well as a higher charge density in BC-rich soils (Liang *et al.*, 2006). Recently, Slavich *et al.* (2013) showed on an Australian acid Ferralsol that the application of BC increased soil pH, but did not affect the pH-dependent CEC.

As a result of its recalcitrance, biochar-C is largely unavailable to soil microbes, but changes in soil physico-chemical properties and biologically available labile C compounds deriving from BC may accelerate microbial biomass and its activity (Warnock *et al.*, 2007; Steinbeiss *et al.*, 2009; Anderson *et al.*, 2011). The increase in microbial biomass may lead to a temporal immobilisation of mineral N, as shown for instance by Bargmann *et al.* (2014). Sradnick *et al.* (2013) reported for the same soil similar microbial biomass C and microbial biomass N. Their findings for microbial biomass C support our results. We detected a significant increase in microbial biomass N that arose from manure application; microbial biomass N decreased with increasing AC enrichment, possibly due to decreased availability of N with a widening treatment CN ratio.

4.3 Effects of AC enriched manure on plant growth, yield, and nutrient concentrations

Under the controlled conditions of this study growth of millet was fast and no tillering occurred likely due to limited light intensity. For the last two measurements of plant height, we observed growth retardation for all plants grown on soils amended with AC-manure as compared with the control group. Often discussed as a reason for plant yield reductions after charcoal applications is N immobilisation that leads to N deficiency in plants (Lehmann *et al.*, 2003; Deenik *et al.*, 2011; Kloss *et al.*, 2014) which may have been the case in the last period of our study. Previous work has shown an increase in plant nutrient availability and growth after biochar incorporation across a range of soils (Glaser *et al.*, 2002; Lehmann *et al.*, 2003; Steiner *et al.*, 2007; Liu *et al.*, 2012). However, differences in the substrate and pyrolysis conditions greatly affect BC characteristics and its effects on key soil parameters such as CEC and nutrient content (Glaser *et al.*, 2002; Gaskin *et al.*, 2008, 2010; Alburquerque *et al.*, 2014) which may thus be study specific. Charcoal-induced reductions or zero effects on plant growth were also reported (Gaskin *et al.*, 2010; Jones *et al.*, 2012; Güerena *et al.*, 2013; Kloss *et al.*, 2014). Also, micronutrient deficiencies after pH-induced decreases in micronutrient availability were found (Bolan *et al.*, 2003).

5 Conclusions

Based on our experiment we conclude that short-term feeding of goats with diets containing up to 9 % AC in the ration is possible without negative effects on animal health, while the excretion of C into manure could be increased. Applied as a low-cost feed additive at this rate, charcoal may have positive effects on animal health (endoparasites and related internal intoxication leading to diarrhoea). Most soil fertility parameters and microbial biomass indices remained unaffected by AC-manure application, except for SOC, pH, and microbial biomass N. WHC tended to increase with AC enrichment of manure. Yield of millet remained unaffected by AC enrichment, whereas K uptake was improved with increasing AC application. Main responsible factors may have been soil type, dosage of manure and AC, properties of AC, and duration of the trial.

Acknowledgements

We gratefully acknowledge the technical assistance of Christian Wagner, Eva Wiegard, Claudia Thieme and Gabi Dormann. We are also thankful to Dr. Herbert Dietz and Royal Court Affairs (Sultanate of Oman) for their essential support and to the German Research Foundation (DFG) for funding this project within the Graduate Research Training Group 1397 'Regulation of Soil Organic Matter and Nutrient Turnover in Organic Agriculture'.

References

Al-Farsi, S. A. (2001). *Classification of soils from an alluvial fan and coastal plain near Sohar, Sultanate of Oman*. Muscat, Oman.

Alburquerque, J. A., Calero, J. M., Barron, V., Torrent, J., Del Campillo, M. C., Gallardo, A. & Villar, R. (2014). Effects of biochars produced from different feedstocks on soil properties and sunflower growth. *Journal of Plant Nutrition and Soil Science*, 177, 16—25.

Alkindi, A., Schlecht, E. & Schiborra, A. (2013). Nitrogen balance and rumen microbial protein synthesis in goats fed quebracho tannin and activated charcoal. *In:* Proceedings of the Society of Nutrition Physiology 2013, March 19–21, Goettingen, Germany.

Anderson, C. R., Condron, L. M., Clough, T. J., Fiers, M., Stewart, A., Hill, R. A. & Sherlock, R. R. (2011). Biochar induced soil microbial community change: implications for biogeochemical cycling of carbon, nitrogen and phosphorus. *Pedobiologia*, 54, 309–320.

Banner, R. E., Rogosic, J., Burritt, E. A. & Provenza, F. D. (2000). Supplemental barley and charcoal increase intake of sagebrush by lambs. *Journal of Range Management*, 53, 415–420.

Bargmann, I., Martens, R., Rillig, M. C., Kruse, A. & Kuecke, M. (2014). Hydrochar amendment promotes microbial immobilization of mineral nitrogen. *Journal of Plant Nutrition and Soil Science*, 177, 59–67.

Bolan, N., Adriano, D. & Curtin, D. (2003). Soil acidification and liming interactions with nutrient and heavy metal transformation and bioavailability. *Advances in Agronomy*, 78, 216–272.

Braendli, R. C., Hartnik, T., Henriksen, T. & Cornelissen, G. (2008). Sorption of native polyaromatic Hydrocarbons (PAH) to black carbon and amended activated carbon in soil. *Chemosphere*, 73, 1805–1810.

Brookes, P. C., Landman, A., Pruden, G. & Jenkinson, D. S. (1985). Chloroform fumigation and the release of soil nitrogen: a rapid direct method to measure microbial biomass nitrogen in soil. *Soil Biology and Biochemistry*, 17, 837–842.

Buck, W. B. & Bratich, P. M. (1986). Activated charcoal: preventing unnecessary death by poisoning. *Veterinary Medicine*, 81, 73–77.

Deenik, J., Diarra, A., Uehara, G., Campbell, S., Sumiyoshi, Y. & Antal Jr., M. (2011). Charcoal ash and volatile matter effects on soil properties and plant growth in an acid ultisol. *Soil Science*, 176, 336–345.

Ding, Y., Liu, Y. X., Wu, W. X., Shi, D. Z., Yang, M. & Zhong, Z. K. (2010). Evaluation of biochar effects on nitrogen retention and leaching in multi-layered soil columns. *Water, Air and Soil Pollution*, 213, 47–55.

Gaskin, J., Speir, R., Harris, K., Das, K., Dewey Lee, R., Morris, L. A. & Fisher, D. S. (2010). Effect of peanut hull and pine chip biochar on soil nutrients, corn nutrient status, and yield. *Agronomy Journal*, 102, 623–633.

Gaskin, J., Steiner, C., Harris, K., Das, K. & Bibens, B. (2008). Effect of low-temperature pyrolysis conditions on biochar for agricultural use. *Transactions of the ASABE*, 51, 2061–2069.

Ghoshal, N. & Singh, K. P. (1995). Effects of farmyard manure and inorganic fertilizer on the dynamics of soil microbial bomass in a tropical dryland. *Agroecosystem. Biology and Fertility of Soils*, 19, 231–238.

Glaser, B., Haumaier, L., Guggenberger, G. & Zech, W. (2001). The 'Terra Preta' phenomenon: a model for sustainable agriculture in the humid Tropics. *Naturwissenschaften*, 88, 37–41.

Glaser, B., Lehmann, J. & Zech, W. (2002). Ameliorating physical and chemical properties of highly weathered soils in the Tropics with charcoal - a review. *Biology and Fertility of Soils*, 35, 219–230.

Güerena, D., Lehmann, J., Hanley, K., Enders, A., Hyland, C. & Riha, S. (2013). Nitrogen dynamics following field application of biochar in a temperate North American maize-based production system. *Plant and Soil*, 365, 239–254.

Huwig, A., Freimund, S., Käppeli, O. & Dutler, H. (2001). Mycotoxin detoxication of animal feed by different adsorbents. *Toxicology Letters*, 122, 179–188.

Ingold, M., Dietz, H., Sradnick, A., Joergensen, R. G., Schlecht, E. & Buerkert, A. (2015). Effects of activated charcoal and quebracho tannin amendments on soil properties in irrigated organic vegetable production under arid subtropical conditions. *Biology and Fertility of Soils*, 51, 367–377.

Ingold, M., Schiborra, A., Schlecht, A. & Buerkert, A. (2011). Influence of biochar and tannin amendments to goat manure on gaseous C and N emissions. *In*: Tielkes, E. (ed.), *Proceedings of Tropentag 2011, October 5–7, "Development on the Margin"*, Bonn, Germany. p. 226.

Jones, D. L., Rousk, J., Edwards-Jones, G., DeLuca, T. H. & Murphy, D. V. (2012). Biochar-mediated changes in soil quality and plant growth in a 3 year field trial. *Soil Biology and Biochemistry*, 45, 113–124.

Kloss, S., Zehetner, F., Wimmer, B., Buecker, J., Rempt, F. & Soja, G. (2014). Biochar application to temperate soils: effects on soil fertility and crop growth under greenhouse conditions. *Journal of Plant Nutrition and Soil Science*, 177, 3–15.

Lehmann, J. & Rondon, M. (2006). Bio-char soil

management on highly weathered soils in the humid Tropics. *In:* Biological Approaches to Sustainable Soil Systems. pp. 517–530, Taylor and Francis, Boca Routon, USA.

Lehmann, J., da Silva Jr., J. P., Steiner, C., Nehls, T., Zech, W. & Glaser, B. (2003). Nutrient availability and leaching in an archaeological anthrosol and a ferralsol of the Central Amazon Basin: fertilizer, manure and charcoal amendments. *Plant and Soil*, 249, 343–357.

Liang, B., Lehmann, J., Solomon, D., Kinyangi, J., Grossman, J., O'Neill, B., Skjemstad, O., Thies, J., Luizao, F. J., Petersen, J. & Neves, E. G. (2006). Black carbon increases cation exchange capacity in soils. *Soil Science Society of America Journal*, 70, 1719–1730.

Liu, J., Schulz, H., Brandl, S., Miehtke, H., Huwe, B. & Glaser, B. (2012). Short-term effect of biochar and compost on soil fertility and water status of a dystric cambisol in NE Germany under field conditions. *Journal of Plant Nutrition and Soil Science*, 175, 698–707.

Lundquist, E. J., Jackson, L. E. & Scow, K. M. (1999a). Wet-dry cycles affect dissolved organic carbon in two California agricultural soils. *Soil Biology and Biochemistry*, 31, 1031–1038.

Lundquist, E. J., Scow, K. M., Jackson, L. E., Uesugi, S. L. & Johnson, C. R. (1999b). Rapid response of soil microbial communities from conventional, low input, and organic farming systems to a wet/dry cycle. *Soil Biology and Biochemistry*, 31, 1661–1675.

Mukherjee, A. & Zimmerman, A. R. (2013). Organic carbon and nutrient release from a range of laboratory-produced biochars and biochar-soil mixtures. *Geoderma*, 193, 122–130.

Murdiati, T. D., McSweeney, C. & Lowry, J. B. (1991). Complexing of toxic hydrolysable tannins of yellowwood (*Terminalia oblongata*) and harendong (*Clidmeia hirta*) with reactive substances: an approach to preventing toxicity. *Journal of Applied Toxicology*, 11, 333–338.

Murphy, J. & Riley, J. P. (1962). A modified single solution method for the determination of phosphate in natural waters. *Analytica Chimica Acta*, 27, 31–36.

Nelson, N. O., Agudelo, S. C., Yuan, W. & Gan, J. (2011). Nitrogen and phosphorus availability in biochar-amended soils. *Soil Science*, 176, 218–226.

Poage, G. W., Scott, C. B., Bisson, M. G. & Hartman, F. S. (2000). Activated charcoal attenuates bittersweet toxicosis in sheep. *Journal of Range Management*, 53, 73–78.

Quaranta, L., Schlecht, E. & Schiborra, A. (2013). Supplementing goats with charcoal: effects on feeding behaviour and faecal nutrient output. *In:* Tielkes, E. (ed.), *Proceedings of Tropentag 2013, September 17–19, "Agricultural development within the rural-urban continuum", Stuttgart-Hohenheim, Germany.* p. 544.

Rawls, W. J., Pachepsky, Y. A., Ritchie, J. C., Sobecki, T. M. & Bloodworth, H. (2003). Effect of soil organic carbon on soil water retention. *Geoderma*, 116, 61–76.

Rogosic, J., Estell, R. E., Ivankovic, S., Kezic, J. & Razov, J. (2008). Potential mechanisms to increase shrub intake and performance of small ruminants in Mediterranean shrubby ecosystems. *Small Ruminant Research*, 74, 1–15.

Rogosic, J., Pfister, J. A., Provenza, F. D. & Grbesa, D. (2006). The effect of activated charcoal and number of species offered on intake of Mediterranean shrubs by sheep and goats. *Applied Animal Behaviour Science*, 101, 305–317.

Siegfried, K., Dietz, H., Gallardo, D. A., Schlecht, E. & Buerkert, A. (2013). Effects of manure with different C/N ratios on yields, yield components and matter balances of organically grown vegetables on a sandy soil of Northern Oman. *Organic Agriculture*, 3, 9–22.

Slavich, P. G., Sinclair, K., Morris, S. G., Kimber, S. W. L., Downie, A. & Van Zwieten, L. (2013). Contrasting effects of manure and green waste biochars on the properties of an acidic ferralsol and productivity of a subtropical pasture. *Plant and Soil*, 366, 213–227.

Sradnick, A., Ingold, M., Marold, J., Murugan, R., Buerkert, A. & Joergensen, R. G. (2013). Impact of activated charcoal and tannin amendments on microbial biomass and residues in an irrigated sandy soil under arid subtropical conditions. *Biology and Fertility of Soils*, 50, 95–103.

Steinbeiss, S., Gleixner, G. & Antonietti, M. (2009). Effect of biochar amendment on soil carbon balance and soil microbial activity. *Soil Biology and Biochemistry*, 41, 1301–1310.

Steiner, C., Teixeira, W. G., Lehmann, J., Nehls, T., Vasconcelos de Macedo, J. L., Blum, W. E. H. & Zech, W. (2007). Long term effects of manure, charcoal and mineral fertilization on crop production and fertility on a highly weathered Central Amazonian upland soil. *Plant and Soil*, 291, 275—290.

Van, D. T. T., Mui, N. T. & Ledin, I. (2006). Effect of method of processing foliage of *Acacia mangium* and inclusion of bamboo charcoal in the diet on performance of growing goats. *Animal Feed Science and Technology*, 130, 242—256.

Vance, E. D., Brookes, P. C. & Jenkinson, D. S. (1987). An extraction method for measuring soil microbial biomass C. *Soil Biology and Biochemistry*, 19, 703—707.

Villalba, J. J., Provenza, F. D. & Banner, R. E. (2002). Influence of macronutrients and activated charcoal on intake of sagebrush by sheep and goats. *Journal of Animal Science*, 80, 2099–2109.

Warnock, D. D., Lehmann, J., Kuyper, T. W. & Rillig, M. C. (2007). Mycorrhizal responses to biochar in soil - concepts and mechanisms. *Plant and Soil*, 300, 9–20.

Wichern, F., Müller, T., Joergensen, R. G. & Buerkert, A. (2004). Effects of manure quality and application forms on soil C and N turnover of a subtropical oasis soil under laboratory conditions. *Biology and Fertility of Soils*, 39 (3), 165—171.

Wu, J., Joergensen, R. G., Pommerening, B., Chaussod, R. & Brookes, P. C. (1990). Measurement of soil microbial biomass C by fumigation extraction - an automated procedure. *Soil Biology and Biochemistry*, 22, 1167—1169.

Zech, W., Senesi, N., Guggenberger, G., Kaiser, K., Lehmann, J., Miano, T. M., Miltner, A. & Schroth, G. (1997). Factors controlling humification and mineralization of soil organic matter in the tropics. *Geoderma*, 79, 117—161.

8

Land use change from rainforests to oil palm plantations and food gardens in Papua New Guinea: Effects on soil properties and S fractions

Richard Alepa , Rajashekhar Rao B. K.[*]

Department of Agriculture, The Papua New Guinea University of Technology, Lae 411, Morobe Province, Papua New Guinea

Abstract

Changes in soil sulfur (S) fractions were assessed in oil palm and food garden land use systems developed on forest vegetation in humid tropical areas of Popondetta in northern Province. The study tested a hypothesis that S in food gardens are limiting nutrient factor and are significantly lower than in plantations and forests. Subsistence food gardens are under long-term slash and burn practice of cropping and such practice is expected to accelerate loss of biomass S from the ecosystem. From each land use, surface soil (0–15 cm) samples were characterised and further pseudo-complete fractionated for S. Conversion of forest to oil palm production decreased ($p < 0.001$) soil pH and electrical conductivity values. The reserve S fraction in soil increased significantly ($p < 0.05$) due to oil palm production ($\sim 28\%$) and food gardening activity ($\sim 54\%$). However, plant available SO_4^{2-}-S was below $15\,mg\,kg^{-1}$ in the food garden soils and foliar samples of sweet potato crop indicating deficiency of plant available S. Soil organic carbon content (OC) was positively and significantly correlated to total S content ($r = 0.533$; $p < 0.001$) among the land use systems. Thus, crop management practices that affect OC status of the soils would potentially affect the S availability in soils. The possible changes in the chemical nature of mineralisable organic S compounds leading to enhanced mineralisation and leaching losses could be the reasons for the deficiency of S in the food garden soils. The results of this study conclude that long-term subsistence food gardening activity enriched top soils with reserve S or total S content at the expense of soluble S fraction. The subsistence cropping practices such as biomass burning in food gardens and reduced fallow periods are apparently threatening food security of oil palm households. Improved soil OC management strategies such as avoiding burning of fallow vegetation, improved fallows, mulching with fallow biomass, use of manures and S containing fertilisers must be promoted to sustain food security in smallholder oil palm system.

Keywords: food garden, land use change, oil palm plantation, organic carbon, reserve sulfur, soil fertility

1 Introduction

Papua New Guinea (PNG) is very diverse with respect to crops grown, cropping systems and their management techniques. Subsistence food production is the most important part of PNG agriculture. It provides most of the food consumed in the country – an estimated 83% of food energy and 76% of protein (Bourke & Harwood, 2009). On the other hand, intensive crop and animal production with commercial interests also feature the agricultural scenario. In PNG, land use conversion to subsistence agriculture is one of the major drivers (46%) of forest cover change while clearing for agricultural plantations contribute to small portion (1%) of forest cover change (Shearman *et al.*, 2009).

Use of forest land for agricultural production is known to change fertility and biological characters of

---
[*] Corresponding author

Department of Agriculture, The Papua New Guinea University of Technology, Private Mail Bag, Lae 411, Morobe Province, Papua New Guinea. Email: rsraobk@rediffmail.com

soil, including cycling of nutrients. Changes in soil properties have implications to the maintenance of productivity of new land use systems and subsequent soil management. Effects of land use conversion on stock of soil C, N, physical properties and microbiological activities, besides loss of biodiversity and greenhouse gas emissions are understood (Acin-Carrera *et al.*, 2013; Havaee *et al.*, 2014; Hunke *et al.*, 2015). Except for few studies (Solomon *et al.*, 2001; Wang *et al.*, 2006), there is paucity of information on the changes in S fractions due to land use change in the tropics.

In many South-East Asian countries prominent forest change took place to oil palm (*Elaeis guineensis* Jacq.) plantations. In PNG area planted to oil palm is around 130,000 ha and further expansion of oil palm production is expected at a rate of 3000 ha yr^{-1} (Nelson *et al.*, 2014b). In 2010, oil palm made up 56 % of the total value of PNGs agricultural exports (i.e., ~ US$ 1.5 billion). Oil palm industry in PNG includes a plantation sector and a smallholder sector. Plantation sector is operated by several large milling companies and the smallholders mostly depend on them for inputs and sale of oil palm fruits. The smallholder sector covers around 40 % area planted with oil palm, but produces only around 32 % of the oil palm (Fisher *et al.*, 2012). Thus, the productivity of smallholder oil palm is relatively poor. Conversion of forest vegetation to oil palm reported to maintain soil structure and major nutrients balance in the desirable range despite a decline in pH and exchangeable cations over medium-to long-term (Nelson *et al.*, 2014a).

In the smallholder oil palm blocks of PNG, apart from oil palms, an adjoining parcel of land is allocated to food gardening. Each smallholder typically owns an oil palm block. Food gardens are an integral part of smallholder oil palm blocks for meeting household food requirements and ensuring food security (Koczberski *et al.*, 2012). Food gardens are cultivated by subsistence agricultural activities and involve slashing the natural or fallow vegetation at varying stages of regeneration and burning the dried biomass to grow a variety of food crops. Soils are worked manually to prepare the land for planting food crops (mostly tubers, vegetables and fruits) in complex pattern to establish mixed food crop gardens (Bourke & Harwood, 2009). Finally, the lands are fallowed as crop productivity in gardens declines. Food gardens are periodically subjected to intentional biomass burning after a short-fallow period (1–2 years). Smallholders maintain small parcel of land under food gardens while other portions of food gardens could be under fallow. Fallow periods

have progressively shortened from 5–10 years due to limited arable land within the blocks and rapid population growth. Food security of smallholder oil palm producers depends largely on food gardening as estimated contribution of purchased food to the total calories consumed by these smallholders is just 16 % in PNG (Gibson & Rozelle, 1998).

The practice of slashing fallow vegetation and burning dried biomass before food crops causes considerable effects on cycling of nutrients and environment. Burning the biomass results in 38–84 % of the biomass carbon and nitrogen loss, including emission of sulfur (S) as SO_2 with adverse environmental consequences (Guild *et al.*, 2004; McCarty, 2011). Besides, the aerial inputs of S are quite small in pristine tropical regions of the World due to a low industrialisation level (Lara *et al.*, 2001). Thus, improper handling of biomass in subsistence cropping systems over long-term can have potential consequence on soil S concentrations (Jez, 2008) which could include extensive deficiency of S in cropping systems (Rajashekhar Rao *et al.*, 2012). The study was aimed to test the hypothesis that change of forest lands to long-term subsistence food gardening activity would lead to the depletion of total S and available S fractions in the soil. Such decline is observed in food gardens in highlands of PNG due to increased land intensification and nutrient mining linked to rapid population growth (Bourke & Harwood, 2009). We also contemplate that such depletion in S fractions would not occur in oil palm production systems due to absence of biomass burning.

2 Materials and methods

The hypothesis was tested by comparing top soil issues of smallholder food gardens against oil palm plantations, and rainforests. Soil samples were collected from four sites (Isivini, Igora, Sorovi and Puhemo villages) in Popondetta, northern Province of PNG. These sites were located between latitudes 08.721° and 08.797° S, and longitudes between 148.117° and 148.263° E. The elevation of the study sites was 65 to 258 m asl, and the geological structure consisted of basaltic and andesitic volcanic materials, brecciated lava, scoria, agglomerate, fanglomerate, tuff and volcanic ash types of materials. The land forms are mostly well preserved constructional land forms in a flat landscape with Vitrand soils (Soil Survey Staff, 2014) formed by alluvial re-deposition of tephra. The climate is a humid tropical with an annual rainfall of ~ 2380 mm. In these study sites, oil palm blocks were established by clearing primary forests under Land Settlement Schemes or Village Oil Palm Scheme beginning from 1967 (Koczberski et al., 2012). Alienated forest land after clearing was sub-divided into smallholding oil palm blocks (2–4 ha in size) for the primary purpose of oil palm production. Using a semi-structured questionnaire, 25 farmers involved in oil palm production were interviewed about cropping history and management practices and also confirmed that prior to oil palm planting the lands were under natural forest. Selection of farmers within the village was based on the road access and proximity of blocks to forest site. Oil palm blocks were established 18 to 52 years before the sampling and had oil palm trees planted in triangular pattern (125–140 palms ha^{-1}).

Apart from oil palms, these farmers practiced food gardening on an adjoining parcel of land. Diversified crops were grown in the food gardens such as sweet potato (Ipomoea batatas (L.) Lam.), taro (Colocasia esculanta Schott), Yam (Dioscorea sp.), cassava (Manihot utilissima Pohl), leafy vegetables such as pumpkins (Cucurbita pepo L.), aibika (Abelmoschus manihot (L.) Medik.), Amaranthus sp., papaya (Carica papaya L.), banana (Musa sp.), snake bean (Vigna unguiculata ssp. sesquipedalis [L.] Verdc.), peanut (Arachis hypogea L.) and many more. The ratio of food garden area to number of persons ranged from 0.06 ha person^{-1} to 0.71 ha person^{-1} with an average food garden size of 3150 m^2 person^{-1}. Sweet potato is the staple crop, providing more than 80 % dietary energy of the households and hence occupied considerable area of the food gardens.

The forest plots utilised for the study consisted of secondary forests, occasionally with stands of tall grass (Saccharum spontaneum L.) interspersed with pioneering low-or medium-crowned trees such as Piper aduncum L., Casuarina sp., Anisoptera sp., Spathodea campanulata P.Beauv., Trichospermum sp., and Macaranga sp..

2.1   Soil and plant sampling

In each site, geo-referenced soil samples from three land use systems (natural forest, oil palm and food gardens) were collected in November 2012. The land use systems were 0.1 to 2 km apart within a site, thus, had similar soil characters before different land use practices were imposed. A sampling unit of 5 m × 5 m was identified in the forest patches and food gardens. In each sampling unit 10 soil samples (0–15 cm depth) were collected randomly by a hand auger and pooled. Study was limited to surface samples because highest changes in soil biological and chemical fertility could be expected in the surface layer. The oil palm plantations

were sampled from the 'between-zones' (Nelson et al., 2014a). About 10 random soil samples were taken in 'between-zone' stretches and composited. A total of 50 soil samples (4 from forest, 24 from oil palm and 22 from food gardens) were studied. Foliar nutrient diagnosis was carried out by collecting leaf samples of 40–50 day old sweet potato crop (O'Sullivan et al., 1997). Air-dried leaf samples were transported to the laboratory and further oven dried at 80 °C until constant weight.

2.2   Soil-plant analysis and S fractionation

Air-dried soil samples (2 mm sieved) were analysed for chemical properties and sulfur fractions at the Unitech Analytical Services Laboratory, PNG. The soil pH and electrical conductivity were measured with a soil to water ratio of 1 : 5 in a potentiometer and conductivity bridge, respectively (Sparks, 1996). Soil particle size analysis was carried out by hydrometer method. Exchangeable cations were extracted in ammonium acetate at pH 7 (ibid.) and analysed for Ca, Mg and K in an ICP-OES (Varian 725ES model). A sub-sample (0.17 mm sieved) was analysed for the soil organic carbon content by Walkley and Black's wet digestion method (ibid.). Total soil N content was determined by modified Kjeldahl method (ibid.). Sulfate-S was extracted by shaking 20 g soil sample in 100 mL Ca(H$_2$PO$_4$)$_2$.H$_2$O for 1 h and S content measured by turbidimetric method in a UV-VIS spectrophotometer (Shimadzu UV-1800) at 420 nm (Chesnin & Yien, 1951). For soluble S determination, 10 g soil was shaken with 25 mL of 0.5 N ammonium acetate + 0.25 N acetic acid mixtures for 3 min and filtered through Whatman 42 paper (Bardsley & Lancaster, 1965). Oxidisable S in the extract was estimated turbidimetrically in a UV-VIS spectrophotometer at 420 nm (Chesnin & Yien, 1951). Soluble S fraction includes adsorbed, water soluble and a part of easily oxidisable organic forms of S. For total soil S analysis, 2.5 g soil (0.17-mm sieved) was ignited with NaHCO$_3$ powder in an electric muffle furnace at 500 °C for 3 h. Cooled off sample was then extracted with 25 ml of NaH$_2$PO$_4$.H$_2$O and filtered through Whatman 1 paper. Sulfur content in the extract was analysed by turbidimetric method (Bardsley & Lancaster, 1960; Ribeiro et al., 2001). The difference between total soil S and soluble S was considered as reserve S. Sweet potato leaf samples milled to < 1 mm particle size were analysed for total N, P, K and S contents. The concentration of total P, K, and S were determined in an ICP-OES (Varian 725ES model) after digesting in a mixture of HClO$_4$ and HNO$_3$. Leaf N content was estimated by micro-Kjeldahl method (Kalra, 1998).

2.3 Statistical Analysis

The data on soil chemical properties and S fractions were tested for normality by Shapiro-Wilk's test and if needed were log transformed prior to statistical analysis. Significant differences between the three land use systems regarding soil properties were determined by ANOVA and LSD tests in Minitab 16 software for windows (McDonald, 2014). Relationships between soil properties and S fractions were examined by calculating Pearson's linear correlation coefficients and simple regression analysis.

3 Results

Land use had significant ($p < 0.05$) effects on pH, electrical conductivity, exchangeable Ca, Mg and K concentrations (Table 1). The pH values of top soils significantly ($p < 0.001$) decreased due to conversion of forests to oil palm plantation. Electrical conductivity values of forest soils significantly ($p < 0.001$) decreased by $\sim 35\%$ due to oil palm production. Food garden soils showed slight enrichment of the basic cations (exchangeable Ca, Mg and K) whilst, the oil palm soils registered their depletion compared to adjacent forest patches.

Food gardening activities significantly ($p < 0.05$) improved the reserve S concentrations of forest soils (Fig. 1a). The soluble S fraction in food garden soils was significantly ($p < 0.05$) lower than that in oil palm soils. There was no significant effect of land use change on status of total S.

The reserve S fraction accounted for 90.6-94.5% of the total S among land use systems. Sulfur status and availability of the soils is mostly influenced by soil OC content and elemental ratios of nutrients in the soil organic matter. The OC: total S ratio and OC: reserve S ratio in oil palm soils were remarkably ($p < 0.05$) greater than in food garden soils (Fig 1b). Conversion of forests to oil palm production increased OC: total S ratio ($\sim 24\%$) and OC: reserve S ratio's ($\sim 22\%$) even as the conversion to food garden did not impact these ratios.

Table 1: *Physical and chemical properties of the surface (0–0.15 m) soil samples in different land use systems (n=50). Values in brackets indicate standard deviation.*

Soil properties		Natural forest (n=4)	Oil palm (n=24)	Food garden (n=22)	ANOVA p value
pH	Range	6.26–6.40	5.13–6.60	5.61–6.54	
	Mean	6.32 (0.06)[a]	5.69 (0.29)[b]	6.09 (0.24)[a]	0.0001
Electrical conductivity (dS m^{-1})	Range	0.54–0.91	0.22–0.71	0.33–0.79	
	Mean	0.73 (0.20)[a]	0.46 (0.14)[c]	0.59 (0.12)[b]	0.0001
Organic carbon (g kg^{-1})	Range	17.2–29.4	14.7–57.5	20.4–53.9	
	Mean	22.8 (5.13)	35.5 (12.7)	32.5 (9.97)	0.688
Total nitrogen (g kg^{-1})	Range	3.60–6.10	3.10–7.90	3.80–10.9	
	Mean	4.80 (1.28)	5.61 (1.30)	6.24 (1.43)	0.604
Exchangeable Ca (cmol kg^{-1})	Range	1.63–4.95	0.56–5.35	1.65–5.26	
	Mean	3.02 (1.39)[ab]	2.27 (1.18)[b]	3.23 (0.82)[a]	0.008
Exchangeable Mg (cmol kg^{-1})	Range	0.16–0.63	0.07–0.74	0.42–0.80	
	Mean	0.49 (0.22)[ab]	0.35 (0.18)[b]	0.57 (0.12)[a]	0.0001
Exchangeable K (cmol kg^{-1})	Range	0.05–0.18	0.02–0.25	0.06–0.26	
	Mean	0.12 (0.06)[ab]	0.09 (0.06)[b]	0.16 (0.07)[a]	0.003
Clay (g kg^{-1})	Range	92–112	29.2–176	51.8–164	
	Mean	102 (9.52)	76.8 (21.8)	77.3 (24.7)	0.131
Silt (g kg^{-1})	Range	136–224	80–284	80–416	
	Mean	166 (40.6)	141 (61.8)	188 (65.5)	0.341
Sand (g kg^{-1})	Range	664–764	624–824	512–864	
	Mean	732 (46.1)	760 (54.4)	734 (72.0)	0.301

[a-c] values within the same row with different superscript are significantly different at $p < 0.05$

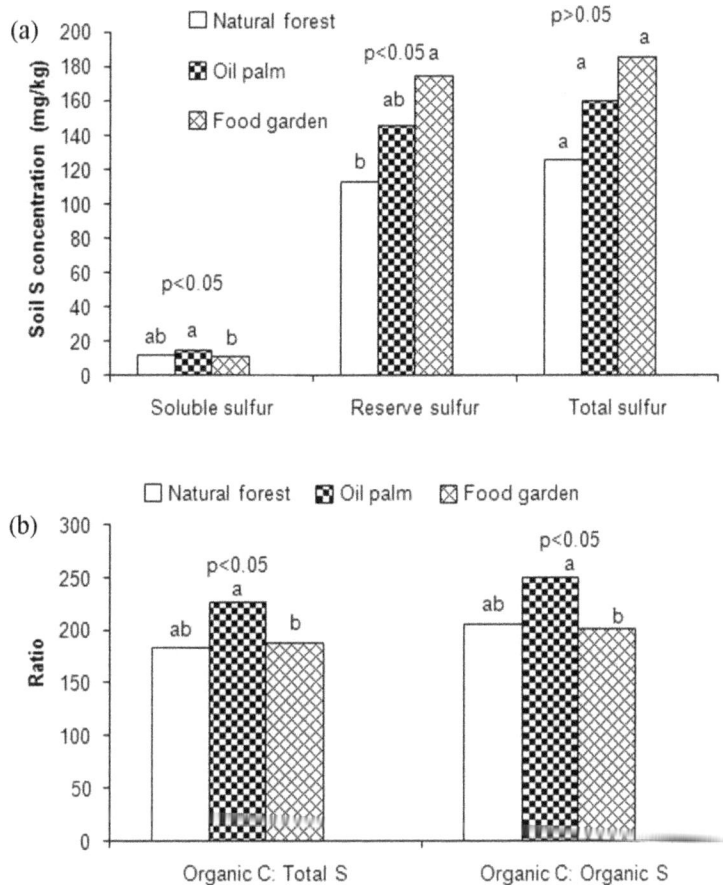

Fig. 1: *(a) Sulfur fractions and (b) Ratio of sulfur fractions in the land use systems. Same lower case letters on a column series are not significantly different ($p > 0.05$).*

Table 2: *Interpretation categories (Hari & Dwivedi, 1994) of plant available SO_4^{2-}-S in soils of food gardens and oil palm plantations.*

		Range of soil SO_4^{2-}-S ($mg\,kg^{-1}$)			
		0–10 Deficient	10–15 Low	15–20 Medium	>20 High
Food garden	Frequency (N)	3	19	0	0
	Mean (standard deviation) ($mg\,kg^{-1}$)	5.27 (1.80)	12.3 (1.10)	0	0
	Percent	13.6	86.4	0	0
Oil palm	Frequency (N)	13	5	2	4
	Mean (standard deviation) ($mg\,kg^{-1}$)	5.60 (2.65)	11.2 (0.80)	16.5 (2.1)	22.8 (2.2)
	Percent	54.2	20.8	8.33	16.7

Table 3: *Foliar nutrient status and critical nutrient concentration (O'Sullivan et al., 1997) of sweet potato crop in food gardens of oil palm blocks.*

Nutrient	Critical nutrient concentration	Observed range in leaf samples	Mean (SD)	% deficient samples
Nitrogen (%)	4.20	2.48–4.63	3.80 (1.10)	72.7
Phosphorous (%)	0.22	0.23–0.41	0.34 (0.05)	0
Potassium (%)	2.60	2.10–3.16	2.72 (0.36)	27.3
Sulfur (%)	0.34	0.10–0.19	0.15 (0.03)	100

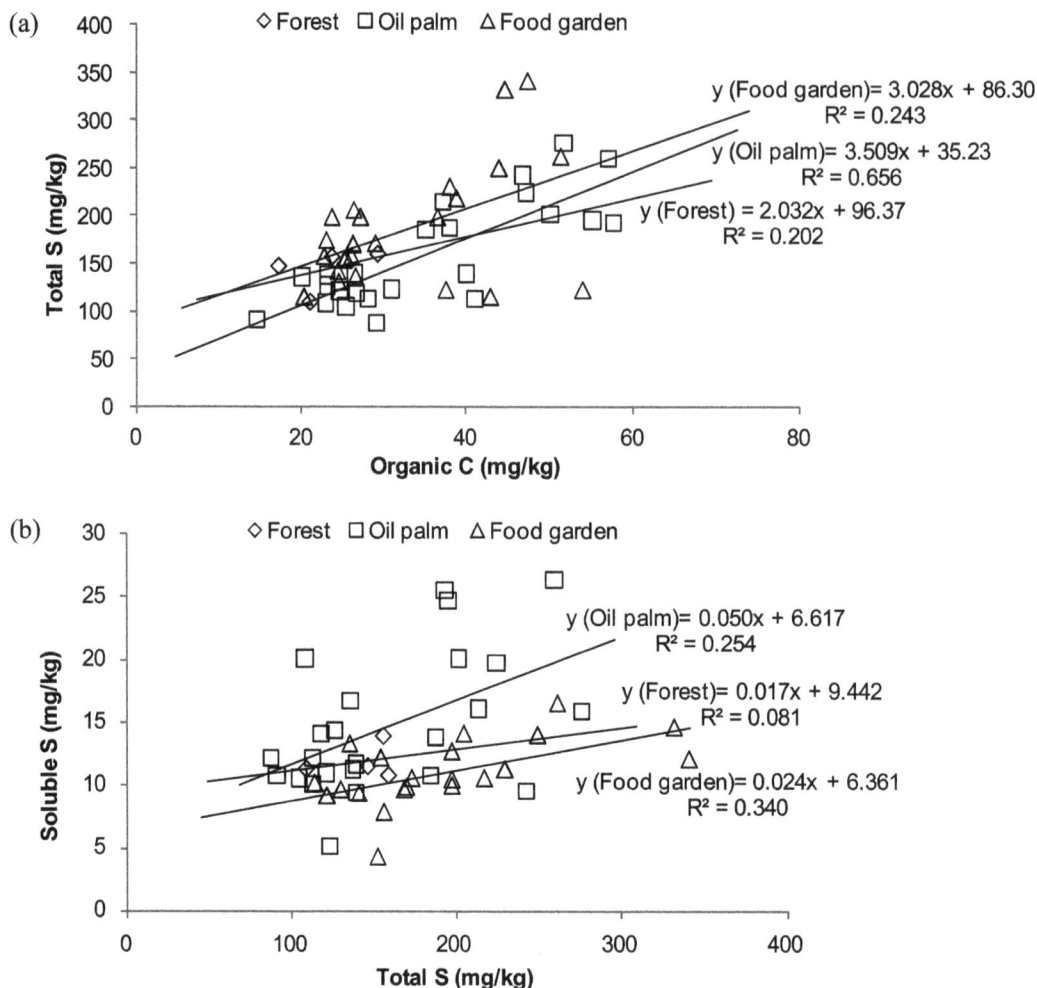

Fig. 2: *(a) Relationship between total S and OC (y = a + bx). y-axis is for total S and x-axis for OC; (b) Relationship between soluble S and total S (y = a + bx). y-axis is for soluble S and x-axis for total S in different land use systems.*

In food gardens 100 % of the topsoil samples had the SO_4^{2-}-S levels below 15 $mg\,kg^{-1}$, thus, categorised into 'deficient' and 'low' fertility (Table 2). On the contrary, ∼ 17 % of the soil samples from oil palm plantations had the SO_4^{2-}-S levels in 'high' S fertility category following Hari & Dwivedi (1994). Foliar properties revealed that 100 % of the sweet potato leaf samples had S content below critical nutrient concentration of 0.34 % (Table 3); followed by the deficiency of N (72.7 %) and K (27.3 %) while the P concentration was 'adequate' in all foliar samples.

The soil soluble S fraction (plant available S) was positively and significantly correlated (r=0.412; $p <$ 0.01) to soil OC content. The soil OC content (r=0.533; $p < 0.001$), was also highly correlated with total S content (r=0.533; $p < 0.001$) (Fig 2a). The soil soluble S content showed a positive relationship with total S, which varied among the land use systems (Fig 2b).

4 Discussion

The results of this study conclude that long-term subsistence food gardening activity enriched top soils with reserve S or total S content at the expense of soluble S fraction. The subsistence cropping practices such as biomass burning in food gardens and reduced fallow periods are apparently threatening food security of oil palm households.

A larger total S pool was expected in oil palm soils than the food garden soils. The repeated burning of fallow vegetation in the food gardens may lead to loss of S from burning biomass. In contrast to the prediction the mean total S content showed an increasing trend in the following order: forest system < oil palm plantations < food gardens. The total S status observed among the three land use soils were comparable to that reported by Itanna (2005) in Ethiopia and slightly greater than that observed by Ribeiro *et al.* (2001) for Oxisols of Brazil. Besides, deep rooted oil palm and forest trees are efficient in pumping-up and recycling S and other nutrients from the lower soil layer to the surface soil (Solomon *et al.*, 2001). However, our results showed enrichment of S in surface soils of food gardens. The reserve S fraction too showed similar pattern of that of total S. The availability of S from total S or reserve S fractions to crops entirely depends on microbial mineralisation which in turn is influenced by the C : S ratio of soils. A net mineralisation is expected in both food gardens and forest soils as in both C : total S ratio is lower than 200 (Blum *et al.*, 2013). The lower mean OC : reserve S ratio of food garden soils indicate greater S mineralisation rates assuming other environmental variables are uniform among the land use systems (Fig. 1b). The biomass charring process in food gardens can create oxidized S species enabling the faster mineralisation of S from the biomass and consequent leaching losses (Blum *et al.*, 2013). Besides, fallow species can contribute nutrients from deeper soil layers through recycling process during fallowing practice of the food gardens (Hartemink, 2003).

In the study there was no significant difference between soluble S status of soils under oil palm and reference forest sites (Fig. 1). Slightly higher S status of oil palm soils could be due to farmers' application of N fertiliser (ammonium sulfate) up to 3 kg palm^{-1} which also contains S (up to 24 % S), mostly in 'between-zone'. Herbaceous, annual crops generate greater biomass and therefore food garden soils had greater soil OC and reserve S status (Aguiar *et al.*, 2014). Although the burning vegetation can cause gaseous emission of C, N and S (Guild *et al.*, 2004), frequent burning promotes an increase of soil OC due to the increased surface crop biomass in the subsequent crop cycle and enhanced OC pool from dead roots (Zhao *et al.*, 2012). Besides, transformations exerted by fire on soil humus, including the accumulation of new particulate C forms bound to S, which are highly resistant to oxidation and biological degradation such as black carbon (González-Pérez *et al.*, 2004).

Except for the pH and electrical conductivity of the soils, other properties did not lend any evidence of chemical degradation due to conversion of forest to either oil palm or food gardens. Oil palm production decreased the pH of forested soils considerably, while, food gardening showed declining electrical conductivity values of soils. Slash and burning of fallow vegetation in Ghana reported to add 1.5–3 t ha^{-1} of Ca and 180 kg ha^{-1} Mg to soil (Schulte & Ruhiyat, 1998).

Conversely, a decline in pH, exchangeable Ca, Mg and electrical conductivity values in oil palm could be due to non-agricultural acidification processes such as N-cycling, nitrification, leaching of nitrates, uptake and sequestration of non-acidic cations in oil palm biomass and harvested fruit bunches (Nelson *et al.*, 2014a). The burning of fallow vegetation and subsequent incorporation of ash material in food garden soils is expected to contribute to enrichment of bases. However, other soil processes such as addition and decomposition of crop biomass and loss of bases through leaching, crop uptake and runoff could have counteracted accretion of basic cations in food gardens.

More land use change is expected in the coming years to expand smallholder oil palm production largely due to socio-economic benefits associated with the crop. Unfortunately, much of the research on oil palm nutrition is directed towards rationalizing use of nutrients such as N, P, K, Mg and B (Comte *et al.*, 2012). There is not much information available on the extent of S deficiency in oil palm plantations except for report from Gerendás *et al.* (2011) in Indonesia. Sulfur status has been found to be severely deficient in several oil palm blocks based on leaf sample analysis. Deficiency of S leads to decreasing use efficiency of N fertilisers in oil palm which is not acceptable as fertilisation costs make up more than 60–65 % of the upkeep cost of plantations. Equally concerning was the fertility status of the food gardens within the oil palm blocks. Besides S, N and K were in deficient supply to the sweet potato crop in food gardens. With the increasing population of oil palm dependents and concomitant intensification of food gardening multi-nutrient deficiencies are emerging. From the results of the study, it appears that sustaining garden food production and oil palm productivity requires use of mineral fertilisers and cheap S sources. For managing S deficiency, mineral S fertilisers such as sulfate of ammonia or kieserite (MgSO$_4$.H$_2$O) could be used.

Further studies are required to evaluate improved soil organic matter management strategies and use of mineral S sources in the food gardens. Suggested strategies are avoiding burning of fallow vegetation, improved fallows, soil mulching with fallow biomass, besides, use of manures and S containing fertilisers. Possible organic matter management strategies may include improved fallows and mulching slashed fallow vegetation. Such strategies may help to improve S cycling in the soil-plant system, besides reducing S losses to atmosphere.

Acknowledgements

The first author is thankful to the Research Committee, Papua New Guinea University of Technology for funding this work.

References

Acin-Carrera, M., Marques, M. J. E., Carral, P., Alvarez, A. M., Lopez, C., Lopez, B. M. & Gonzalez, J. A. (2013). Impacts of land-use intensity on soil organic carbon content, soil structure and water-holding capacity. *Soil Use and Management*, 29, 547–556.

Aguiar, M. I., Fialho, J. S., Campanha, M. M. & Oliveira, T. S. (2014). Carbon sequestration and nutrient reserves under different land use systems. *Revista Árvore*, 38 (1), 81–93.

Bardsley, C. E. & Lancaster, J. D. (1960). Determination of reserve sulfur and soluble sulfates in soils. *Soil Science Society of America Proceedings*, 24, 265–268.

Bardsley, C. E. & Lancaster, J. D. (1965). Sulphur. *In:* Black, C. A. (ed.), *Methods of Soil Analysis, Part 2: Chemical and microbiological properties*. pp. 1102–1116, American Society of Agronomy, Madison, USA.

Blum, S. C., Lehmann, J., Solomon, D., Caires, E. F. & Alleoni, L. R. F. (2013). Sulfur forms in organic substrates affecting S mineralization in soil. *Geoderma*, 201, 156–164.

Bourke, R. M. & Harwood, T. (2009). *Food and agriculture in Papua New Guinea*. ANU E-press, The Australian National University, Canberra.

Chesnin, L. & Yien, C. H. (1951). Turbidimetric determination of available sulfates. *Soil Science Society of America Proceedings*, 15, 149–151.

Comte, I., Colin, F., Whalen, J. K., Grunberger, O. & Caliman, J. (2012). Agricultural practices in oil palm plantations and their impact on hydrological changes, nutrient fluxes and water quality in Indonesia: A review. *Advances in Agronomy*, 116, 71–124.

Fisher, H., Winzenried, C. & Sar, L. (2012). Oilpalm pathways: an analysis of ACIARs oil palm projects in Papua New Guinea. ACIAR Impact Assessment Series Report No. 80. Australian Centre for International Agricultural Research: Canberra, 63 pp.

Gerendás, J., Donough, C. R., Oberthür, T., Lubis, A., Indrasuara, K., Dolong, T., Abdurrohim, G. & Rahmadsyah (2011). Function and nutrient status of sulphur in oil palm in Indonesia. Conference on International Research on Food Security, Natural Resource Management and Rural Development. Tropentag 2011 Oct 5–7, 2011. University of Bonn, Germany. URL http://www.tropentag. de/2011/abstracts/full/301.pdf

Gibson, J. & Rozelle, S. (1998). *Results of the household survey component of the 1996 poverty assessment for Papua New Guinea*. Population and Human Resources Division, The World Bank. Washington DC, USA.

González-Pérez, J. A., González-Vila, F. J., Almendros, G. & Knicker, H. (2004). The effect of fire on soil organic matter-a review. *Environment International*, 30, 855–870.

Guild, L. S., Kauffman, J. B., Cohen, W. B., Hlavka, C. A. & Ward, D. E. (2004). Modeling biomass burning emissions for Amazon forest and pastures in Rondonia, Brazil. *Ecological Applications*, 14 (4), S232–S246.

Hari, R. & Dwivedi, K. N. (1994). Delineation of sulfur-deficient soil groups in the central alluvial tract of Uttar Pradesh. *Journal of Indian Society of Soil Science*, 42 (2), 284–286.

Hartemink, A. E. (2003). Sweet potato yields and nutrient dynamics after short-term fallows in the humid low lands of Papua New Guinea. *Netherlands Journal of Agricultural Sciences*, 50, 297–320.

Havaee, S., Ayoubi, S., Mosaddeghi, M. R. & Keller, T. (2014). Impacts of land use on soil organic matter and degree of compactness in calcareous soils of central Iran. *Soil Use and Management*, 30, 2–9.

Hunke, P., Roller, R., Zeilhofer, P., Schröder, B. & Mueller, E. (2015). Soil changes under different landuses in the Cerrado of Mato Grosso, Brazil. *Geoderma Regional*, 4, 31–43.

Itanna, F. (2005). Sulfur distribution in five Ethiopian Rift Valley soils under humid and semi-arid climate. *Journal of Arid Environment*, 62, 597–612.

Jez, J. (2008). *Sulfur: A Missing Link between Soils, Crops and Nutrition*. ASA-CSSA-SSSA Madison, Wisconsin, USA.

Kalra, Y. P. (1998). *Handbook of Reference Methods for Plant Analysis*. CRC press, Taylor and Francis Group, New York, USA.

Koczberski, G., Curry, G. N. & Bue, V. (2012). Oil palm, food security and adaptation among smallholder households in Papua New Guinea. *Asia Pacific Viewpoint*, 53, 288–299.

Lara, L. B. L. S., Artaxo, P., Martinelli, L. A., Victoria, R. L., Camargo, P. B., Krusche, A., Ayers, G. B., Ferraz, E. S. B. & Ballester, M. V. (2001). Chemical composition of rainwater and anthropogenic influences in the Piracicaba River Basin, Southeast Brazil. *Atmospheric Environment*, 35, 4937–4945.

McCarty, J. L. (2011). Remote sensing-based estimates of annual and seasonal emissions from crop residue burning in the contiguous United States. *Journal of Air and Waste Management Association*, 61, 22–34.

McDonald, J. H. (2014). *Handbook of biological statistics*. Sparky House Publishing, Baltimore, Maryland. USA.

Nelson, P. N., Banabas, M., Nake, S., Goodrick, I., Webb, M. J. & Gabriel, E. (2014a). Soil fertility changes following conversion of grassland to oil palm. *Soil Research*, 52, 698–705.

Nelson, P. N., Gabriel, J., Filer, C., Banabas, M., Sayer, J. A., Curry, G. N., Koczberski, G. & Venter, O. (2014b). Oil Palm and deforestation in Papua New Guinea. *Conservation Letters*, 7, 188–195.

O'Sullivan, J. N., Asher, C. J. & Blamey, F. P. C. (1997). *Nutrient disorders of sweet potato, ACIAR monograph*. ACIAR, Canberra, Australia.

Rajashekhar Rao, B. K., Krishnappa, K., Srinivasarao, C., Wani, S. P., Sahrawat, K. L. & Pardhasaradhi, G. (2012). Alleviation of multinutrient deficiency for productivity enhancement of rain-fed soybean and finger millet in the semi-arid region of India.

Communications in Soil Science and Plant Analysis, 43, 1427–1435.

Ribeiro, E. S., Dias, L. E., Alvarez, V. H., Mello, J. W. V. & Daniels, W. L. (2001). Dynamics of Sulfur Fractions in Brazilian Soils Submitted to Consecutive Harvests of Sorghum. *Soil Science Society of America Journal*, 65 (3), 787–794.

Schulte, A. & Ruhiyat, D. (1998). *Soils of Tropical Forest Ecosystems*. Springer-Verlag, Heidelberg, Berlin, Germany.

Shearman, P., Ash, J., Mackey, B., Bryan, J. E. & Lokes, B. (2009). Forest Conversion and Degradation in Papua New Guinea between 1972–2002. *Journal of Biotropica*, 41, 379–390.

Soil Survey Staff (2014). *Keys to Soil Taxonomy*. (12th ed.). USDA Natural Resources Conservation Service, Washington DC, USA.

Solomon, D., Lehmann, J., Tekalign, M., Fritzsche, F. & Zech, W. (2001). Sulfur fractions in particle-size separates of the sub-humid Ethiopian highlands as influenced by land use changes. *Geoderma*, 102, 41–59.

Sparks, D. L. (1996). *Methods of Soil Analysis Part 3-Chemical Methods. SSSA Book Series 5*. SSSA and ASA, Madison, Wisconsin, USA.

Wang, J., Solomon, D., Lehmann, J., Zhang, X. & Amelung, W. (2006). Soil organic sulfur forms and dynamics in the Great Plains of North America as influenced by long-term cultivation and climate. *Geoderma*, 133, 160–172.

Zhao, H., Tong, D., Lin, Q., Lu, X. & Wang, G. (2012). Effect of fires on soil organic carbon pool and mineralization in a Northeastern China wetland. *Geoderma*, 189, 532–539.

9

Adoption of an improved bean seed variety and consumption of beans in rural Madagascar

Christine Bosch *, Manfred Zeller , Domenica Deffner

Hans-Ruthenberg Institute, University of Hohenheim, 70593 Stuttgart, Germany

Abstract

This paper studies access to, and adoption of improved seed, as well as the diffusion of improved seed information in a remote area of central Madagascar. The analysis is based on panel data gathered from 2012 to 2014 from 390 households in three villages. In 2013, a randomised control trial was applied. Half of the 390 households were randomly assigned to receive improved lima bean seed (*Phaseolus lunatus*), which were specifically bred for dry regions. Of the seed-receiving households, 50 % were randomly assigned to receive information on how to store, plant, and cultivate the improved seed, as the variety was unfamiliar in the region. The control group and the two treatment groups are compared with respect to baseline characteristics, bean cultivation, information exchange with other farmers, legume consumption, and willingness to pay (WTP) for improved bean seed. To account for non-compliance, contamination and spillover effects, local average treatment effects (LATE) are estimated. Of the seed-receiving households, 54 % cultivated the seed, reaping an average yield of 6.3 kg per kg of seed obtained. Seed information did not lead to higher yields. A small significant positive impact of seed distribution on legume consumption is found. WTP is 171 % of the local market price for bean seed, free provision of seed and information did not result in a higher WTP.

Keywords: information dissemination, legumes, local average treatment effects, technology adoption, willingness to pay

1 Introduction

Agricultural productivity in Madagascar is low due to climate hazards and limited adoption of improved agricultural technologies. This limited adoption is attributed to: labour and liquidity constraints at planting time (Moser & Barrett, 2003, 2006); increased prices and high transaction costs for inputs due to remoteness and poor transport infrastructure (Stifel & Minten, 2008; Minten *et al.*, 2013); and risk aversion, social conformity, and customs (Moser & Barrett, 2003; Barrett *et al.*, 2004; Barrett, 2008; Stifel *et al.*, 2011). In addition to

the low demand from farmers, low supply and the resulting limited access to agricultural inputs are constraints to adoption (Minten *et al.*, 2013).

This low supply and uptake of technologies has also been studied in the seed market and through the lens of seed aid as a disaster response (Sperling *et al.*, 2008). Several authors (Alemayehu, 2009; Sperling & McGuire, 2010; Katungi *et al.*, 2011b) argue that informal seed markets, while not fully understood, present a potential for more, higher quality, and more diversified seed. Establishing links between variety innovators and those who can multiply and distribute seed at affordable prices, is suggested. Newly created seed material could be delivered directly to important community-

---
* Corresponding author
Email: christine.bosch@uni-hohenheim.de

based nodes, instead of solely to parastatal and commercial entities (Sperling *et al.*, 2008; Gibson, 2013).

Randomised control trials to study adoption and diffusion of improved agricultural technologies are increasingly popular (Banerjee & Duflo, 2008; Duflo *et al.*, 2008; Barrett & Carter, 2010). Some of the experiments showed that rates of return of improved agricultural technologies, like improved seed or fertiliser, are not as positive in real world situations as they are in demonstration plots or controlled conditions (Vandercasteelen *et al.*, 2013; Bulte *et al.*, 2014). One study found positive impacts on yields, but not on profits (Beaman *et al.*, 2013). Fixed costs, including the psychological costs of changing habits, might be substantial (Duflo *et al.*, 2011). A growing number of studies is testing how information can best be disseminated among farmers (Hotz *et al.*, 2012; Vasilaky, 2013; Culbertson *et al.*, 2014).

Under- and malnutrition is prevalent in Madagascar, where 33 % of the population is undernourished (FAOSTAT, 2015). Calories are mainly obtained from staple foods, such as rice and cassava. Given the poor diets, hidden hunger is widespread. The share of cereals, roots and tubers in Madagascar's dietary energy supply was 79 % in 2011, which is by far the highest value globally. At 48 g per capita per day, protein intake is very low and is less than the average of all least developed countries (FAOSTAT, 2011). Higher dietary diversity among Malagasy children is highly correlated with micronutrient intake (Moursi *et al.*, 2008). Legumes improve diets by adding essential vitamins and minerals, especially iron, and are high in protein and dietary fibre (Aykroyd & Doughty, 1982).

The lima bean (*Phaseolus lunatus / kabaro* in Malagasy) is a perennial plant that achieves highest yields in the hot and humid tropics. Lima beans are tolerant to mild drought, high temperatures, and poor soils. In Madagascar, mean yield is one ton per hectare (Ministry of Agriculture, 2004). To raise yield and quality, a research station of FOFIFA (National Centre of Applied Research and Rural Development) in Toliara, southwestern Madagascar, is developing an improved variety (personal interview with FOFIFA representative, 2014).

Willingness to pay (WTP) can be defined as the amount of money an individual assigns to the benefits or costs of a particular product or service. WTP surveys have often been used to assess social benefits of environmental policies or projects. The application to private goods, like agricultural products, is rather un-

common, as these goods are traded in markets and have observable prices. However, when it comes to the assessment of non-traded goods or value components that are not (yet) reflected by real market data, WTP surveys are a useful tool. Recent studies assess WTP for improved or certified seed (Dalton *et al.*, 2011; Kaguongo *et al.*, 2014; Kassie *et al.*, 2014), traditional and locally produced foods (Chelang'a *et al.*, 2013), fertiliser (Minten *et al.*, 2007), and extension services (Ulimwengu & Sanyal, 2011).

This paper explores the impact of seed and seed information distribution on yield, willingness to pay and consumption, for the case of lima beans. The following research questions are addressed: (1) If seed is distributed for free, do households plant or consume it? (2) Does the inclusion of agronomic information with the distributed seed increase seed utilisation and bean yield? (3) How much are farmers willing to pay for improved seed? (4) Does the inclusion of agronomic information with the distributed seed increase the willingness to pay for improved seed? (5) Does seed receipt and bean cultivation increase legume consumption?

2 Materials and methods

2.1 Randomised control trial: sampling strategy and study design

The study was carried out within the framework of a household panel that ran from 2009 to 2014 in three villages in the community of Fenoarivo, which belongs to the district of Ambalavao in the Haute Matsiatra region (Fig. 1). Fenoarivo is the local centre of administration, is connected with transport, and hosts a weekly market. The other two villages in the sample, Maroilo and Sakafia, are eight and twelve kilometres away from Fenoarivo, respectively.

Baseline characteristics originate from a household survey that took place between December 2012 and February 2013. Bean seed was obtained from the research station of FOFIFA in Toliara and distributed during a second survey from September to November 2013. Of the 390 eligible households in the panel, 196 (50 %) were randomly assigned to receive 0.6 kg of bean seed[1]. Of those, 84 (43 %) were randomly assigned to receive detailed information on how to store and cultivate the seed, following recommendations by the Ministry of

---

[1] Households received 1.5 *kapoaka* of beans. *Kapoaka* is a common expression for milk tins used to measure amounts of various items. For beans, 1 *kapoaka* is equivalent to approximately 390 g.

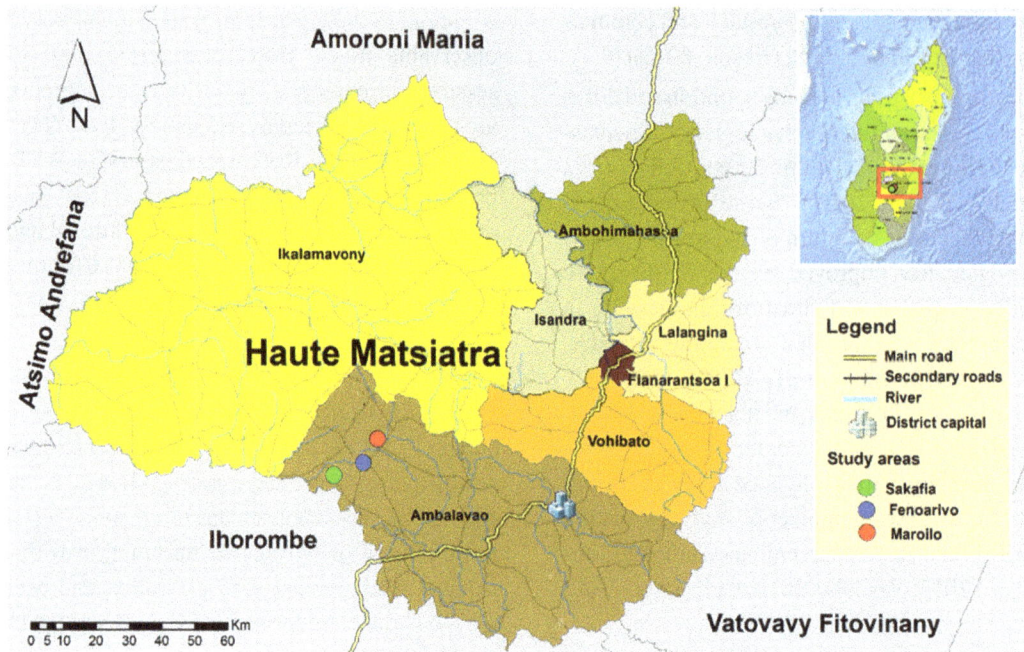

Fig. 1: *Map of the study area.*

Agriculture (2013). It was recommended to plant the seed in April 2014 and to harvest in September/October 2014.

From November to December 2014, after the harvest of the beans, a follow-up survey and additional focus group discussions were carried out. Net-maps, a participatory, interview-based mapping technique developed by Schiffer & Hauck (2010), were used to enable participants to visualize and discuss the bean seed market, the actors involved, their linkages, their importance and influence, and their individual objectives, as well as existing knowledge about and attitudes towards improved seed. Because of the remoteness of the research area, not all actors involved could be present. Representatives of the agricultural extension service and the research station of FOFIFA in Toliara were interviewed separately.

2.2 Descriptive statistics

The baseline characteristics used in this study are variables that are expected to influence the adoption of lima bean seed and related production and consumption outcomes. In addition to demographic and agricultural characteristics, this includes information on innovations and social capital of the households. Innovations were elicited for five years prior to the interview and include dummies for five categories, namely crop diversification, technology adoption, access to new markets and traders, as well as innovations in work organisation, resulting in a score ranging from 0 to 5 innovations (Hart-

mann & Arata, 2011). The index on social capital comprises seven questions on trust, honesty, and willingness to help in the villages. Answers range from full agreement (1) to no agreement at all (5) on a Likert scale. The index is the mean of the seven answers. Attitudes towards work are elicited using the question "Can hard work improve your living standard?".

Some villagers have cultural taboos (*fady* in Malagasy) concerning certain bean types that prohibit consumption, cultivation, or talking about the beans, since they are believed to inhibit rainfall or successful prevention of cattle rustling. Common beans (*Phaseolus vulgaris* / *tsaramaso* in Malagasy) are widely cultivated in the area and not considered as taboo for the village fields. Apart from a climbing variety grown in home gardens, lima beans were an unknown bean species in the villages and it was unknown whether they are assigned with a taboo when grown in fields. Therefore, as a proxy, we use taboos for bambara groundnut (*Vigna subterranea* / *voanjobory* in Malagasy), a legume introduced from West Africa and widely believed to inhibit rainfall if cultivated in village fields.

In the follow-up survey, the following issues were examined in the control and the two treatment groups (seed-only and seed-and-information-receiving households): seed utilisation, problems during cultivation, the importance and diffusion of information, potential spillover effects regarding this information, evaluation of and WTP for improved seed, and bean consumption. For yield the seed multiplication rate is used as an operational proxy variable, given in kg per kg of seed. Lima bean and legume consumption data were elicited for different recall periods. Ravallion (2008) and Deaton (2010) recommend the use of intermediate indicators, in addition to outcome indicators, to understand the processes determining impacts. Descriptive statistics are based on the initial assignment of the households to the three groups. Statistical differences for categorical variables were determined with the help of chi-square tests, and for ordinal and interval variables with Kruskal-Wallis tests. The Wilcoxon rank-sum test served as the post-hoc test. All tests were done with STATA.

The contingent valuation method (CVM) is a survey-based method to elicit WTP which does not rely on experimental or real purchase decisions (Whittington, 1998; Bateman et al., 2002). Potential buyers are asked how much they would be willing to pay for the product contingent on a description of an alternative or a hypothetical scenario. Following Haab & McConnell (2002), enumerators described the benefits of improved bean seed (yield roughly twice that of locally available beans, higher pest and disease resistance, and higher drought tolerance) and explained the need to pay. Due to the fact that lima beans were assigned with a taboo, households were given the choice between lima and common beans. Enumerators then showed the household a so-called payment card with a list of price ranges, ordered from lowest to highest. The lower bound was set roughly double the price of bean seed available at the local market. The household was asked to pick the range that included the maximum amount they were willing to pay. WTP is estimated by taking the mean value of these price ranges. Compared with open questions, the payment card has the advantage that it offers respondents a visual aid for the choice.

To check for the reliability of stated WTP, additional questions were included: "Do you think the seed would be available at the market?", "What amount of seed would you buy for the indicated price?", and "Would you be able to afford the seed at the indicated price?".

2.3 Empirical strategy

Impact evaluation generally aims to assess a program's effect against a counterfactual, showing the situation in the absence of the program (Rubin, 1974; Ravallion, 2008). Random assignment of households to a treatment group seeks to ensure that the control group is a valid counterfactual and allows simple comparisons of outcomes. If there are no differences in household characteristics between the control and treatment groups

at baseline, any changes of outcomes can be attributed solely to the program. Significant differences in baseline characteristics could indicate a problem in the random assignment of treatment.

The average treatment effect (ATT) is the average gain of households from having received the seed, whether they received them from an enumerator or from another household, ignoring random assignment to treatment groups. By adding control variables, heterogeneity of impacts for observed control variables can be estimated as:

$$y_i = \beta_0 + \beta_1 treated_1 + \beta_2 treated_2 + \beta_i x_i + \varepsilon_i,$$

where y_i are the outcome indicators (lima bean yield, consumption, and WTP) $treated_1$ and $treated_2$ are dummy variables for seed and information received, β_1 and β_2 are the respective treatment effects, x_i are household characteristics at baseline, β_i the respective coefficients, and ε_i the error term (Ravallion, 2008). Ceteris paribus, the regression model predicts how a unit change in an explaining variable would increase (or decrease) the outcome variable. ATT is likely to be overestimated as it is subject to self-selection. Control households that received seed might differ from the average household, for example in bean production experience. The intention-to-treat estimate (ITT) approximates the average treatment effect on those intended to treat with random assignment:

$$y_i = \beta_0 + \beta_1 treat\_intended_1 + \beta_2 treat\_intended_2 + \beta_i x_i + \varepsilon_i,$$

where $treat\_intended_1$ and $treat\_intended_2$ are dummy variables for the assignment to seed and information receipt. ITT is likely to be underestimated, as not all households intended to treat actually received, kept, and cultivated the seed. In the latter case, outcomes do not just depend on random assignment, but also on purposive assignment of others. Selective compliance and contamination into the control group can lead to biased estimates of the impacts of treatment. Imbens & Angrist (1994) showed that an average treatment effect under mild restrictions (local average treatment effect – LATE) can still be identified, even when there is no subpopulation for whom the probability of treatment is zero. Using assignment to treatment in a randomised trial as an instrument variable, LATE requires three conditions to be held: (1) eligibility for the treatment group has to be exogenous that is held under random assignment by the design of the study, (2) the use of an instrument requires an exclusion restriction, meaning that random assignment to treatment only affects outcomes through ac-

Table 1: *Household characteristics at baseline in 2012/2013.*

	Seed-and-information (*n* = 112)	Seed-only (*n* = 84)	Control group (*n* = 194)	Total (*n* = 390)
Age of household head (years)	44.3 (12.5)	47.1 (15.4)	46.6 (14.0)	46.0 (14.0)
Education of household head (years)	4.5 (3.2)	4.1 (3.3)	4.2 (3.1)	4.3 (3.2)
Maximum education among household members (years)	6.3 (3.3)	6.1 (3.1)	6.8 (3.3)	6.5 (3.3)
Household size (n)	6.0 (2.4)	6.0 (2.5)	6.3 (2.8)	6.1 (2.6)
Household members in working age (n)	4.0 (1.8)	3.8 (2.0)	4.4 (2.4)	4.1 (2.1)
Dependents in household (n)	2.2 (1.4)	2.3 (1.3)	2.3 (1.4)	2.3 (1.4)
Legume consumption (number of days, past 7 days)	1.9 (1.7)	2.0 (2.0)	2.0 (2.0)	1.9 (1.9)
Dietary diversity (7 days)	43.1 (13.4)	44.7 (13.4)	44.0 (14.0)	43.9 (13.4)
Cultivated land per capita (ha)	0.4 (0.5)	0.3 (0.4)	0.3 (0.4)	0.4 (0.4)
Seasonally flooded land (dummy)	0.5 (0.5)	0.6 (0.5)	0.6 (0.5)	0.6 (0.5)
Wealth self-assessment (1–10)	4.0 (1.7)**	3.5 (1.4)	3.7 (1.7)	3.7 (1.6)
Cattle per capita (n)	1.3 (2.7)*	0.8 (1.3)	1.0 (2.2)	1.0 (2.2)
Agricultural equipment (dummy)	0.7 (0.5)	0.6 (0.5)	0.6 (0.5)	0.6 (0.5)
Bean production (kg per household)	27.6 (67.0)	29.0 (75.7)	28.3 (54.8)	28.2 (63.4)
Legume types cultivated (n)	1.5 (1.0)	1.5 (1.1)	1.5 (1.0)	1.5 (1.0)
Selling crops to trader (dummy)	0.9 (0.3)	0.9 (0.4)	0.9 (0.4)	0.9 (0.3)
Legume sales revenue (year, in EUR)	11.2 (37.1)	8.4 (17.7)	9.4 (34.6)	9.7 (32.3)
Innovations (number, last 5 years)	1.9 (1.0)	1.7 (1.1)	1.8 (1.1)	1.8 (1.1)
Taboo for at least one bean species (dummy)	0.5 (0.5)	0.5 (0.5)	0.4 (0.5)	0.4 (0.5)
Cultivation of bambara groundnut until 2013 (dummy)	0.4 (0.5)	0.3 (0.5)	0.4 (0.5)	0.4 (0.5)
Attitude towards work (mean agreement, 1–5)	4.0 (1.0)	4.1 (0.8)	3.9 (1.0)	4.0 (1.0)
Social capital (mean agreement, 1–5)	2.8 (0.5)	2.8 (0.5)	2.7 (0.5)	2.8 (0.5)

Numbers in parenthesis indicate standard deviations. ** (*) indicates differences at the 5 % (10 %) significance level.

tual participation in the program, (3) anyone who would take the treatment if assigned to the control group would also take treatment if assigned to the treatment group. If these conditions are met, LATE is the average treatment effect for those households that always comply with their assignment and for those whose treatment status is changed by the instrument (Angrist *et al.*, 1996; Ravallion, 2008). Instrumental-variable regressions are estimated with the help of the ivreg2 command in Stata (Baum *et al.*, 2007).

3 Results

3.1 Baseline characteristics

Table 1 compares household characteristics at baseline between the two treatments (seed-only and seed-and-information-receiving households) and the control group. Because of randomization, we expect that there are no significant differences between the groups. This holds true for all variables, except for subjective wealth and the possession of cattle. Wealth is significantly correlated with cattle, an important status

symbol in the region. The significant difference is based on two verified outliers with 20 cattle per capita that were assigned to control and seed-and-information-receiving group, respectively. In the 2012/2013 season, 83 % of households planted one or more types of legume, and for the 34 % that sold legumes in 2013, average sales amounted to 10.2 EUR[2]. Legumes are also important for consumption: in September 2013, they were consumed two days per week on average. The households that bought legumes at the market (27 %) spent an average of 0.4 EUR per week. Almost half of the households reported a taboo for bambara groundnut. However, by 2013, 38 % of households were growing it.

3.2 Utilisation and cultivation of bean seed

Of the 390 panel households, 354 were revisited to evaluate the seed distribution. The remaining 36 households had moved away, or were not available in the survey period (Table 2).

---

[2] Euro (EUR) values in this paper are converted from Malagasy Ariary (MGA) using official yearly averages: 1 EUR = 2,945 MGA (2013) and 3,273 MGA (2014).

Table 2: *Household attrition after baseline survey and seed distribution.*

	Seed-and-information	Seed-only	Control group	Total
Households in baseline survey (2013)	112	84	194	390
Households dropping out	6	12	18	36
Households in follow-up survey (2014)	106	72	176	354

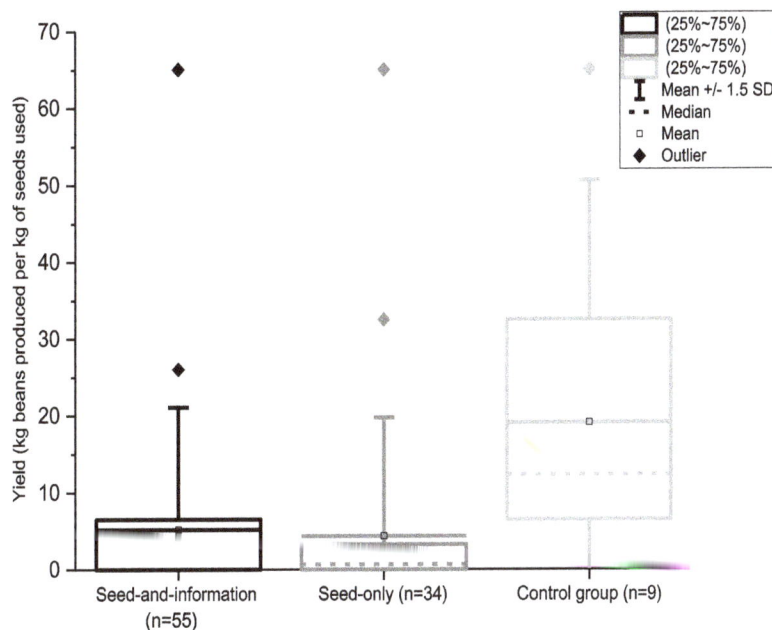

Fig. 2: *Yield of lima beans, 2014.*

At baseline in 2013, 88 % of the 390 households stated planting the seed, if given, 1.2 % would give the seed to another person, and 1.8 % rejected the seed due to taboos or would cook them. In 2014, 98 out of the 354 revisited households reported having cultivated the received seed, mostly out of curiosity and with the objective of home consumption or as food for agricultural labourers. Seed-and-information-receiving households ($n = 55$, 52 %) were not significantly more likely to cultivate the seed than seed-only-receiving households ($n = 34$, 45 %). Nine control households (5 %) reported having received lima bean seed from other sources (neighbours, family, friends) and cultivated these. Insect damage (50 %), consumption of seed (41 %), or taboos were reported as main reasons for not planting, with no significant differences between the groups. Three households reported having replaced other legumes, the remaining households said they cultivated the seed in addition to existing legumes. Women were more involved in bean cultivation than man, with no significant differences between the groups.

Seed quality, seeding, cultivation, and yield of lima beans were evaluated as better than average and bet-

ter than other legumes. Control households planting lima beans perceived cultivation compared with other legumes to be significantly easier than treatment households. The average yield was 6.3 kg beans per kg of seed used. Taking out those households that did not achieve any yield (48 %), gives an average yield of 12.2 kg per kg of seed. Control households planting lima beans achieved a significantly higher yield, whereas information provision did not result in a higher yield (Fig. 2).

The most cited reasons for low yield were drought and destruction of the plants by cattle or insects, with no significant difference between the groups. Of the legume-cultivating households, 63 % rated climatic conditions for legumes in 2014 "much worse" or "worse" than in the past five years.

3.3 Information dissemination

Of the information-receiving households, 83 % rated the given information as useful. For 80 % of those households, the information was sufficient. Out of the seed-and-information-receiving group, significantly more households (20 %) also received information from others, compared with 12.5 % of the seed-only-

Table 3: *Local average treatment effect (LATE) of information provision on lima bean yield.*

	LATE	LATE with controls
Seed-only	3.87 (2.1)*	4.29 (2.0)**
Seed-and-information	−1.45 (2.1)	−1.18 (2.0)
Age of household head (years)		0.07 (0.05)
Maximum education among household members (years)		−0.07 (0.4)
Dependency ratio (dependent members/members in working age)		3.40 (1.0)***
Gender of household head (dummy)		−1.14 (1.7)
Willingness to take risks (self-assessment, 1–10)		0.36 (0.3)
Cultivated land per capita (ha)		−0.80 (1.5)
Legume types cultivated (n)		0.28 (0.6)
Selling crops to trader (dummy)		−4.15 (1.9)**
Innovations (number in the last 5 years)		1.29 (0.6)**
Adjusted R^2		0.12
N		354

Numbers in parenthesis indicate standard deviations. *** (**) (*) indicates significance at the 1 % (5 %) (10 %) level.

Table 4: *Consumption of lima beans and legumes, differentiated by recall period, 2014.*

	Seed-and-information ($n=106$)	Seed-only ($n=76$)	Control group ($n=172$)	Total ($n=354$)
Lima bean consumption (dummy, 12 months)	0.6 (0.5)***	0.6 (0.5)***	0.3 (0.4)	0.5 (0.5)
Legume consumption (dummy, 12 months)	0.9 (0.4)	0.8 (0.4)	0.8 (0.4)	0.8 (0.4)
Legume consumption (n, 7 days)	3.2 (2.4)	3.4 (2.5)	3.2 (2.5)	3.3 (2.4)
Legume consumption from own production (dummy, 7 days)	0.7 (0.5)	0.8 (0.4)	0.7 (0.5)	0.7 (0.5)
Expenditures for legumes (EUR, 7 days)	0.1 (0.3)	0.1 (0.3)	0.1 (0.3)	0.1 (0.3)

Numbers in parenthesis indicate standard deviations. *** indicates difference at the 1 % significance level.

receiving households. Households receiving additional information from sources other than the enumerators, achieved a significantly higher yield. Most reported information sources were family (49 %), neighbours (43 %), and friends (8 %). Only 7 % of the control households were informed about the new bean variety.

Significantly more seed-and-information-receiving households (74 %) planted the bean seed on seasonally flooded fields next to a river or rice fields, as advised in the included information. Seed-only-receiving households planted mostly on other fields (62 %), and the control households mostly next to the house (56 %). Yet the planting location had no significant impact on reported bean yield.

Table 3 shows the impact of seed and information distribution on lima bean yield. The regression models do not predict a significant impact of information on yield. Adding control variables shows that next to seed dis-

tribution, a household's willingness to take risks and innovations are positive and significant predictors of yield. Dependency ratio and access to traders are negative predictors of yield.

3.4 Consumption

Almost all households (98 %) reported consuming their harvested beans. Seven households saved seed for the next cultivation period and two households sold parts of their harvest. No significant differences with respect to the use of the harvested crop between the three groups could be detected.

Of the 354 households, 45 % stated having eaten lima beans in the past year. Control households were significantly less likely to consume lima beans than treatment households. No significant differences could be detected when looking at legume consumption in general in the week prior to the interview (Table 4).

Table 5: *Local average treatment effect (LATE) of seed distribution on legume consumption.*

	LATE	LATE with controls
Seed-only	0.63 (0.47)	0.89 (0.5)**
Seed-and-information	−0.49 (0.47)	−0.67 (0.5)
Education of household head (years)		0.03 (0.1)
Dependency ratio (dependent members/members in working age)		−0.28 (0.2)
Gender of household head (dummy)		−1.04 (0.4)***
Agricultural equipment (dummy)		−0.57 (0.3)*
Crop diversity (number of different crops)		0.11 (0.04)***
Selling crops to trader (dummy)		−0.47 (0.4)
Livestock sales (dummy)		0.48 (0.3)*
Income from own business (dummy)		−0.79 (0.3)***
Income from agricultural labour (dummy)		−0.85 (0.3)***
Access to mutual help (dummy)		0.50 (0.3)
Adjusted R^2		0.05
N		354

Numbers in parenthesis indicate standard deviations. *** (**) (*) indicates significance at the 1 % (5 %) (10 %) level.

Table 5 shows the impacts of seed distribution on the frequency of legume consumption in the past seven days before the interviews. When controlling for household characteristics, a significant positive impact of seed distribution on legume consumption is observed.

3.5 Willingness to pay

More than half (58 %) of the farmers stated they usually produce their own legume seed, 30 % of the farmers said they usually buy their seed at the market, 10 % buy them from other farmers, and the rest mostly receive seed from family members. In total, at least once in the last five years, 49 % of all households cultivating legumes bought seed at the market, 64 % used their own seed, and 21 % bought from other farmers in the village. The net-maps compiled during focus group discussions in 2014 (Fig. 3) show the most important seed and information sources.

The Malagasy agricultural extension service, Centre de Services Agricoles (CSA), an NGO funded by the European Union and managed by local officials, is distributing improved seed to farmers. It aims to increase agricultural productivity by linking service demands of farmers with appropriate service providers, operating in the rural districts, and training village representatives (Ministry of Agriculture, 2015).

In the district capital Ambalavao, CSA provides free samples of improved rice and bean seed, for which farmers can apply directly or indirectly through local administration with a written contract. In 2013, very few farmers knew about this possibility and none had received seed. There are also problems on the supply side: those who had applied for improved seed were told that reserves had already been exhausted. On the local market, there is no improved seed. Due to a breeding and dissemination project on improved common beans, a farmer who is being trained as village representative of CSA understood yield and quality advantages of improved bean seed.

When asked about the importance of seed traits, farmers listed yield and potential for sale, followed by taste and ease of cultivation as most important. Pest and disease resistance and drought tolerance were not highly ranked by most households. Results from focus groups showed that the majority of households were unaware of the possibility of breeding resistant and tolerant seed. It was stated that improved seed is properly sorted, thus the biggest and least damaged seed.

According to focus group discussions and information from traders at the local market, a reference value of 0.2 EUR was chosen, the average price for local bean seed. Mean WTP of all households amounts to 0.3 EUR/*kapoaka*, which is 171 % of the reference value (premium of 42 % compared with the average bean price). Interviewees stated a higher mean WTP for common bean seed than for lima bean seed, the differences between species and between treatment and control households were not significant.

Net-Map of the bean seed market in Fenoarivo

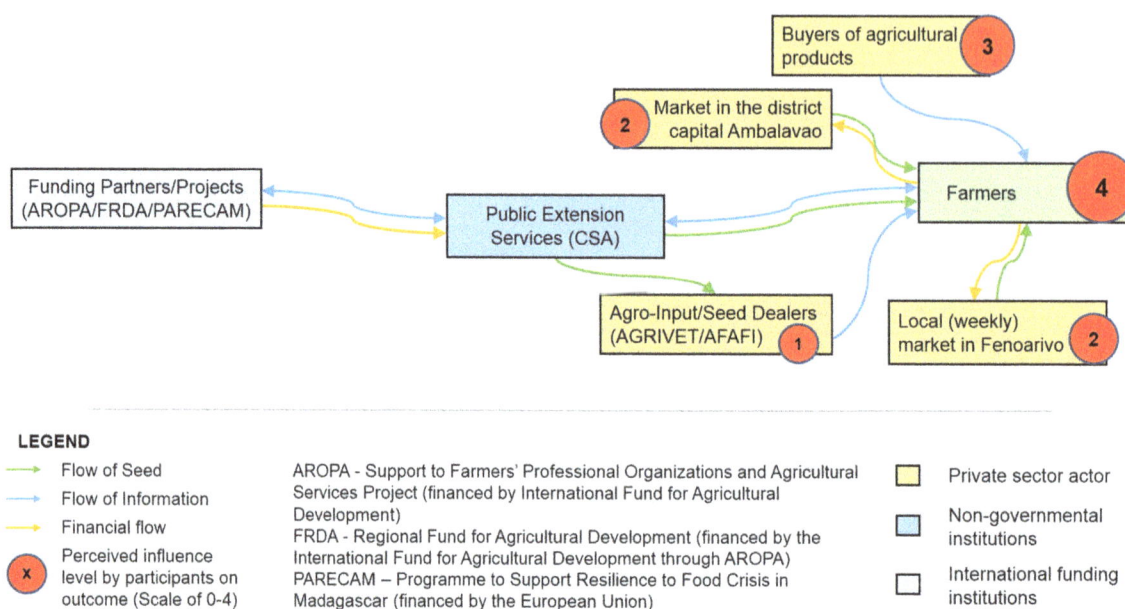

Fig. 3: *Exemplary net-map of the bean seed market in Fenoarivo.*

4 Discussion

Almost two-thirds (63 %) of respondents thought it was likely that improved bean seed would be offered at the market in Fenoarivo, and 60 % that they could be offered at their stated WTP. The majority (86 %) of households thought it was likely that they could afford the bean seed at their stated WTP. No differences between treatment and control households could be detected.

No significant impact of either seed or information on WTP could be observed in the regression models (Table 6). Positive predictors of WTP were: additional information sources on improved seed, cultivated land area, social capital, willingness to take risks, and household wealth Legume consumption was a negative predictor.

Of all households, 40 % stated that they would purchase lima beans for consumption and 49 % said being interested in cultivating the beans. Seed-and-information-receiving households are significantly more likely to buy and cultivate lima beans in the future than the two other groups. The majority of households would prefer to buy improved seed in March (217), February (54), and April (31), and at the local market rather than from other farmers or the extension service.

4.1 Utilisation and cultivation of lima bean seed

Although seed utilisation rates (55 %) seem low, they are comparable to other areas in Madagascar. Snoeck (2016) reported for the northeast of Madagascar that about one out of two freely distributed improved clove seed has been planted. Moser & Barrett (2003) reported an adoption rate of only 25 % for improved rice technology (SRI) over a period of five years for different sites in the island's central and southern highlands. Utilisation rates were affected negatively by limited storage possibilities and long storage time, as seed was distributed six months before sowing time. This might have led to the high number of households reporting that insects destroyed their seed. Taboos against beans were a significant hindrance to utilisation. Some households that were intended to receive seed rejected it, and were not willing to cultivate, eat, or talk about them, while others wanted to keep them for consumption. This confirms that social conformity in Madagascar limits adoption of new varieties and technologies (Moser & Barrett, 2003). The adoption process of bambara groundnut shows that it might take some years, but that if beliefs are updated, taboos can be overcome. After its introduction at a nearby large-scale *Jatropha* plantation and by some innovative farmers, villagers realised that rainfall had been unaffected and started planting the beans.

Table 6: *Local average treatment effect (LATE) of seed distribution on willingness-to-pay for improved seed.*

	LATE	LATE with controls
Seed-only	−62.25 (42.8)	−48.40 (38.0)
Seed-and-information	3.89 (42.9)	−32.73 (37.7)
Information from other sources		123.85 (34.4)***
Age of household head (years, squared)		0.01 (0.01)
Maximum education among household members (years)		−1.46 (7.1)
Gender of household head (dummy)		−41.4 (31.6)
Dependency ratio (dependent members/members in working age)		5.42 (18.5)
Willingness to take risk (self-assessment, 1–10)		35.29 (4.9)***
Subjective wealth assessment (1–10)		11.82 (7.1)*
Cultivated land per capita (ha)		59.91 (27.2)**
Mutual help (dummy)		49.70 (28.1)
Selling crops to trader (dummy)		48.41 (35.2)
Frequency of legume consumption (number of days, past 7 days)		−10.94 (5.8)*
Social capital (mean agreement)		24.92 (13.0)*
Positive attitude towards work (mean agreement)		4.09 (11.0)
Adjusted R^2		0.20
N		350

Numbers in parenthesis indicate standard deviations. *** (**) (*) indicates significance at the 1 % (5 %) (10 %) level.

Seed was planted mostly by older family members in home gardens with the objective of contributing to household consumption. The seed quantity distributed was low and unlike in the Toliara region, farmers do not have access to lima bean markets. According to focus group discussions, reliable access to markets would increase investments in agricultural production. The maximum production obtained was 65 kg, which is very close to the yield expectation of FOFIFA (43–69 kg of beans per kilogram of seed). Households with an above-average lima bean yield were more experienced in bean production at baseline and also cultivated bambara groundnut. Seed-receiving households unwilling to cultivate might have given their seed to people they knew were experienced in bean cultivation. Prior to the seed distribution a climbing lima bean variety was known in the villages and grown by some farmers in home gardens, and in home gardens people might spend more time weeding and watering and therefore achieve a higher yield.

4.2 Information dissemination

Information-receiving households did not have significantly higher cultivation rates or lima bean yields than their seed-only-receiving counterparts. Seed-only-receiving households, however, were more likely to consume the seed at the beginning of the lean season and not take the risk of keeping them until the planting season. Information-receiving households might have kept the seed with the intention to plant them, pointing to the importance of information dissemination.

As illiteracy is still widespread in the region, information was given verbally only, thus households were not able to look up information that they might not have remembered at sowing time. Lima beans had not been planted before in the villages, there was no local contact person to consult, and the nearest place to access information was in the district capital. Households with above-average yield recalled or knew more information concerning lima bean cultivation. One reason for the rather low information dissemination from information-receiving households to others might have been that lima beans were considered taboo by some households, and it is common to avoid talking about taboos. Furthermore, the distributed information did not target the most mentioned problems, drought and destruction of the plants by cattle or insects. Interestingly, out of the seed-and-information-receiving households, significantly more households also received information from others, thus were more likely to discuss and to be consulted by others. FOFIFA Toliara is training technicians from farmers' organisations in lima bean cultivation, who are then responsible for disseminating this expertise to farmers in their organisations. Hotz *et al.* (2012) found that group-level trainings in nutrition and cultivation of orange sweet potato had a significant impact on production and consumption of rural households in Mozambique. One-year and three-year interventions were found having the same significant impact on vitamin A intake, suggesting that group training could be limited to the first year of intervention.

4.3 Consumption

Bean seed distribution did lead to a small, but significantly higher legume consumption for the seed-only-receiving households. Additional information provision did not lead to higher legume consumption, which might be due to the lower achieved yield of this group, and due to the possibility that households had already consumed their supply by the time of the interview. A trader at the market confirmed that after the seed had been distributed, he sold lima beans originating from the region of Toliara. This might explain the high percentage of control households having consumed lima beans, also because only two households sold part of their harvest. Larochelle *et al.* (2015) found that adoption of improved bean seed in Uganda led to increased dietary diversity through the channel of home consumption, and indirectly through farm income, productivity, and empowerment of women. Because beans are mainly cultivated by women, the authors hypothesize a positive effect for dietary diversity as women might have control over bean sales and, therefore, might be in a better position influencing household nutrition outcomes. Similarly, Kabunga *et al.* (2014) showed that fruit and vegetable production led to significant improvements in household nutrition. Although many studies support the hypothesis that household agricultural production is correlated with household and individual consumption, the evidence for a pathway from agriculture to nutrition is mixed. Types of food (especially when comparing crops and dairy products), context, and location cause the effects to vary greatly in size (Carletto *et al.*, 2015). According to Sibhatu *et al.* (2015) the effect of crop and livestock diversification on dietary diversity diminishes with farm size, probably because of foregone benefits from specialisation for already diversified farms.

4.4 Willingness to pay

Mean WTP for improved bean seed was estimated at 171 % of the market price for traditional seed or at a premium of 42 % compared to traditional seed. This result is comparable to other studies. Kaguongo *et al.* (2014) valued certified potato seed in Kenya, where farmers on average are willing to pay 190 % of the price

of farmer seed for certified seed, and 170 % for clean seed. Chelang'a *et al.* (2013) found that consumers prefer African leafy vegetables to exotic ones and are willing to pay an average premium of 79 %.

The FOFIFA research station sells its improved lima bean seed to farmers at a price of 1.2–1.5 EUR per kilogram (0.5–0.6 EUR/*kapoaka*). Seed originating from farmers' own seed production ranges from 0.5–0.6 EUR per kilogram (0.2 EUR/*kapoaka*) with regional and seasonal differences (personal interview with representative of FOFIFA, 2014). Comparing these prices with WTP from this study, farmers cannot or do not want to afford improved seed directly from research stations, but would be willing to pay a premium for farmer's own produced seed. FOFIFA had a similar experience in the Toliara region when marketing their improved bean seed to farmers' organisations (personal interview with representative of FOFIFA, 2014). As the case of AfricaRice in the Antsirabe region shows, bean seed could be marketed through participatory varietal selection (personal interview with representative of AfricaRice, 2015).

An interesting result is that 87 % of households answering the WTP question chose the familiar common bean over the lima bean, when asked which of the two they prefer to buy. An explanation for this can be that households selecting lima beans were significantly more willing to take risks, regardless of whether they were successful in cultivating the seed or might have been disappointed by their experience. Risk aversion was also associated with a significantly lower WTP for weather-index insurance (Hill *et al.*, 2013). Common beans are not taboo in the village fields, therefore taboos for lima beans and consequent social pressure might have played a role in this decision.

Households that received additional information on improved lima bean seed reported a significantly higher WTP. This aligns with the study of Kaguongo *et al.* (2014) where farmers' awareness of seed quality had a positive impact on WTP. Cultivated land per capita and subjective household wealth increase WTP significantly. Ulimwengu & Sanyal (2011) also indicated that the amount of land owned and the income level increased farmers' WTP for agricultural services. Legume consumption is a negative predictor, suggesting that households cultivating legumes mostly for home consumption are willing to invest less in inputs. Similarly to Kaguongo *et al.* (2014), remoteness to markets does not result in a lower WTP neither does proximity to markets increase WTP. Contrary to Chelang'a *et al.* (2013), gender, education, and dependency ratio have no significant influence on WTP.

Whittington (1998) summarizes potential biases of WTP studies. Both overstatements due to prestige effects and understatements due to consumers trying to influence the final price can occur. As in this study no data on the purchasing of improved beans is available, these biases cannot be ruled out. To minimize hypothetical bias, which is the difference between stated and revealed values, lower bounds were set for the payment card, but with an effort to keep a realistic range of prices from which households could choose (Murphy *et al.*, 2005).

Contrary to the results of Bates *et al.* (2012) that receiving a product for free increased peoples' WTP for it later, in this study neither seed distribution nor information provision increased WTP. Insect damage, low yield due to climatic conditions, and not following cultivation recommendations may be to blame. Households planting the bean seed might have had higher expectations or might have been disappointed by the performance of the seed. CRS *et al.* (2013) discuss the downside risk of free distribution of seed in Madagascar. Institutions are still buying the largest amount of seed, which is hindering the development of a sustainable private sector serving smallholder farmers.

5 Conclusions and recommendations

The amount of seed distributed to households was small and only 54 % of the seed-receiving households cultivated the lima bean seed. Of those, 48 % reported zero yield, mainly due to drought. Seed information did not increase cultivation. Taboos played a role, as did insect damage due to the long storage time. A timely distribution of packaged seed could avoid those problems. Local seed could be tested in comparison to improved seed on demonstration plots of farmer field schools, so that farmers could experience the differences. Similarly to FOFIFA Toliara, some farmers could be trained in the use of improved seed and other agricultural technologies and given incentives to disseminate this knowledge to farmers in their organisations and villages. CRS *et al.* (2013) suggest participatory varietal selection, decentralised seed production (by farmers or farmers' associations), low-cost delivery mechanisms (e.g. through village committees or with the help of radio programs), and technologies to minimize storage losses. These results also strengthen importance and need for public extension services in rural Madagascar. We recommend participatory community-level interventions that match farmers' needs and focus not only on technological issues in introducing new varieties, but also allow to consider social processes that hinder innovations. Net-maps compiled during the focus group discussions turned out

to be a helpful and easy-to-implement tool, as participants learned about the breeding program of the Ministry of Agriculture and the possibility of obtaining improved seed from the extension service.

Seed information did not increase yield. Reasons for low yield are climatic conditions as well as crop losses due to insects and cattle. The explanatory power of the factors determining lima bean yield is low, suggesting that unobserved factors, like cow dung application, irrigation, plot-specific rainfall during critical times of bean growth, or time invested played a role. If cultivation is more closely monitored, inputs used or problems during cultivation could be used as intermediate indicators to better explain yield variances. Local representatives could be trained in storage and cultivation techniques and serve as contact persons for households in case they have questions or encounter problems. This would allow for information sharing on topics that initial information provision cannot address. Group training sessions or demonstration plots could give impetus to information dissemination. As mostly women are responsible for bean production, female contribution and knowledge sharing at the demonstration plot and effects on cultivation and decision making could be tested. Information diffusion via videos or visual aids containing pictures and short sentences could be tested. While targeting farmers, the extension service could cooperate with schools. Educating children in agricultural production processes and the agricultural market might lead to long-term benefits and to more participation of the rural population in development programs.

A small positive effect on legume consumption was found. As unobserved factors, like food or cooking preferences or taboos might have played a role, cooking and nutrition information should be included during the dissemination of new varieties (Katungi *et al.*, 2011a).

Neither seed distribution nor information provision increased WTP. The training of a local farmer in the commune has shown to be an effective means of disseminating knowledge about improved inputs from the extension service to the village; however, no improved seed has been supplied yet. Given the poor infrastructure and high poverty rates, as well as market failures in a remote and underdeveloped setting lacking cash crops, continued state and non-governmental organisations support might be justified.

Acknowledgements

We thank the foundations *Stiftung Energieforschung Baden-Wuerttemberg* and *fiat panis* for funding this research. We are grateful to the respondents for their continued participation in the survey, to the field assistants and enumerators for their efforts to collect high-quality data. We thank Albert Randrianasolo of FOFIFA Toliara, CSA Ambalavao and Romaine and Sylvain Ramananarivo for useful discussions. Moreover, we thank two anonymous reviewers for making suggestions, leading to considerable improvements in the article.

References

Alemayehu, M. (2009). Reinvigorating bean seed system in Ethiopia: role of farmers and social institutions. *Seed Info*, 36, 11–16.

Angrist, J. D., Imbens, G. W. & Rubin, D. B. (1996). Identification of causal effects using instrumental variables. *Journal of the American Statistical Association*, 91 (434), 444–455.

Aykroyd, W. R. & Doughty, J. (1982). *Legumes in human nutrition*. Food and Agriculture Organisation of the United Nations, Rome, Italy.

Banerjee, A. V. & Duflo, E. (2008). The experimental approach to development economics. *Annual Review of Economics*, 1, 151–178.

Barrett, C. B. (2008). Poverty traps and resource dynamics in smallholder agrarian systems. *In:* Dellink, R. B. & Ruijs, A. (eds.), *Economics of poverty, environment and natural-resource use*. pp. 17–40, Springer, Wageningen, The Netherlands.

Barrett, C. B. & Carter, M. R. (2010). The power and pitfalls of experiments in development economics: some non-random reflections. *Applied Economic Perspectives and Policy*, 32 (4), 515–548.

Barrett, C. B., Moser, C. M., McHugh, O. V. & Barison, J. (2004). Better technology, better plots, or better farmers? Identifying changes in productivity and risk among Malagasy rice farmers. *American Journal of Agricultural Economics*, 86 (4), 869–888.

Bateman, I. J., Carson, R. T., Day, B., Hanemann, M., Hanley, N., Hett, T., Jones-Lee, M., Loomes, G., Mourato, S. & Oezdemiroglu, E. (2002). *Economic valuation with stated preference techniques: a manual*. Edward Elgar, Cheltenham, UK.

Bates, M. A., Glennerster, R., Gumede, K. & Duflo, E. (2012). The price is wrong. *Field Actions Science Reports*, Special Issue 4: Fighting Poverty, between market and gift, 30–37. Available at: http://factsreports.revues.org/1554

Baum, C. F., Schaffer, M. E. & Stillman, S. (2007).

ivreg2: Stata module for extended instrumental variables/2SLS, GMM and AC/HAC, LIML and k-class regression. Boston College Department of Economics, Statistical Software Components S425401.

Beaman, L., Karlan, D., Thuysbaert, B. & Udry, C. (2013). Profitability of fertilizer: experimental evidence from female rice farmers in Mali. *The American Economic Review*, 103 (3), 381–386.

Bulte, E., Beekman, G., Di Falco, S., Hella, J. & Lei, P. (2014). Behavioral responses and the impact of new agricultural technologies: evidence from a double-blind field experiment in Tanzania. *American Journal of Agricultural Economics*, 96 (3), 813–830.

Carletto, G., Ruel, M., Winters, P. & Zezza, A. (2015). Farm-Level pathways to improved nutritional status: introduction to the special issue. *The Journal of Development Studies*, 51 (8), 945–957.

Chelang'a, P., Obare, G. & Kimenju, S. (2013). Analysis of urban consumers' willingness to pay a premium for African Leafy Vegetables (ALVs) in Kenya: a case of Eldoret Town. *Food security*, 5 (4), 591–595.

CRS, CARE, Caritas, CIAT, UEA/DEV, Tranoben'ny Tantsaha Nasionaly (2013). *Seed system security assessment, East and South Madagascar*. CRS, CIAT and UEA, Antananarivo, Madagascar.

Culbertson, M. J., McCole, D. & McNamara, P. E. (2014). Practical challenges and strategies for randomised control trials in agricultural extension and other development programmes. *Journal of Development Effectiveness*, 6 (3), 284–299.

Dalton, T. J., Yesuf, M. & Muhammad, L. (2011). Demand for drought tolerance in Africa: selection of drought tolerant maize seed using framed field experiments. Annual Meeting of the Agricultural and Applied Economics Association, Pittsburgh, Pennsylvania.

Deaton, A. (2010). Instruments, randomization, and learning about development. *Journal of Economic Literature*, 48 (2), 424–455.

Duflo, E., Glennerster, R. & Kremer, M. (2008). Using randomization in development economics research: a toolkit. *In:* Schultz, T. & Strauss, J. (eds.), *Handbook of Development Economics*. Vol. 4, pp. 3895–3962, Elsevier, Amsterdam and New York: North Holland.

Duflo, E., Kremer, M. & Robinson, J. (2011). Nudging farmers to use fertilizer: theory and experimental evidence from Kenya. *American Economic Review*, 101 (6), 2350–2390.

FAOSTAT (2011). Suite of Food Security Indicators. Available at: http://www.fao.org/faostat/en/#data/FS (accessed on: 24.09.2017).

FAOSTAT (2015). Country Indicators – undernourishment. Available at: http://www.fao.org/faostat/en/#country/129 (accessed on: 24.09.2017).

Gibson, R. (2013). How sweet potato varieties are distributed in Uganda: actors, constraints and opportunities. *Food Security*, 5 (6), 781–791.

Haab, T. C. & McConnell, K. E. (2002). *Valuing environmental and natural resources: the econometrics of non-market valuation*. Edward Elgar Publishing, Cheltenham, UK.

Hartmann, D. & Arata, A. (2011). *Measuring social capital and innovation in poor agricultural communities: the case of Cháparra, Peru*. FZID discussion papers 30, Forschungszentrum Innovation und Dienstleistung, Universität Hohenheim, Germany.

Hill, R. V., Hoddinott, J. & Kumar, N. (2013). Adoption of weather-index insurance: learning from willingness to pay among a panel of households in rural Ethiopia. *Agricultural Economics*, 44 (4–5), 385–398.

Hotz, C., Loechl, C., de Brauw, A., Eozenou, P., Gilligan, D., Moursi, M., Munhaua, B., van Jaarsveld, P., Carriquiry, A. & Meenakshi, J. (2012). A large-scale intervention to introduce orange sweet potato in rural Mozambique increases vitamin A intakes among children and women. *British Journal of Nutrition*, 108 (1), 163–176.

Imbens, G. W. & Angrist, J. D. (1994). Identification and estimation of local average treatment effects. *Econometrica*, 62 (2), 467–475.

Kabunga, N., Ghosh, S. & Griffiths, J. K. (2014). Can smallholder fruit and vegetable production systems improve household food security and nutritional status of women? Evidence from rural Uganda. IFPRI Discussion Paper 01346.

Kaguongo, W., Maingi, G., Barker, I., Nganga, N. & Guenthner, J. (2014). The value of seed potatoes from four systems in Kenya. *American Journal of Potato Research*, 91 (1), 109–118.

Kassie, G. T., Abdulai, A., MacRobert, J., Abate, T., Shiferaw, B., Tarekegne, A. & Maleni, D. (2014). Willingness to pay for drought tolerance (DT) in maize in communal areas of Zimbabwe. 88th Annual Conference of the Agricultural Economics Society, Paris, France.

Katungi, E., Sperling, L., Karanja, D., Farrow, A. & Beebe, S. (2011a). Relative importance of common bean attributes and variety demand in the drought areas of Kenya. *Journal of Development and Agricultural Economics*, 3 (8), 411–422.

Katungi, E., Wozemba, D. & Rubyogo, J. C. (2011b). A cost benefit analysis of farmer based seed production for common bean in Kenya. *African Crop Science Journal*, 19 (4), 409–415.

Larochelle, C., Alwang, J., Norton, G. W., Katungi, E. & Labarta, R. A. (2015). Impacts of improved bean varieties on poverty and food security in Uganda and Rwanda. *In:* Walker, T. S. & Alwang, J. (eds.), *Crop Improvement, Adoption and Impact of Improved Varieties in Food Crops in Sub-Saharan Africa*. pp. 314–337, CGIAR and CAB International, Croydon, UK.

Ministry of Agriculture (2004). Filière grains secs, fiche 106. MAEP UPDR – VALY Agridéveloppement, Madagascar. Available at: http://inter-reseaux.org/IMG/pdf_106_Filiere_Grains_secs.pdf (accessed on: 24.09.2017).

Ministry of Agriculture (2013). Fiches techniques de base destinées aux techniciens agricoles: Pois du cap. Ministère auprès de la Présidence chargé de l'Agriculture et de l'Élevage, Madagascar. Available at: http://www.mpae.gov.mg/donnees/fiches-techniques (accessed on: 24.09.2017).

Ministry of Agriculture (2015). Le processus de mise en place des Centres de Service Agricole. Ministère auprès de la Présidence chargé de l'Agriculture et de l'Élevage, Madagascar. Available at: http://www.mpae.gov.mg/csa (accessed on: 24.09.2017).

Minten, B., Koru, B. & Stifel, D. (2013). The last mile(s) in modern input distribution: pricing, profitability, and adoption. *Agricultural Economics*, 44 (6), 629–646.

Minten, B., Randrianarisoa, J. & Barrett, C. B. (2007). Productivity in Malagasy rice systems: wealth-differentiated constraints and priorities. *Agricultural Economics*, 37 (s1), 225–237.

Moser, C. M. & Barrett, C. B. (2003). The disappointing adoption dynamics of a yield-increasing, low external-input technology: the case of SRI in Madagascar. *Agricultural Systems*, 76 (3), 1085–1100.

Moser, C. M. & Barrett, C. B. (2006). The complex dynamics of smallholder technology adoption: the case of SRI in Madagascar. *Agricultural Economics*, 35 (3), 373–388.

Moursi, M. M., Arimond, M., Dewey, K. G., Treche, S., Ruel, M. T. & Delpeuch, F. (2008). Dietary diversity is a good predictor of the micronutrient density of the diet of 6- to 23-month-old children in Madagascar. *The Journal of Nutrition*, 138 (12), 2448–2453.

Murphy, J. J., Allen, P. G., Stevens, T. H. & Weatherhead, D. (2005). A meta-analysis of hypothetical bias in stated preference valuation. *Environmental and Resource Economics*, 30 (3), 313–325.

Ravallion, M. (2008). Evaluating anti-poverty programs. *In:* Schultz, T. & Strauss, J. (eds.), *Handbook of Development Economics*. Vol. 4, pp. 3787–3846, Elsevier, Amsterdam and New York: North Holland.

Rubin, D. B. (1974). Estimating causal effects of treatments in randomized and nonrandomized studies. *Journal of Educational Psychology*, 66 (5), 688–701.

Schiffer, E. & Hauck, J. (2010). Net-Map: collecting social network data and facilitating network learning through participatory influence network mapping. *Field Methods*, 22 (3), 231–249.

Sibhatu, K. T., Krishna, V. V. & Qaim, M. (2015). Production diversity and dietary diversity in smallholder farm households. *Proceedings of the National Academy of Sciences of the United States of America*, 112 (34), 10657–10662.

Snoeck, D. (2016). *Agroforestry for Food Security. Final Narrative Report*. CIRAD, Montpellier, France.

Sperling, L., Cooper, H. D. & Remington, T. (2008). Moving towards more effective seed aid. *The Journal of Development Studies*, 44 (4), 586–612.

Sperling, L. & McGuire, S. (2010). Understanding and strengthening informal seed markets. *Experimental Agriculture*, 46 (2), 119–136.

Stifel, D., Fafchamps, M. & Minten, B. (2011). Taboos, agriculture and poverty. *Journal of Development Studies*, 47 (10), 1455–1481.

Stifel, D. & Minten, B. (2008). Isolation and agricultural productivity. *Agricultural Economics*, 39 (1), 1–15.

Ulimwengu, J. & Sanyal, P. (2011). Joint estimation of farmers' stated willingness to pay for agricultural services. IFPRI Discussion Paper 1070, International Food Policy Research Institute (IFPRI), Washington, D.C.

Vandercasteelen, J., Dereje, M., Minten, B. & Taffesse, A. S. (2013). Scaling-up adoption of improved technologies: the impact of the promotion of row planting on farmers' teff yields in Ethiopia. LICOS Discussion Paper Series 344, LICOS Centre for Institutions and Economic Performance, KU Leuven.

Vasilaky, K. (2013). Female Social Networks and Farmer Training: Can Randomized Information Exchange Improve Outcomes? *American Journal of Agricultural Economics*, 95 (2), 376–383.

Whittington, D. (1998). Administering contingent valuation surveys in developing countries. *World Development*, 26 (1), 21–30.

10

Towards an assessment of on-farm niches for improved forages in Sud-Kivu, DR Congo

Birthe K. Paul [a,b,*], Fabrice L. Muhimuzi [c], Samy B. Bacigale [c,d],
Benjamin M. M. Wimba [e], Wanjiku L. Chiuri [a],
Gaston S. Amzati [c], Brigitte L. Maass [a,f]

[a] *International Center for Tropical Agriculture (CIAT), Nairobi, Kenya*
[b] *Wageningen University & Research (WUR), Wageningen, the Netherlands*
[c] *Université Evangélique en Afrique (UEA), Bukavu, DR Congo*
[d] *International Institute of Tropical Agriculture (IITA), Bukavu, DR Congo*
[e] *Institut National pour l'Etude et la Recherche Agronomiques (INERA), Bukavu, DR Congo*
[f] *Department for Crop Sciences, University of Göttingen, Germany*

Abstract

Inadequate quantity and quality of livestock feed is a persistent constraint to productivity for mixed crop-livestock farming in eastern Democratic Republic of Congo. To assess on-farm niches of improved forages, demonstration trials and participatory on-farm research were conducted in four different sites. Forage legumes included *Canavalia brasiliensis* (CIAT 17009), *Stylosanthes guianensis* (CIAT 11995) and *Desmodium uncinatum* (cv. Silverleaf), while grasses were Guatemala grass (*Tripsacum andersonii*), Napier grass (*Pennisetum purpureum*) French Cameroon, and a local Napier line. Within the first six months, forage legumes adapted differently to the four sites with little differences among varieties, while forage grasses displayed higher variability in biomass production among varieties than among sites. Farmers' ranking largely corresponded to herbage yield from the first cut, preferring Canavalia, Silverleaf desmodium and Napier French Cameroon. Choice of forages and integration into farming systems depended on land availability, soil erosion prevalence and livestock husbandry system. In erosion prone sites, 55–60 % of farmers planted grasses on field edges and 16–30 % as hedgerows for erosion control. 43 % of farmers grew forages as intercrop with food crops such as maize and cassava, pointing to land scarcity. Only in the site with lower land pressure, 71 % of farmers grew legumes as pure stand. When land tenure was not secured and livestock freely roaming, 75 % of farmers preferred to grow annual forage legumes instead of perennial grasses. Future research should develop robust decision support for spatial and temporal integration of forage technologies into diverse smallholder cropping systems and agro-ecologies.

Keywords: mixed crop-livestock systems, tropical forages, Napier grass, farming system research, participatory research

1 Introduction

In the Sud-Kivu province of eastern Democratic Republic of Congo (DRC), farmers traditionally practice mixed crop-livestock production. Since 1996, cattle

have become a target of war, so that mixed farming is threatened with complete breakdown, lacking manure to sustain crop cultivation (Cox, 2012). Especially due to the "quasi-disappearance of cattle" (Vlassenroot, 2005), overall livestock holdings have been severely reduced to 0.2–0.5 Tropical Livestock Units (TLU) per household, which is too low to satisfy subsistence or allow for

* Corresponding author
Email: B.Paul@cgiar.org

regular sale (Maass *et al.*, 2012; Ouma & Birachi, 2012). Despite their low productivity, existing smallholder production systems can still provide a steady source of animal protein and manure for household consumption and sale. With a certain level of intensification, these systems could provide a pathway out of poverty (Maass *et al.*, 2013). In addition to a general lack of knowledge and skills in animal husbandry and poor access to veterinary services, farmers consider scarcity of (quality) livestock feed, especially in the dry season, as one of the main constraints for livestock production (Zozo *et al.*, 2010). Currently, grazing on natural pastures and collection of roadside grasses and herbs constitute the main source of feed, while 37 % of farmers cultivate forages on small plots contributing only 6 % to the livestock diet (Bacigale *et al.*, 2014).

Improved forages can play an important role in enhancing livestock production, while generating additional environmental co-benefits. Direct benefits for crop production include weed suppression, pest and disease reduction, and soil fertility improvement through N fixation, while environmental effects could be soil organic matter and carbon increase, reduced greenhouse gas emission intensities and soil erosion control. Increased feed production on agricultural land less suitable for cropping and recuperation of degraded land also contribute to increased land use efficiency (Peters *et al.*, 2013). However, despite the range of potential benefits, the cultivation of forages and especially forage legumes in sub-Sahara Africa remains low (Thomas & Sumberg, 1995; Muir *et al.*, 2014). One of the reasons for the low adoption of technologies is the high heterogeneity of farming systems and agro-ecologies in sub-Sahara Africa. Instead of trying to find silver bullets to development, provision of multiple technologies and targeting is key (Giller *et al.*, 2011). Sown forages are knowledge intensive technologies, and they can play various roles in farming systems, determined by market access, population density, and agro-ecological potential (from low to high): (1) as pasture in grazing systems, e.g. through over-sowing of natural grasslands; (2) a niche role in semi-intensive mixed crop-livestock systems which rely on a diverse feed basket, e.g., planted as live barriers on farm and field boundaries, under-story in plantation, or as cover crop, green manure or intercrop with food crops; (3) as pure stand on arable land in intensive mixed crop-livestock systems, e.g., grass in sole stand or forage legume rotations with grasses and food crops. Regarding these different roles of forage technologies, it is key to develop multiple forage options to-

gether with farmers, using participatory methods (Peters *et al.*, 2003).

This study is based on two approaches that have been underlined in targeting interventions to specific socio-economic and agro-ecological environments: (1) The concept of socio-ecological niches which was initially developed to match legume technologies with the heterogeneity of agro-ecologies and farming systems in terms of soil fertility, rainfall, socio-economic status and resource endowment (Ojiem *et al.*, 2007); (2) Participatory on-farm research which is essential to integrate farmers' priorities into technology development and dissemination. Evidence from participatory trials in Sud-Kivu shows that such research approaches can increase farmers' learning and technology uptake while contributing to the scientific evidence base for improving technologies (Paul *et al.*, 2014). For the present study, researcher-managed demonstration plots as well as farmer-managed on-farm niche trials with selected forage legumes and grasses were established in four sites with contrasting agro-ecological conditions in order to (i) locally confirm agro-ecological adaptation of improved forage grasses and legumes; (ii) appraise farmers' preferences for improved forage varieties; and (iii) evaluate on-farm niches for forage grasses and legumes integrated in the cropping systems.

2 Materials and methods

2.1 Study area

The research was conducted in four sites with different agro-ecological conditions representative of Sud-Kivu: Muhongoza (02°04′S, 028°54′E – Kalehe territoire), Nyacibimba (02°29′S, 028°47′E – Kabare Territoire), Tubimbi (02°44′S, 028°35′E – Walungu Territoire), and Kamanyola (02°44′S, 029°01′E – Walungu Territoire) (Figure 1). Territoire refers to the administrative unit used in Sud-Kivu province which are, from superior to inferior: 'Territoire', 'Collectivité', 'Groupement', 'Localité', 'Village'. Sites differ in altitude, average land sizes, soil fertility, and erosion potential (Table 1).

All sites are characterised by a long rainy season from September to December, a short rainy season from February to April, and a dry season from June to August. Figure 2 shows the monthly rainfall for Nyacibimba which was chosen as most central site. Total rainfall during 2011–2013 ranged between 1345 and 1597 mm year^{-1}, with 2014 being a drier year (922 mm year^{-1}) (Figure 2).

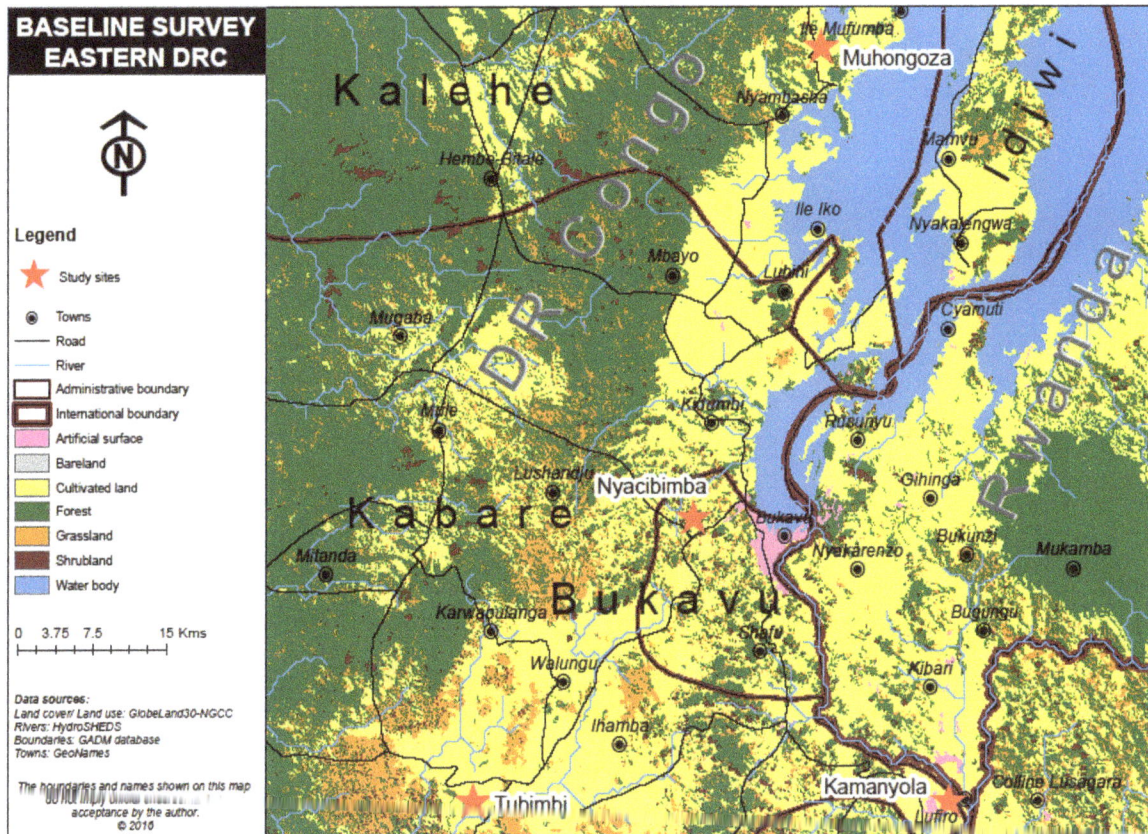

Fig. 1: *Map of study sites*

Table 1: *Study site characteristics. Land size is divided into small (<0.5 ha), medium (0.5–1 ha) and large (>1 ha). Soil fertility was scored in general terms, taking into account pH, organic matter content and nutrient availability. Soil erosion was classified as none, medium, or strong, depending on topography and rainfall.*

Site	Territoire*	Altitude (m asl)[†]	Land size	Soil fertility	Slope (%)	Soil erosion
Muhongoza	Kalehe	1548	Medium	Medium	5–10	Medium
Nyacibimba	Kabare	1955	Medium	Low	>10	Strong
Kamanyola	Walungu	940	Small	High	<5	None
Tubimbi	Walungu	1100	Medium	Low	<5	None

* In Sud-Kivu province, administrative units used are, from superior to inferior, 'Territoire', 'Collectivité', 'Groupement', 'Localité', and 'Village'; [†] asl, above sea level.

Table 2: *Physical and chemical soil quality in study sites.*

Site	Clay (%)	Sand (%)	Silt (%)	pH (water)	K (%)	Olsen P (ppm)	Total N (%)	Total C (%)
Muhongoza	58.41	27.25	14.33	4.03	0.09	13.16	0.25	3.61
Nyacibimba	40.52	39.12	20.36	4.45	0.33	16.32	0.30	2.97
Kamanyola	34.51	29.15	36.35	5.41	0.20	7.86	0.25	3.26
Tubimbi	34.50	33.16	32.34	3.89	0.10	28.71	0.22	2.77

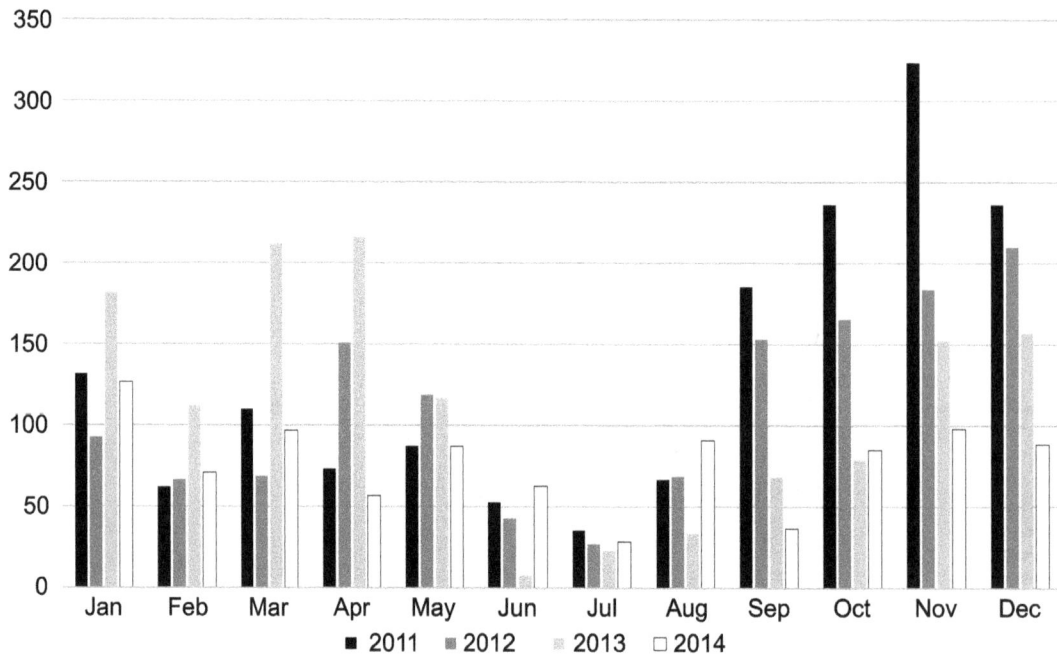

Fig. 2: *Monthly rainfall (mm) in Nyacibimba, DR Congo. Data retrieved from NASA for 01/01/2011–31/12/2014.*

Soil fertility was determined through soil sampling in August 2014 and analysed for texture, pH, and soil nutrients at the CIAT (International Center for Tropical Agriculture) soil laboratory in Nairobi, Kenya. Fertility differs among sites, mainly restricted by the low pH limiting nutrient availability and increasing Al toxicity. Only in Kamanyola, pH was close to the critical level of 5.5 for plant production (Table 2). Farmers in all four sites were organised in Innovation Platforms (IP) since end of August 2012 that were established according to the principles of Integrated Agricultural Research for Development (IAR4D) (Chiuri *et al.*, 2013) in order to improve the cavy (*Cavia porcellus*) value chain. Each IP appointed four different technical committees and held regular monthly meetings to plan common activities which included forage experimentation to improve livestock feeding.

2.2 Trial establishment, management and agronomic measurements

Two demonstration trials were established in each of the four study sites. The demonstration plots were planted, managed and harvested by the IP farmers themselves, however, directed by a researcher from UEA (Université Evangélique en Afrique) who also used the field events for training both farmers and agronomy students from the university. The first trial tested the forage legumes *Canavalia brasiliensis* CIAT 17009 (Canavalia), *Stylosanthes guianensis*

CIAT 11995 (Stylo) and *Desmodium uncinatum* cv. Silverleaf (Silverleaf desmodium). These legumes had previously performed well in the study area (Katunga *et al.*, 2014). Seeds were obtained from Karama Research station of RAB (Rwanda Agriculture Board) in Eastern Rwanda. Silverleaf desmodium is already naturalised in the study area and was used in this trial as local control. The second trial tested the forage grasses Napier grass (*Pennisetum purpureum*) cv. French Cameroon, a local Napier variety as well as Guatemala grass (*Tripsacum andersonii*). Cuttings of cv. French Cameroon and Guatemala were acquired from INERA (Institut National pour l'Etude et la Recherche Agronomiques) Mulungu, while those of local Napier grass were obtained in the respective sites and used as local control. The two trials were established in a completely randomized block design with three replications and lasted six months from October 2012 to April 2013. Planting dates were 2, 4, 6 and 19 October in Tubimbi, Muhongoza, Nyacibimba and Kamanyola, respectively. In the legume trial, each plot measured 3 m × 1.5 m with sowing spacing of 50 cm × 25 cm, while in the grass trial, plots measured 4 m × 3 m with planting spacing of 1 m × 1 m. Blocks and plots were separated by 1 m. Neither chemical fertilisers nor pesticides were applied, but two hands of cow manure per hole were applied in both trials before planting. Fresh biomass was harvested only once after six months between 26 March and 4 April 2013. For the legume trial, biomass was assessed from an area of 1 m² by cutting any green material 15 or 10 cm above soil surface (for Canavalia or Stylo/Silverleaf desmodium, respectively); whereas for the grasses, fresh biomass was cut inside an area of 6 m^2 and 50 cm above soil surface in order to allow plants to regenerate. A homogeneous sample of at least 100 g fresh matter was collected from each of the three replications for all forage varieties. Samples were oven-dried at INERA in Mulungu at 75°C during 48 hours to obtain dry matter (DM) content. Percentage (%) soil cover was estimated before biomass harvest. This helped to evaluate the potential to reduce soil erosion and suppress weed growth. Average plant height was taken on five plants in each plot from the plant bottom up to top of the highest leaf. Analysis of variance (ANOVA) was carried out with R Studio Version 0.97.449 (R Core Team, 2013). Grass and legume trials were analysed separately for DM yield, plant height and soil cover, with site, variety and block as treatment factors. Local Napier in Kamanyola was disregarded from the statistical analysis since there was no yield from any replicate. Means are presented with standard errors. A P-value ≤ 0.05 was considered significant.

2.3 Participatory evaluation

A participatory evaluation was carried out in each site before forage biomass was harvested. Across all sites, a total of 63 farmers (43 women) were involved in the forage appraisal. After explaining the methodology, farmers were arranged in two groups by gender. In every site, one field replication was randomly chosen for the exercise. From each group, participants went individually to appraise the three plots within the forage grass and legume trials, together with a scientist who took note of their ranking. Farmers ranked forages from 1 (most preferred) to 3 (least preferred) according to their own selection criteria; data in this paper are presented as percentages. Farmers were also individually asked which of the criteria were most important for ranking the different forage varieties.

2.4 On-farm niche planting and data collection

An on-farm study was carried out from October to December 2013 to assess niches for forages. A total of 79 farmers of which 73 % women volunteered to test improved forages in their own fields. Before sowing, farmers were trained on importance and management of forages: planting, maintenance, harvesting biomass and seed, feeding animals as well as about the potential spatial niches on their farms. Two packages were availed to each farmer to be planted during the wet season: a package of forage legume seeds (Canavalia, Stylo, *Lablab*

purpureus CIAT 22759 and Silverleaf desmodium) of about 60–100 g for each variety and a pack of at least 40 cuttings per variety of grasses including Guatemala and Napier grass (French Cameroon and the local variety). Farmers were free to choose where and how to integrate these forages in their farming systems. After successful establishment, farmers were also encouraged to share any planting material or seeds produced in the future with other IP members or neighbours. The IP technical committees recorded all farmers who had planted forages and shared these data in December 2013. A second assessment of on-farm niches was carried out in September and October 2014 using the same methodology.

3 Results

3.1 Demonstration trials

Agro-ecological conditions of the different sites significantly influenced DM yields of the legumes ($P < 0.001$), with DM yields ranging from 6.1–6.9 t ha^{-1} in fertile soils (Kamanyola), 2.6–4.1 t ha^{-1} in medium soils (Muhongoza) and 0.3–2.1 t ha^{-1} in poor soils (Nyacibimba and Tubimbi) with the exception of Silverleaf desmodium which performed well (5.4 t ha^{-1}) in Nyacibimba (Table 3). There were no significant differences among the legume varieties across all sites ($P = 0.63$), but a significant interaction effect between site and variety ($P = 0.002$) (Table 3). Plant height was closely related to DM biomass production. Soil cover depended both on site conditions ($P < 0.001$) and the legume varieties ($P < 0.001$) tested. Stylo covered consistently less soil than Silverleaf desmodium and Canavalia (Table 3).

The three forage grass varieties performed differently in each site with regard to DM yield ($P = 0.006$), but the effect was weaker than in legumes. The effect of site on plant height ($P = 0.13$) and soil cover ($P = 0.08$) was not significant, in contrary to legumes (Table 4). Differences between the grass varieties were strong in terms of DM yield ($P < 0.001$), plant height ($P < 0.001$) and soil cover ($P < 0.001$). French Cameroon DM yields ranged from 3.5–10.1 t ha^{-1}, being the highest yielding variety in Kamanyola and Tubimbi, and the second highest in Nyacibimba and Muhongoza. Guatemala grass DM yields varied between 1.3–2.6 t ha^{-1} and were the lowest in all sites except Kamanyola where local Napier did not yield any biomass at all. This was probably due to the fact that local Napier grass in Kamanyola originated from swampy lowland areas and it was not adapted to the drier conditions of the site where the demonstration plot was established.

Table 3: *Dry matter (DM) yield (t ha^{-1}), plant height (cm) and estimated soil cover (%) of three forage legume varieties tested at four sites in Sud-Kivu, DRC.*[†]

	Variety	Site			
		Muhongoza	Nyacibimba	Kamanyola	Tubimbi
DM yield (t ha^{-1})	Canavalia	3.2 (0.3)	2.2 (0.5)	6.9 (1.0)	1.5 (0.2)
	Silverleaf desmodium	2.6 (0.2)	5.4 (0.5)	6.1 (0.6)	0.3 (0.1)
	Stylo	4.2 (0.5)	1.6 (0.6)	6.7 (0.7)	2.1 (1.0)
ANOVA	Variety	$P = 0.63$ ^{ns}			
	Site	$P < 0.001$ ***			
	Variety*Site	$P = 0.002$ **			
Plant height (cm)	Canavalia	48.4 (4.1)	36.9 (2.1)	45.7 (4.4)	40.3 (1.1)
	Silverleaf desmodium	27.6 (5.6)	61.7 (6.2)	53.3 (4.0)	15.1 (2.3)
	Stylo	42.3 (2.3)	30.9 (1.6)	53.1 (2.5)	41.5 (3.7)
ANOVA	Variety	$P = 0.32$ ^{ns}			
	Site	$P < 0.001$ ***			
	Variety*Site	$P < 0.001$ ***			
Soil cover (%)	Canavalia	86.7 (0.1)	66.7 (0.3)	96.7 (1.0)	68.3 (0.2)
	Silverleaf desmodium	66.7 (0.5)	100.0 (0.2)	68.3 (0.6)	16.7 (0.1)
	Stylo	65.0 (0.6)	37.5 (0.5)	73.3 (0.7)	63.3 (1.0)
ANOVA	Variety	$P < 0.001$ ***			
	Site	$P < 0.001$ ***			
	Variety*Site	$P < 0.001$ ***			

[†] Values are means with standard error (N=3). Levels of significance: * <0.05, ** <0.01, *** <0.001, ^{ns} not significant

Table 4: *Dry matter (DM) yield (t ha^{-1}), plant height (cm) and estimated soil cover (%) of three forage grass varieties tested at the study sites in Sud-Kivu.*[†]

	Variety	Site			
		Muhongoza	Nyacibimba	Kamanyola	Tubimbi
DM yield (t ha^{-1})	French Cameroon	3.5 (0.6)	5.6 (1.5)	10.1 (1.7)	3.6 (1.7)
	Local Napier	6.2 (1.0)	7.5 (1.8)	NA	1.9 (1.0)
	Guatemala	2.4 (0.4)	1.3 (0.1)	2.6 (0.8)	2.4 (0.5)
ANOVA	Variety	$P < 0.001$ ***			
	Site	$P = 0.006$ **			
	Variety*Site	$P = 0.02$ *			
Plant height (cm)	French Cameroon	239.1 (1.7)	228.9 (12.0)	280.8 (24.8)	238.1 (24.1)
	Local Napier	278.8 (6.3)	270.1 (22.4)	NA	183.3 (35.7)
	Guatemala	116.9 (5.8)	143.9 (7.3)	149.4 (10.7)	156.1 (4.9)
ANOVA	Variety	$P < 0.001$ ***			
	Site	$P = 0.13$ ^{ns}			
	Variety*Site	$P = 0.01$ *			
Soil cover (%)	French Cameroon	64.0 (1.5)	80.0 (0.6)	83.3 (1.7)	73.3 (1.7)
	Local Napier	70.0 (1.8)	66.7 (1.0)	NA	40.0 (1.0)
	Guatemala	78.3 (0.1)	66.7 (0.4)	86.7 (0.8)	86.7 (0.5)
ANOVA	Variety	$P < 0.001$ ***			
	Site	$P = 0.08$ ^{ns}			
	Variety*Site	$P < 0.001$ ***			

[†] Values are means with standard error (N=3). Levels of significance: * <0.05, ** <0.01, *** <0.001, ^{ns} not significant

3.2 Participatory evaluation

Generally, farmers' preferences for forages were guided by the following criteria in decreasing order of importance: biomass production, leaf size, animal preference, recovery and drought tolerance (legumes); biomass production, animal preference, adaptation, use in erosion control, tillering capacity, and novelty in the area (for grasses only). Women and men expressed the same preference criteria. For legumes, the overall highest priority in high altitude (Muhongoza and Nyacibimba) was given to Silverleaf desmodium (40% and 70%), while second choice was Canavalia (47% and 55% respectively). In mid altitude (Tubimbi) and low altitude (Kamanyola), first choice was Canavalia (61% and 70%, respectively) and second Stylo (56%) for Tubimbi and Silverleaf desmodium (50%) for Kamanyola (Table 5). For the grasses, cv. French Cameroon clearly was the first choice across all sites (87% in Muhongoza, 70% in Nyacibimba and 61% in Tubimbi) except in Kamanyola, where it was ranked second (60%) after Guatemala grass (Table 5). Gender had no influence on the choice of forages (data not presented).

3.3 On-farm planting

Interest of farmers varied across sites, and in 2013 participating farmers ranged from 32 in Muhongoza to 8 in Kamanyola. Overall, women were more active in on-farm experimentation of improved forages across all the sites, with 58 out of 79 participating farmers, corresponding to 73% (Figure 3a). In Muhongoza, Nyacibimba and Tubimbi 54% of farmers decided to plant grasses together with legumes, 45% grasses alone and

1% legumes alone; while in Kamanyola 25% of farmers planted grasses and legumes together, 0% grasses alone and 75% legumes alone (Figure 3a). Compared to 2013, the overall number of farmers who planted forages was the same in 2014; however, percentage of women decreased to 47%. While in Muhongoza more farmers planted forages in 2014 than in 2013 (+40%, with all additional farmers being men), in Nyacibimba, they almost halved (−45%, with all dropping out farmers being women). In Nyacibimba the farmers growing grasses decreased to 9%, while in Kamanyola 43% of farmers started growing grasses (Figure 3b).

Farmers' integration of forages into their farming system also varied according to site as well as forage type. In 2013, forage legumes were intercropped with other crops such as maize and cassava across all sites by 43% of the farmers, while 40% grew them as pure stand. Especially in Tubimbi, more farmers grew legumes as pure stand (71%) in small plots around the homesteads and later mixed for feeding with natural forages (Figure 4a). Across sites, grasses were mainly planted on field edges (56% of farmers), followed by pure stands hedges for erosion control (19%) and pure stand (10%). In the high altitude sites of Muhongoza and Nyacibimba, more farmers planted grasses on field edges (55% and 60%, respectively) and in hedgerows for erosion control (16% and 30%). Only in Tubimbi, 47% of farmers initially planted grasses as pure stand (Fig. 4c). In 2014, there was more niche diversity for grasses in Tubimbi in 2014, where farmers experimented with planting forages in banana plantations, under trees as well as on field edges and for erosion control (Figure 4d). Overall, the integration of both forage legumes and grasses into farming systems in 2014 was similar to 2013 (Figure 4).

Table 5: *Farmers' choice (%) among three forage grasses and three forage legumes according to site in Sud-Kivu, DRC.*

Site	Grass variety	Choice (%)			Legume variety	Choice (%)		
		1st	*2nd*	*3rd*		*1st*	*2nd*	*3rd*
Muhongoza (N=15)	French Cameroon	86.7	6.7	6.7	Canavalia	33.3	46.7	20.0
	Local Napier	6.7	66.7	26.7	Silverleaf desmodium	40.0	13.3	46.7
	Guatemala	6.7	26.7	66.7	Stylo	26.7	40.0	33.3
Nyacibimba (N=20)	French Cameroon	70.0	15.0	15.0	Canavalia	30.0	55.0	15.0
	Local Napier	15.0	40.0	45.0	Silverleaf desmodium	70.0	30.0	0.0
	Guatemala	15.0	45.0	40.0	Stylo	0.0	15.0	85.0
Kamanyola (N=10)	French Cameroon	20.0	60.0	20.0	Canavalia	70.0	10.0	20.0
	Local Napier	10.0	10.0	80.0	Silverleaf desmodium	0.0	50.0	50.0
	Guatemala	70.0	30.0	0.0	Stylo	30.0	40.0	30.0
Tubimbi (N=18)	French Cameroon	61.1	38.9	0.0	Canavalia	61.1	27.8	11.1
	Local Napier	0.0	5.6	94.4	Silverleaf desmodium	11.1	16.7	72.2
	Guatemala	38.9	55.6	5.6	Stylo	27.8	55.6	16.7

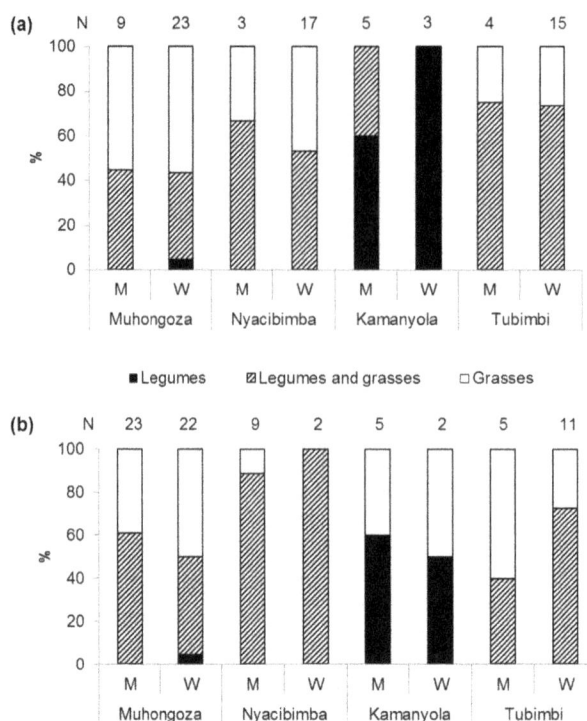

Fig. 3: *Forage types cultivated by farmers on their own farms in four different sites of Sud-Kivu, DRC in (a) December 2013 and (b) September/October
2014. M refers to men, W to women. All sites are described in Tables 1 and 2.*

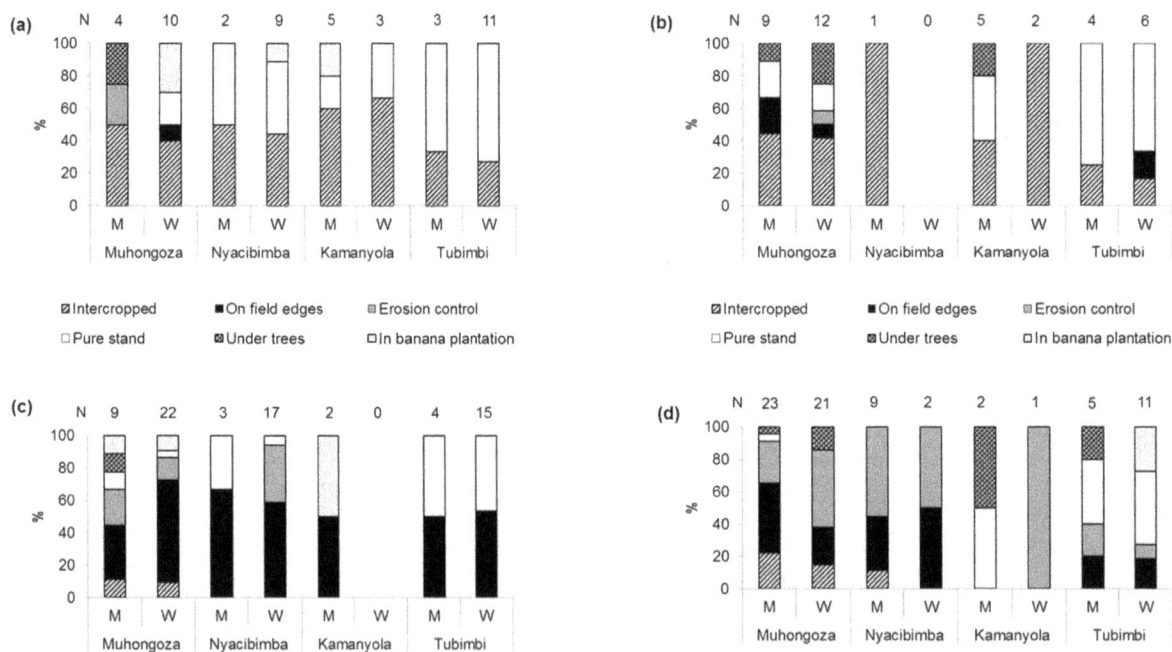

Fig. 4: *Integration of (a, b) forage legumes and (c, d) grasses into farming systems by farmers on their own land in four
different sites of Sud-Kivu, DRC in (a, c) December 2013 and (b, d) September/October 2014. M refers to men, W to women.
All sites are described in Tables 1 and 2.*

4 Discussion

4.1 Improved forage legume and grass options across different agro-ecologies

Herbage yield of the first cut of both legumes and grasses was dependent on agro-ecological conditions. Biomass production was highest in Kamanyola, which has the least acidic soils of all sites. Tubimbi showed lowest and most variable biomass production, which is probably due to most acidic soils (pH 3.9) limiting P availability and increasing Al toxicity as well as plant susceptibility for diseases and pests. Forage legumes adapted differently to the four sites with no differences among varieties, confirming that forage legumes are highly site specific. Stylo appeared less suitable for soil protection due to relatively low above-ground soil cover, although roots also contribute to reduction of soil erosion. Forage grasses displayed higher variability in biomass production among varieties than among sites, underlining their broader adaptability. French Cameroon showed consistently higher biomass yield than Guatemala grass, as did local Napier except for its failure in Kamanyola.

However, it is important to note that the agronomic data from the demonstration trials presented in this study can only be considered as indicative for agro-ecological adaptation. The main reason is that only one herbage cut was done, and it was carried out later than normally (six months after planting). This does not only limit comparability with other agronomic forage trials, but also rather gives insights into the establishment of the forages tested, not their performance over several seasons. For example, farmers in Nyacibimba abandoned French Cameroon in the second year (2014) as it was susceptible to pests (small flies) in most farms that caused drying of the leaves (F. L. Muhimuzi, unpublished observation). Nevertheless, the trials confirmed performance of the best bet forages evaluated in a previous agronomic study in the same sites (Katunga et al., 2014). The study season (October 2012 – April 2013) fell into a normal rainfall season when comparing with previous and following seasons (Figure 2). Thus the results can still be interpreted as indicative for establishment and agro-ecological adaptation.

Farmers in the region have very minor tradition of cultivating improved forages (Maass et al., 2012; Bacigale et al., 2014), and if so, they are rather familiar with grasses (Napier, Guatemala) and fodder shrubs. Particularly, the offered annual legumes were new to most farmers so that they had to choose new selection criteria during focus group discussions. Farmers prioritized biomass production and animal preference, which resulted in highest rankings for herbage yield. The same priorities had been previously observed for forage legumes (Katunga et al., 2014) and cassava-legume intercropping in the same area (Pypers et al., 2011). However, visual assessment of biomass production corresponded to apparent and not necessarily real biomass production, which may have resulted in farmers favouring Canavalia with its large leaves. Two additionally important criteria for grasses were novelty in the area and capacity for erosion control, which made French Cameroon the most popular choice especially in new research sites (Muhongoza and Nyacibimba), where farmers have not previously been in contact with this variety. It is known that Guatemala grass is well adapted to the higher areas of Sud-Kivu (Compere, 1960), although it is not prevalent in the region. In this study, farmers of most sites selected Guatemala second after French Cameroon. As Guatemala grass is fairly drought-tolerant, Guatemala can still offer digestible forage during the dry season when Napier grass is losing its forage quality.

4.2 On-farm niches for forages in Sud-Kivu

Farmers' interest to participate in forage research and planting varied among sites, also reflecting livestock husbandry systems and traditions. In Kamanyola, the relatively lower interest can be partly explained by the predominant free livestock roaming system, which is the most common practice once crops have been harvested. In such areas, there is no tradition of planting forages for livestock feeding, which has been confirmed by a previous feed assessment showing the low percentage of farmers currently growing fodder (Bacigale et al., 2014).

Initially, women were more interested in forage cultivation than men across research sites. This might have two reasons: Firstly, women are often in charge of livestock feeding (Maass et al., 2012), and planting forages close to the homestead can considerably reduce time and effort for collecting forages from far. In South-East Asia, the labour saving effect for women was an important entry point for forage technology adoption (Phengsavanh & Stür, 2006). Women are likely to favour forage crops, especially when they have multiple uses such as food crop, soil erosion and weed control, as well as soil fertility maintenance and/or farm or field boundary demarcation. Secondly, the IPs through which the experimentation was conducted focussed on cavies that are predominantly tended by women and boys (Zozo et al., 2010; Maass et al., 2012). One year later, men also became interested in planting forages to feed goats, resulting into almost equal participation of men and women.

The choice of forage type as well as their integration into cropping systems depended on various factors, mostly land availability, land tenure, soil erosion prevalence and livestock husbandry system. While in most sites, farmers chose to cultivate both legumes and grasses, Kamanyola was the only site with a majority of farmers initially choosing to grow (annual) legumes only. The high percentage of rented land and free livestock roaming in the dry season on fields does not favour cultivation of perennial grass varieties that remain when the annual food crops have been harvested. Farmers in high altitude and sloping sites of Muhongoza and Nyacibimba had the highest preference for grass varieties planted as hedgerows due to the erosion control benefits. In general, farmers preferred cropping system integration over monocropping, especially for legumes. This can be explained with low land availability in Sud-Kivu, where average farm sizes are as low as 0.4 ha (Ouma & Birachi, 2012), although livestock farmers were shown to have larger land sizes of 1.5 ha (Maass *et al.*, 2012). Only in Tubimbi, more forage grasses and legumes are grown as sole stand due to the comparably larger average land sizes and lower pressure on land. Furthermore, men work in nearby artisanal mines. Thus women can more easily decide to allocate more land to forage cultivation since normally their access to land is limited by their ability to negotiate with the male household head (Zozo *et al.*, 2010).

4.3   Future opportunities and further research needs

These results underline the importance of targeting forages well to specific agro-ecological conditions, but also to conduct adaptive research to fine-tune matching forages to the farmers' interests, needs and production systems. It has been suggested before to provide farmers with a basket of forage options to choose from (Stür *et al.*, 2002). A well-functioning, robust and convincing forage innovation is most critical to adoption (Stür *et al.*, 2013). As shown in this study, forage crops' adaptability can be highly site-specific and integration into cropping systems depends on various factors, especially in heterogeneous smallholder farming systems. Rigorous, long-term multi-locational system agronomy trials are needed to test spatial and temporal integration of annual and perennial forage crops with other food and forage crops. Good examples are provided by the work of Naudin *et al.* (2011), who have experimented with integrating different forages (*Vigna unguiculata*, *Vicia villosa*, *Lablab purpureus*, *Stylosanthes guianensis*) into maize, cassava and rice systems in Madagascar for feed, soil fertility, and protection purposes.

Our study underlines the usefulness of the concept of on-farm niches for better understanding why and how farmers choose to grow forage crops. However, the socio-ecological niche approach needs more systematic operationalisation in order to be useful for targeting of technologies to agro-ecologies and farming systems for maximum impact on farmers' livelihoods and environmental quality. Formalising previous research and expert knowledge into simple decision support tools could help to match contexts (e.g. constraints in terms of land, labour, capital, input availability, knowledge and markets) and farmers' objectives and needs (e.g. in terms of food, feed, soil protection, income) with suitable forage technologies. These are needed to assist scientists, development and extension workers in recommending and promoting most suitable technologies with maximum potential impact.

Moreover, better understanding of forage technology uptake and adoption is needed for sub-Sahara Africa. Although the principle of IAR4D proved useful in linking farmers to a value chain, further research from the area is needed on economic and labour benefits of forage cultivation and their (potential) contribution to smallholders' incomes and livelihoods. Developed value chains and functioning extension services can provide crucial market pulls and expertise stimulating behavioural change towards farm-grown fodder and stall feeding, such as recently shown in Vietnam (Stür *et al.*, 2013).

Acknowledgements

Our gratitude goes to farmers, members of the cavy Innovation Platforms from Muhongozi, Nyacibimba, Kamanyola and Tubimbi, for their unreserved collaboration. We are grateful to the Australian Agency for International Development (AusAID) for funding this research under the Africa Food Security Initiative and through the partnership between the Commonwealth Science and Industrial Research Organisation (CSIRO) and the Biosciences eastern and central Africa (BecA) Hub at the International Livestock Research Institute (ILRI), who was leading the project. Special thanks to Caroline Sibomana, Valence Bwana, Pyame Balemirwe and Dieudonné Katunga Musale for assistance in implementing the research activities. Nicholas Koech (CIAT) mapped the study sites (Figure 1). Staff time by the CIAT authors was partly supported by the CGIAR Research Program on Integrated Systems for the Humid Tropics (Humidtropics).

References

Bacigale, S. B., Paul, B. K., Muhimuzi, F. L., Mapenzi, N., Peters, M. & Maass, B. L. (2014). Characterizing feeds and feed availability in Sud-Kivu province, DR Congo. *Tropical Grasslands – Forrajes Tropicales*, 2, 9–11.

Chiuri, W., Birachi, E., Buruchara, R., Adekunle, W., Fatunbi, O., Pali, P. N., Wimba, B., Bizosa, A., Nyamurinda, B., Nyamwaro, S. O., Habumugisha, P., Tuyisenge, J., Bonabana-Wabbi, J., Karume, K., Kaenge, V. & Kamugisha, R. (2013). Market access for agro-enterprise diversity in the Lake Kivu Pilot Learning Site of the sub-Saharan Africa Challenge Programme. *African Journal of Agricultural and Resource Economics*, 8 (3), 120–134.

Compere, R. (1960). Introduction into Kivu (Congo) of *T. laxum*, a forage plant for dairy cows in the dry season. *Bulletin agricole du Congo belge*, 51 (5), 1085–1103.

Cox, T. P. (2012). Farming the battlefield: The meanings of war, cattle and soil in South Kivu, Democratic Republic of Congo. *Disasters*, 36 (2), 233–248.

Giller, K. E., Tittonell, P., Rufino, M. C., van Wijk, M. T., Zingore, S., Mapfumo, P., Adjei-Nsiah, S., Herrero, M., Chikowo, K. R., Corbeels, M., Rowe, E. C., Baijukya, F., Mwijage, A., Smith, J., Yeboah, E., van der Burg, W. J., Sanogo, O. M., Misiko, M., de Ridder, M., Karanja, S., Kaizzi, C., K'ungu, J., Mwale, M., Nwaga, D., Pacini, C. & Vanlauwe, B. (2011). Communicating complexity: Integrated assessment of trade-offs concerning soil fertility management within African farming systems to support innovation and development. *Agricultural Systems*, 104 (2), 191–203.

Katunga, M. M. D., Muhigwa, J. B. B., Kashala, K. J. C., Ipungu, L., Nyongombe, N., Maass, B. L. & Peters, M. (2014). Testing agro-ecological adaptation of improved herbaceous forage legumes in South-Kivu, D.R. Congo. *American Journal of Plant Sciences*, 5, 1384–1393.

Maass, B. L., Chiuri, W. L., Zozo, R., Katunga, D., Metre, K. T. & Birachi, E. (2013). Using the "livestock ladder" to exit poverty for poor crop-livestock farmers in South Kivu province, eastern DR Congo. *In:* Vanlauwe, B., Blomme, G. & Van Asten, P. (eds.), *Agro-ecological Intensification of Agricultural Systems in the African Highlands*. pp. 145–155, New York: Routledge.

Maass, B. L., Musale, D. K., Chiuri, W. L., Gassner, A. & Peters, M. (2012). Challenges and opportunities for smallholder livestock production in post-conflict South Kivu, eastern DR Congo. *Tropical Animal Health and Production*, 44 (6), 1221–1232.

Muir, J. P., Pitman, W. D., Dubeux Jr, J. C. & Foster, J. L. (2014). The future of warm-season, tropical and subtropical forage legumes in sustainable pastures and rangelands. *African Journal of Range & Forage Science*, 31 (3), 187–198.

Naudin, K., Scopel, E., Andriamandroso, A. L. H., Rakotosolofo, M., Andriamarosoa Ratsimbazafy, N. R. S., Rakotozandriny, J. N., Salgado, P. & Giller, K. E. (2011). Trade-Offs Between Biomass Use and Soil Cover. the Case of Rice-Based Cropping Systems in the Lake Alaotra Region of Madagascar. *Experimental Agriculture*, 48 (2), 194–209.

Ojiem, J. O., Vanlauwe, B., de Ridder, N. & Giller, K. E. (2007). Niche-based assessment of contributions of legumes to the nitrogen economy of Western Kenya smallholder farms. *Plant and Soil*, 292 (1-2), 119–135.

Ouma, E. & Birachi, E. (eds.) (2012). *CIALCA Baseline Survey. CIALCA Technical Report 17.* CIALCA (Consortium for Improving Agriculturebased Livelihoods in Central Africa). Nairobi, Kenya and Ibadan, Nigeria: CIAT, IITA and Bioversity International. URL http://www.cialca.org/files/files/cialca%20baseline%20survey%20report_print.pdf (last accessed: 25.10.2014).

Paul, B. K., Pypers, P., Sanginga, J. M., Bafunyembaka, F. & Vanlauwe, B. (2014). ISFM Adaptation Trials: Farmer-to Farmer Facilitation, Farmer-Led Data Collection, Technology Learning and Uptake. *In:* Vanlauwe, B., Van Asten, P. & Blomme, G. (eds.), *Challenges and Opportunities for Agricultural Intensification of the Humid Highland Systems of Sub-Saharan Africa*. pp. 385–397.

Peters, M., Herrero, M., Fisher, M., Erb, K., Rao, I., Subbarao, G., Castro, A., Chara, J., Murgueitio, E., van der Hoek, R., Läderach, P., Hyman, G., Tapasco, J., Strassburg, B., Paul, B., Rincon, A., Schultze-Kraft, R., Fote, S. & Searchinger, T. (2013). Challenges and opportunities for improving eco-efficiency of tropical forage-based systems to mitigate greenhouse gas emissions. *Tropical Grasslands - Forrajes Tropicales*, 1, 137–149.

Peters, M., Lascano, C. E., Roothaert, R. & De Haan,

N. C. (2003). Linking research on forage germplasm to farmers: the pathway to increased adoption–a CIAT, ILRI and IITA perspective. *Field Crops Research*, 84 (1), 179–188.

Phengsavanh, P. & Stür, W. (2006). The use and potential of supplementing village pigs with Stylosanthes guianensis in Lao PDR. *In:* Preston, R. & Ogle, B. (eds.), *Workshop on Forages for Pigs and Rabbits.* pp. 1–6, Phnom Penh, Cambodia.

Pypers, P., Sanginga, J.-M., Kasereka, B., Walangululu, M. & Vanlauwe, B. (2011). Increased productivity through integrated soil fertility management in cassava-legume intercropping systems in the highlands of Sud-Kivu, DR Congo. *Field Crops Research*, 120 (1), 76–85.

R Core Team (2013). *R: A language and environment for Statistical Computing.* R Foundation for Statistical Computing, Vienna, Austria. URL http://www.R-project.org/

Stür, W., Horne, P. M., Gabunada Jr, F. A., Phengsavanh, P. & Kerridge, P. C. (2002). Forage options for smallholder crop–animal systems in Southeast Asia: working with farmers to find solutions. *Agricultural Systems*, 71 (1), 75–98.

Stür, W., Khanh, T. T. & Duncan, A. (2013). Transformation of smallholder beef cattle production in Vietnam. *International Journal of Agricultural Sustainability*, 11 (4), 363–381.

Thomas, D. & Sumberg, J. E. (1995). A review of the evaluation and use of tropical forage legumes in sub-Saharan Africa. *Agriculture, Ecosystems & Environment*, 54 (3), 151–163.

Vlassenroot, K. (2005). Households land use strategies in a protracted crisis context: land tenure, conflict and food security in eastern DRC. Conflict Research Group, University of Ghent, Belgium. URL http://www.fao.org/3/a-ag306e.pdf (last accessed: 07.05.2016).

Zozo, R., Chiuri, W. L., Katunga, D. M. & Maass, B. L. (2010). Report of a Participatory Rural Appraisal (PRA) in the Groupements of Miti-Mulungu and Tubimbi, South Kivu/DR Congo. CIAT Working Document No. 211, 32 pp. Nairobi, Kenya: Centro Internacional de Agricultura Tropical (CIAT). URL http://ciat-library.ciat.cgiar.org/articulos_ciat/CIAT_WD211_PRA_DRC.pdf (last accessed: 25.10.2014).

11

Development of mechanical methods for cell-tray propagation and field transplanting of dwarf napiergrass (*Pennisetum purpureum* Schumach.)

Renny Fatmyah Utamy [a], Yasuyuki Ishii [b,*], Sachiko Idota [b],
Lizah Khairani [c], Kiichi Fukuyama [b]

[a]*Interdisciplinary Graduate School of Agriculture and Engineering, University of Miyazaki, Miyazaki, Japan*
[b]*Faculty of Agriculture, University of Miyazaki, Japan*
[c]*Graduate School of Agriculture, University of Miyazaki, Miyazaki, Japan*
(present: Faculty of Animal Husbandry, University of Padjadjaran, Indonesia)

Abstract

Since dwarf napiergrass (*Pennisetum purpureum* Schumach.) must be propagated vegetatively due to lack of viable seeds, root splitting and stem cuttings are generally used to obtain true-to-type plant populations. These ordinary methods are laborious and costly, and are the greatest barriers for expanding the cultivation area of this crop. The objectives of this research were to develop nursery production of dwarf napiergrass in cell trays and to compare the efficiency of mechanical versus manual methods for cell-tray propagation and field transplanting. After defoliation of herbage either by a sickle (manually) or hand-mowing machine, every potential aerial tiller bud was cut to a single one for transplanting into cell trays as stem cuttings and placed in a glasshouse over winter. The following June, nursery plants were trimmed to a 25–cm length and transplanted in an experimental field (sandy soil) with 20,000 plants ha^{-1} either by shovel (manually) or Welsh onion planter. Labour time was recorded for each process. The manual defoliation of plants required 44 % more labour time for preparing the stem cuttings (0.73 person-min. stem-cutting^{-1}) compared to using hand-mowing machinery (0.51 person-min. stem-cutting^{-1}). In contrast, labour time for transplanting required an extra 0.30 person-min. m^{-2} (14 %) using the machinery compared to manual transplanting, possibly due to the limited plot size for machinery operation. The transplanting method had no significant effect on plant establishment or plant growth, except for herbage yield 110 days after planting. Defoliation of herbage by machinery, production using a cell-tray nursery and mechanical transplanting reduced the labour intensity of dwarf napiergrass propagation.

Keywords: establishment, labour time, mechanical transplanting, nursery production, vegetative propagation

1 Introduction

Dwarf napiergrass (*Pennisetum purpureum* Schumach.) of a late-heading type (dwarf late, DL), which was bred in Florida, USA (Hanna *et al.*, 1993) and brought to the Thai Dairy Promotion Organization (DPO) in Thailand, was introduced to Japan in 1996 (Ishii *et al.*, 1998). The plant's features are quite similar to 'Mott' dwarf napiergrass (Sollenberger *et al.*, 1988) and it has the potential to spread in southern Kyushu (Ishii *et al.*, 1998). This grass can be perennial in the low-altitude areas of Kyushu Island (Mukhtar *et al.*, 2003; Ishii *et al.*, 2005, 2008; Wadi *et al.*, 2008), if the lowest minimum temperature is maintained above –6.2°C in winter (Ishii *et al.*, 2008, 2013).

*Corresponding author
Faculty of Agriculture, University of Miyazaki,
Miyazaki 889-2192, Japan
Email: yishii@cc.miyazaki-u.ac.jp

In 2007, this dwarf napiergrass was spread to 12 extension sites including isolated islands in the Miyazaki, Kumamoto and Kagoshima prefectures of southern Kyushu (Utamy *et al.*, 2011), where the current area under cultivation is more than 10 ha. Dwarf napiergrass can yield sufficient dry matter (DM) at 13 Mg ha^{-1} of a superior herbage quality (Khairani *et al.*, 2013) compared to ordinary tropical grasses under an additional fertiliser application of at least 100 kg N ha^{-1} year^{-1} after establishment (Utamy *et al.*, 2011). This grass has been examined for responses to cutting frequency and plant density (Mukhtar *et al.*, 2003; Wadi *et al.*, 2004), chemical fertiliser and manure application (Hasyim *et al.*, 2008, 2010), rotational grazing management (Ishii *et al.*, 2005, 2009) and weed control at the establishment and early growth stage (Utamy *et al.*, 2012) in southern Kyushu.

However, there remained an agronomic problem to be solved for establishing napiergrass. Napiergrass should be propagated vegetatively by root splitting, stem cuttings or shoot tips due to lack of viable seeds (Burton, 1989). However, it is very laborious to prepare nursery plants and transplant them by hand (manually), and manual methods seem to be inefficient compared with systems using machinery. Although vegetative propagation efficiency was improved by the shoot-cutting method (Kang *et al.*, 2011), mechanical planting is a well-known method for the establishment of rice (Sharma & Singh, 2008), Job's tears (*Coix lacryma-jobi* L. var. *mayuen* (Romman.) Stapf) (Inoue *et al.*, 1984), and fruits and vegetables (Choon, 1999; Sarif *et al.*, 1996). A prototype semi-automated napiergrass planter, on which a worker must ride to set out nursery plants, has been developed (Sollenberger *et al.*, 1987). Transplanting of pangolagrass (*Digitaria eriantha* Steud.) cv. Transvala was developed using a hand transplanter in Okinawa Prefecture (Okinawa Prefectural Livestock Experiment Station, 2005), but it is not available for napiergrass due to over-sizing of nursery plants. An automated mechanical transplanting method was applied to guineagrass (*Panicum maximum* Jacq.) by Nakahara *et al.* (2001) and bahiagrass (*Paspalum notatum* Flügge) by the Kumamoto National Livestock Breeding Center (2006). However, nursery plant production of dwarf napiergrass using cell trays as applied for vegetables, has not yet been tested for napiergrass, and a mechanical planter for vegetables has not been applied yet for nursery plants of this grass species.

Therefore, our objectives were to develop a method for nursery plant production of dwarf napiergrass using cell trays and to compare the efficiency of mechanical versus manual methods for preparing single-nodal stem pieces for cell-tray propagation and in-field transplanting.

2 Materials and methods

2.1 Grass species and site description

The grass species used was dwarf napiergrass of the late-heading type (dwarf late, DL). This grass grows to a height of 1.3–2.0 m (depending on the growing conditions) with an average internode length < 5 cm. This research was composed of two experiments, one for the preparation of stem cuttings and nursery production in a glasshouse (Experiment 1), and another for transplanting nursery plants to the field and establishment of the plants (Experiment 2). Experiment 1 was conducted first at Sumiyoshi Livestock Science Station (SLSS), University of Miyazaki (31° 59′ N, 131° 27′ E, 10 m above sea level) for the preparation of stem cuttings and then at Kibana Agricultural Science Station (KASS), University of Miyazaki (31° 50′ N, 131° 25′ E, 31 m above sea level) for nursery production in the glasshouse from November 2008 to May 2009. Experiment 2 was conducted at SLSS for the transplanting and establishment of nursery plants from June 2009 to October 2009. The soil at SLSS is sandy Regosols up to at least a 1-m depth with a pH (H$_2$O) of 6 and electrical conductivity of 10.4 mS m^{-1} for the topsoil (0–15 cm; Wadi *et al.*, 2008).

2.2 Nursery production (Experiment 1)

In November 2008, nursery production was conducted to record the labour time at SLSS for both manual and mechanical treatments, with four and three replications, respectively. In each manual treatment, 40 stalks of dwarf napiergrass were defoliated by sickle, engaging three persons. In each machinery treatment, 120 stalks were first defoliated by a hand-mowing machine (Model, Makita Co. Ltd.) operated by gasoline, engaging two persons. The number of labourers was reduced from three to two in the machinery treatment because there was no need to cut stalks manually. After gathering the cut herbage, every tiller bud positioned on an aerial node was cut by scissors to a single-nodal stem cutting, which consists of an internode, stem and axillary bud (Moser & Jennings, 2007), engaging 6.0 persons on average in manual treatments and 4.8 persons in machinery treatments, and the labour time was recorded. The labourers for both manual and machinery treatments were randomly chosen from a total of ten available. Stem cuttings were transported to KASS and transplanted to cell-tray beds (sized 3.0 cm × 3.0 cm × 4.5 cm deep per cell) filled with peat moss (Sakae Transportation Service Ltd., Mimata, Miyazaki), engaging 1.8 persons on average in manual and 6.8 people in machinery treatments, with labour time recorded. The cell trays were placed in a glasshouse and watered once or twice a day (depending on growth conditions) by one person to record labour time from November 2008 to May 2009. The surviving percentage of plants showing regrowth from stem cuttings, judged by the presence of green leaves on plants, was measured at monthly intervals from early January to early May in 2009. Air temperature was monitored by thermometer (Thermo Leaf, Taisei E & L Co. Ltd., Tokyo, Japan) from 3 February to 12 May 2009. The temperature in the glasshouse averaged 17.1, 17.7, 23.0 and 25.4 °C, respectively, for February, March, April and May.

2.3 Transplanting, establishment and field management (Experiment 2)

In June 2009, 2000 tillers, which formed 4–5 leaves each and had a basal diameter around 2–3 cm, were prepared as nursery plants of dwarf napiergrass. For the manual treatments, 600 nursery plants were prepared, and for the machinery treatments, 1,200 plants were prepared (for a total of 1,800 plants), engaging 2.0 persons for both plots to record the labour time. Then, green tops and roots below the cell trays were removed by scissors to a fixed 25-cm length, engaging 2.0 persons for both treatments to record labour time. Adjusting the plants to this length was necessary in order to use a Welsh onion planter (Negi-Nira Ishokuki, PNF–3, Katakura-kiki Co. Ltd., Nagano, Japan) in this research. The device is a self-moving planting machine, which must be held by a walking operator, and the capacity of the machine is 580 m hr^{-1} at low-speed operation and 1,210 m hr^{-1} at high-speed operation. In the manual treatment, a shovel was used for transplanting. The field was cultivated by rotary tilling once and no basal fertiliser was supplied before transplanting. The fields were divided into plots 10 m × 10 m for manual treatments and 10 m × 20 m for machinery treatments, each with 3 replications. The inter- and intra-row spacing was 1.0 and 0.5 m, respectively, to maintain a density of 2 plants m^{-2} for both treatments. Plants were transplanted on 5 June 2009 by 4 persons in the manual and on 9 June 2009 by 3 persons in the machinery treatment, and the labour time was recorded.

Plants were transplanted just at the start of the rainy season, before the hot summer season in this area; no irrigation was conducted. Based on data from Miyazaki Meteorological Observatory (Japan Meteorological Agency, 2010), the number of rainy days in the total period, the total (and daily mean) precipitation and the daily mean temperature from transplanting to the observed date for establishment were respectively 17 out of 25 days, 231.0 mm (9.2 mm day^{-1}) and 24.0 °C in the manual plot, and 14 out of 21 days, 230.5 mm (11.0 mm day^{-1}) and 24.3 °C in the machinery plot. These conditions of precipitation and temperature for 3 weeks after transplanting were enough to determine the establishment of this grass. Weeding was conducted on 1 July 2009 by cutting the inter-row space with a hand-mowing machine. Fertiliser was supplied by 3 split applications on 19 June (10–14 days after planting), 11 July (5 weeks after planting) and 3 October 2009 (end of the first defoliation) at 5 g m^{-2} each for N, P$_2$O$_5$ and K$_2$O per application by a compound fertiliser.

2.4 Determination of plant growth characters and dry matter yield

The percentage of established plants was determined 3 weeks after transplanting on 1 July 2009 when new leaf growth appeared to suggest the onset of rooting. After this date, no dead plants were observed in any plot. About 110 days after planting, on 26 September 2009, plant height and tiller number were measured for 10 plants per replicated plot (total of 30 plants per treatment) by the line transect method and the herbage was cut at 10 cm above the ground for 2 plants per replicated plot (total of 6 plants per treatment) by the line transect method to measure fresh weight, and about 400 g of subsample was dried at 70 °C for 4 days in an air-forced oven to determine percentage of dry matter, following Ishii et al. (2005).

2.5 Statistical analysis

The mean values of each variable affecting required labour time for cell-tray stem cuttings in two cutting methods, growth characters (plant height, tiller density, and percentage of leaf blade to whole plant weight) and herbage mass in two transplanting methods of dwarf napiergrass were compared by Student's t-test at the 5 % significance level (SPSS for Windows ver. 16.0, Chicago, IL, USA). The means in the percentage of plants surviving from stem cuttings in the glasshouse and the percentage of plants established in the field were analysed by t-test at the 5 % level after arcsine transformation (McDonald, 2009).

3 Results

3.1 Labour time for nursery production in Experiment 1

In dwarf napiergrass, 160 stalks produced 10,286 stem cuttings, averaging 64.3 stem cuttings stalk^{-1} (Table 1). Required labour for gathering napiergrass stalks was significantly less using machinery (0.06 person-min stem-cutting^{-1}) than manually (0.23 person-min stem-cutting^{-1}). However, the labour time for preparing stem cuttings (averaging 0.30 person-min stem-cutting^{-1}) and for planting stem cuttings into cell trays (averaging 0.16 person-min stem-cutting^{-1}) did not differ significantly between treatments (Table 1). Mechanical harvest of leafage and stubble (the aboveground stem part left after cutting leafage) in dwarf napiergrass was effective in reducing labour time when preparing stubble for production of stem cuttings. The number of stem cuttings produced per stalk was 45.4 manually and 70.6 for the machinery treatment.

3.2 Labour time for watering and surviving percentage of plants showing regrowth from stem cuttings in the glasshouse

Nursery plants of dwarf napiergrass, watered in glasshouse, required labour time at 10.7 person-min application^{-1}. For both treatments (manual and machinery), the surviving percentage of plants showing regrowth tended to increase marginally from early January, reaching a peak in early February at 93.0 and 91.5 % for manual and machinery treatment, respectively, followed by a slight decline through early May to 90.3 and 83.9 %, while significant differences between treatments were not detected on either date.

3.3 Labour time for transplanting nursery plants in Experiment 2

The number of nursery plants transplanted successfully using machinery was 398 (99.5 %) per replication (Table 2). Labour time required for manually preparing nursery plants was 0.03 person-min m^{-2} (2 %) greater than for the machinery treatment, while labour time for manually transplanting in the field was by 0.34 person-min m^{-2} (41 %) lower than for mechanical transplanting (Table 2). Therefore, the entire transplanting process was significantly shorter for the manual operation than for the mechanical operation by 0.30 person-min m^{-2} (12 %), as shown in Table 2. In the present study, manually transplanting 200 nursery plants to the pasture site required 2.20 person-min m^{-2} (367 person-hr ha^{-1}), and machinery transplanting required 2.50 person-min m^{-2} (416 person-hr ha^{-1}).

3.4 Plant survival in the field, growth characters and herbage yield

The percentage of established plants 3 weeks after planting on 1 July 2009 were 99.7 % and 99.5 % in the manual and machinery treatments, respectively, showing no statistical difference (Table 3). The growth characters and herbage yield of dwarf napiergrass were determined on 26 September 2009; plant height reached 114–116cm, tiller density 53–67 m^{-2} and percentage of leaf blade 61–63 %, with no statistical differences between the methods, except for herbage yield, which was significantly lower for mechanical than for manual transplanting, by 2,880 kg DM ha^{-1} (37 %).

Table 1: *Required labour time for two methods for preparing cell-tray stem cuttings of dwarf napiergrass (Experiment 1).*

Method	Replications	Stalks (no.)	Total stem cuttings (no.)	Required labour time (person-min. stem-cutting^{-1})			
				Cut and gather herbage, cut stubble and collect tillers	Prepare stem cuttings	Transplant stem cuttings into cell trays	Total
Manual	4	40	1816	0.226* ±0.023 (3.0, 3.0–3.0)‡	0.319^{ns} ±0.037 (6.0, 5.0–7.0)‡	0.185^{ns} ±0.027 (1.8, 1.0–2.0)‡	0.730* ±0.038
Machinery	3	120	8470	0.055 ±0.012 (0.009 ±0.002 †) (2.0, 2.0–2.0)‡	0.297 ±0.063 (4.8, 4.0–6.0)‡	0.154 ±0.027 (6.8, 5.0–8.0)‡	0.506 ±0.047

Data are presented as means ± standard deviation.
* P < 0.05; ^{ns}, P > 0.05 within each activity by Student's t-test.
† Only for cutting and gathering herbage by machinery.
‡ Mean and range of number of persons engaged.

Table 2: *Required labour time for two methods of transplanting dwarf napiergrass (Experiment 2).*

Method	Replications	Nursery plants (no.)	Plot area (m²)	Required labour time (person-min. m⁻²)		
				Prepare nursery plants	Transplant nursery plants	Total time
Manual	3	200	100	$1.717^* \pm 0.0058$ $(2.0, 2.0–2.0)^\ddagger$	$0.480^* \pm 0.1058$ $(4.0, 4.0–4.0)^\ddagger$	$2.198^* \pm 0.1097$
Machinery	3	$398^\dagger$	200	1.683 ± 0.0000 $(2.0, 2.0–2.0)^\ddagger$	0.815 ± 0.0624 $(3.0, 3.0–3.0)^\ddagger$	2.498 ± 0.0624

Data are presented as means ± standard deviation.
* $P < 0.05$ within each time by Student's t-test.
† Two plants lacked in transplanting on average.
‡ Mean and range of number of persons engaged.

Table 3: *Percentage of established dwarf napiergrass plants 3 weeks after planting, and growth characters and yield 3.5 months after planting.*

Method	Replications	Character				
		Percentage of established plants	Plant height (cm)	Tiller density (No. m⁻²)	Percentage of leaf blade	Herbage mass (g DM m⁻²)
Manual	3	$99.7^{ns} \pm 0.58$	$116.0^{ns} \pm 2.31$	$67.3^{ns} \pm 21.9$	$60.6^{ns} \pm 4.87$	$783.0^* \pm 238.0$
Machinery	3	99.5 ± 0.00	114.0 ± 3.00	53.0 ± 19.1	63.0 ± 2.65	495.0 ± 103.0

Data are presented as mean ± standard deviation.
* $P < 0.05$ within each character, ^{ns}, $P > 0.05$ by Student's t-test.

4 Discussion

4.1 Labour time for nursery production, plant maintenance and transplanting

Yokoyama (1996) revealed that a mechanical system reduced labour time for transplanting Welsh onion with a minimum failure of transplanting ($<5\%$) using the same machinery used in the present study. Mechanical transplanting of guineagrass (*Panicum maximum*) required less labour time as compared to manual transplanting, about 92 % (368 person-hr ha⁻¹) by using the transplanter for paddy rice (Seibyou Pot Transplanter, I. Co. Ltd.) (Nakahara *et al.*, 2001). A 49 % (63.3 person-hr ha⁻¹) reduction in labour time was recorded for mechanically transplanting *Brachiaria brizantha* cv. MG₅ by the Okinawa Prefectural Animal Husbandry Research Center (2010) and a 28 % reduction was recorded for mechanically transplanted rice (Sharma & Singh, 2008). The machinery system reduced labour intensity, with no need for a labourer to stoop when transplanting pangolagrass (*Digitaria eriantha*) cv. Transvala using a hand transplanter (Okinawa Prefectural Livestock Experiment Station, 2005), which was the same situation as in the present study. Boonman (1993) transplanted 20,000 root splits of napiergrass manually into 1 ha of pasture, which required 445 person-hr ha⁻¹. This result was slightly higher than obtained in the present study (367 person-hr ha⁻¹) using the manual method.

The power required to drive the applied transplanting machinery must be considered, because it should be suitable for the target crop species, cultivation system and soil properties. The machinery currently used for Welsh onion, usually operated on Andosols under ridge cultivation, could not operate on the sandy Regosols in this study without establishing ridges due to lack of adequate power of the machinery, resulting in an increase in labour time. The number of people engaged in machine transplanting was three people in the present study, while it could be reduced from three to two people. In the present study, holding the machine, picking up individual nursery plants from the basket and setting the plants on a conveyer were carried out by different people, but if a two-person operation could be established by combining the latter two processes under one person's operation, the labour time at transplanting could be reduced to 0.54 person-min m⁻², which nearly corresponds with that in the present manual transplanting. In addition, this machine should be more effective and efficient when used over a wider area, because the time loss when changing direction at the end of each row could be reduced a great deal with longer rows under cultivation.

4.2 Plant survival in the field, growth characters and yield

Kipnis & Bnei-Moshe (1988) determined the established plant density of napiergrass (N23) and a napiergrass × pearl millet hybrid by mechanical transplanting at 5.8 plants m^{-2} 3 weeks after establishment. This planting density was almost 3 times higher than the density of the present study (2 plants m^{-2}), however, the present 2 plants m^{-2} was determined as the optimum density to achieve the highest range of herbage yield in the year of establishment (Mukhtar et al., 2003; Ishii et al., 2005).

Regarding the growth characters and yield of dwarf napiergrass determined 3.5 months after planting, the lower herbage yield in the machinery treatment may be due to a shallower planting depth using this low-powered machinery (rated output of 1.6 kilowatts) than the manual treatment. It is expected that using a high-powered transplanter could plant dwarf napiergrass deeper than in the current study. The examined site had sandy Regosols, where drought stress severe enough to suppress plant growth happened frequently, especially hampering the development of shallower planted stem cuttings.

4.3 Further application of the present nursery production and transplanting methods of napiergrass

Cultivation of napiergrass at a low-altitude site of Kyushu Island has several merits due to its perennial property (Ishii et al., 2005, 2008; Wadi et al., 2008), high biomass yield with high-quality leafy herbage (Ito et al., 1988; Sunusi et al., 1999), which results in high digestibility and palatability to livestock (Tudsri & Ishii, 2007; Ishii et al., 2009) and dual use as a bioenergy crop (Rengsirikul et al., 2011, 2012, 2013; Khairani et al., 2013) and a biodigester crop (Hasyim et al., 2014). Especially, dwarf napiergrass is suitable for the nutrition of small-holder beef cows in the area, since it can over-winter in the low-altitude areas of Kyushu Island (Ishii et al., 2008, 2013) and it can be fit to the rotational grazing even in abandoned orchard fields (Utamy et al., 2014). Therefore, the present nursery propagation and transplanting practices has a potential to apply to the establishment of napiergrass for a multipurpose use as forage, biodigester and biofuel resource.

5 Conclusion

In the present study, entire processes for nursery production and establishment of dwarf napiergrass were completed, from preparation of stem cuttings based on aerial tiller buds, to cell-tray nursery culture in a glasshouse over winter, to mechanical transplanting by a Welsh onion planter. Even though the manual planting of the stem cutting of dwarf napiergrass needed less labour time per unit area, achieving higher yield after establishment, the entire mechanical processes could be beneficial to reduce labour intensity at transplanting and to expand the cultivation area of this grass species. Considering the actual situation in Japanese agriculture, in which the average age of producers is 65.8 years (MAFF, 2010), this planting machine could be very useful in reducing farmers' work load, and can be maintained for at least 7 years with depreciation costs of less than 100,000-yen yr^{-1} (Agri Business, 2000). Even though farmers have additional costs in mechanical transplantation, it should be a key issue to expand the cultivation area for this plant species. At the moment, plant propagation by the farmers themselves is not so popular, therefore, nursery selling by cell-tray propagation is a next step to expand the production of dwarf napiergrass in the region.

Acknowledgements

The authors would like to express sincere thanks to Mr. Tsuguhiko Yuge for his assistance with the field operation. This work was partly supported by grants-in-aid for scientific research from the Japan Society for the Promotion of Science (JSPS Kiban Research-(C) No. 21580332).

References

Agri Business (2000). Welsh onion. URL http://agri-biz.jp/item/detail/1582?page=7, (last accessed: 28 April 2015).

Boonman, J. G. (1993). *East Africa's Grasses and Fodders: Their Ecology and Husbandry*. Kluwer Academic Publishers, London.

Burton, G. W. (1989). Registration of 'Merkeron' napiergrass. *Crop Science*, 29 (5), 1327.

Choon, Y. K. (1999). Mechanization in chili cultivation. National Engineering Conference on Smart Farming for the Next Millennium, University Putra Malaysia, Serdang, Malaysia, 14–16 March 1999. URL http://eng.upm.edu.my/msae/machinery.htm, (last accessed: 9 June 2009).

Hanna, W. W., Monson, W. G. & Hill, G. M. (1993). Evaluation of dwarf napiergrass. *In:* Proc. 17th Intern. Grassl. Congr., Palmerston North, New Zealand. pp. 402–403.

Hasyim, H., Ishii, Y., Wadi, A. & Idota, S. (2014). Effect of digested effluent of manure on soil nutrient content and production of dwarf napier grass in southern Kyushu, Japan. *Journal of Agronomy*, 13, 1–11.

Hasyim, H., Ishii, Y., Wadi, A., Idota, S. & Sugimoto, Y. (2010). Growth response of dwarf napiergrass to manure application originated from digested effluent. *Journal of Warm Regional Society of Animal Science, Japan*, 53, 115–126.

Hasyim, H., Ishii, Y., Wadi, A., Utamy, R. F., Wang, Y., Idota, S. & Fukuyama, K. (2008). Effect of digested effluent of manure on the growth and yield of dwarf napiergrass in southern Kyushu, Japan. *In:* Organizing Committee of 2008 IGC/IRC Congress (ed.), *Multifunctional Grasslands in a Changing World*. Vol. II, ch. Grasslands/Rangelands Production Systems – Integration of Crops, Forage and Forest Systems, p. 228, Guangdong People's Publishing House, China.

Inoue, T., Izawa, T., Nakajima, Y., Kutsuna, Y., Kato, Y. & Taniguchi, M. (1984). Studies on the mechanical cultivation of Job's tears (*Coix lacryma-jobi* L.). – Nursery period, transplanting density and time. *Research Bulletin of the Aichi ken Agricultural Research Center*, 16, 69–72.

Ishii, Y., Fukuyama, K., Iwakiri, T., Wadi, A. & Idota, S. (2009). Establishment of rotational grazing system on dwarf napiergrass pasture oversown with Italian ryegrass by smallholder farmers of Japanese-Black breeding beef cows in southern Kyushu, Japan. *In:* New Paradigm for Diversity of forage Production in the East Asian Region – Proceedings of the 3rd Korea-China-Japan Joint Symposium on Grassland Agriculture and Livestock Production. pp. 218–219.

Ishii, Y., Hamano, K., Kang, D., Rengsirikul, K., Idota, S., Fukuyama, K. & Nishiwaki, A. (2013). C4-Napier grass cultivation for cadmium phytoremediation activity and organic livestock farming in Kyushu, Japan. *Journal of Agricultural Science and Technology*, A3, 321–330.

Ishii, Y., Mukhtar, M., Idota, S. & Fukuyama, K. (2005). Rotational grazing system for beef cows on dwarf napiergrass pasture oversown with Italian ryegrass for 2 years after establishment. *Grassland Science*, 51 (3), 223–234.

Ishii, Y., Tudsri, S. & Ito, K. (1998). Potentiality of dry matter production and overwintering ability in dwarf napiergrass introduced from Thailand. *Bulletin of the Faculty of Agriculture - Miyazaki University*, 45, 1–10.

Ishii, Y., Wadi, A., Utamy, R. F., Wang, Y., Fukagawa, S., Idota, S. & Fukuyama, K. (2008). Adaptability of dwarf napiergrass to small holders of beef cows in southern Kyushu, Japan. *In:* Organizing Committee of 2008 IGC/IRC Congress (ed.), *Multifunctional Grasslands in a Changing World*. Vol. II, ch. Grasslands/Rangelands Production Systems – Integration of Crops, Forage and Forest Systems, p. 230, Guangdong People's Publishing House, China.

Ito, K., Murata, Y., Inanaga, S., Ohkubo, T., Takeda, T., Numaguchi, H., Miyagi, E. & Hoshino, M. (1988). Studies on the dry matter production of napiergrass II. Dry matter productivities at six sites in southern area of Japan. *Japanese Journal of Crop Science*, 57 (3), 424–430.

Japan Meteorological Agency (2010). Previous climatic data. URL http://www.data.jma.go.jp/obd/stats/etrn/index.php, (last accessed. 28 December 2010).

Kang, D.-J., Ishii, Y. & Nishiwaki, A. (2011). Effects of the shoot-cutting method on field propagation in napiergrass (*Pennisetum purpureum* Schum.). *Journal of Crop Science and Biotechnology*, 14 (2), 139–142.

Khairani, L., Ishii, Y., Idota, S., Utamy, R. F. & Nishiwaki, A. (2013). Variation in growth attributes, dry matter yield and quality among 6 genotypes of napiergrass used for biomass in year of establishment in southern Kyushu, Japan. *Asian Journal of Agricultural Research*, 7, 15–25.

Kipnis, T. & Bnei-Moshe, S. (1988). Improved vegetative propagation of napiergrass and pearl millet × napiergrass interspecific hybrids. *Tropical Agriculture (Trinidad)*, 65, 158–160.

Kumamoto National Livestock Breeding Center (2006). Transplanting of bahiagrass. Kumamoto. URL http://www.nlbc.go.jp/kumamoto, (last accessed: 18 February 2010).

MAFF (2010). Census of Agriculture, Forestry and Fisheries in 2010. Department of Statistics, Ministry of Agriculture, Forestry and Fisheries (MAFF), Japan. URL http://www.maff.go.jp/j/tokei/sokuhou/census10_zantei/index.html, (last accessed: 22 October 2010).

McDonald, J. H. (2009). *Handbook of Biological Statistics (2nd edition)*. Sparky House Publishing, Baltimore, Maryland. URL http://udel.edu/~mcdonald/, (last accessed: 2 November 2010).

Moser, L. E. & Jennings, J. A. (2007). Grass and legume structure and morphology. *In:* Barnes, R. F., Nelson, C. J., Moore, K. J. & Collins, M. (eds.), *Forages Volume II (6th edition)*. pp. 15–35, Blackwell Publishing, Iowa, USA.

Mukhtar, M., Ishii, Y., Tudsri, S., Idota, S. & Sonoda, T. (2003). Dry matter productivity and overwintering ability of the dwarf and normal napiergrasses as affected by the planting density and cutting frequency. *Plant Production Science*, 6 (1), 65–73. doi: 10.1626/pps.6.65.

Nakahara, Y., Kobashi, K. & Matsuoka, H. (2001). Development of seed harvesting system of guineagrass cv. Natsukomaki. *Bull. Kyushu Branch Japan. Grassl. Sci.*, 31 (1), 54–56.

Okinawa Prefectural Animal Husbandry Research Center (2010). Vegetative reproduction of tropical grass, *Brachiaria brizantha* cv. MG$_5$ by cell-tray nursery plant. Okinawa. URL http://www.affrc.go.jp/research/seika/data_kyusyu/h19, (last accessed: 18 February 2010).

Okinawa Prefectural Livestock Experiment Station (2005). Transplanting method of pangolagrass cv. transvala. Okinawa Green Grass, 2: 1-4.

Rengsirikul, K., Ishii, Y., Kangvansaichol, K., Pripanapong, P., Sripichitt, P., Punsuvon, V., Vaithanomsat, P., Nakamanee, G. & Tudsri, S. (2011). Effects of inter-cutting interval on biomass yield, growth components and chemical composition of napiergrass (*Pennisetum purpureum* Schumach) cultivars as bioenergy crops in Thailand. *Grassland Science*, 57 (3), 135 − 141.

Rengsirikul, K., Ishii, Y., Kangvansaichol, K., Sripichitt, P., Punsuvon, V., Vaithanomsat, P., Nakamanee, G. & Tudsri, S. (2013). Biomass yield, chemical composition and potential ethanol yields of 8 cultivars of napiergrass (*Pennisetum purpureum* Schumach.) harvested 3-monthly in central Thailand. *Journal of Sustainable Bioenergy Systems*, 3, 107–112.

Rengsirikul, K., Tudsri, S., Sugimoto, Y., Idota, S. & Ishii, Y. (2012). Dry matter production and energy efficiency of dwarf napiergrass (*Pennisetum purpureum* Schumach) under supply of animal manures, legume

clippings and chemical fertilizer. *Bulletin of the Faculty of Agriculture, Miyazaki University*, 58, 69–77.

Sarif, H. S. H., Kassim, B. & Mohd, S. I. S. (1996). The future of mechanized high density fruit cultivation in Malaysia. Smart Farming 99, Malaysia. URL http://eng.upm.edu.my/msae/machinery.htm, (last accessed: 9 June 2009).

Sharma, S. C. & Singh, T. P. (2008). Development and performance evaluation of a mat type nursery raising device. *AMA, Agricultural Mechanization in Asia, Africa and Latin America*, 39 (2), 64–70.

Sollenberger, L. E., Prine, G. M., Ocumpaugh, W. R., Hanna, W. W., Jones Jr., C. S., Schank, S. C. & Kalmbacher, R. S. (1988). 'Mott' Dwarf Elephantgrass: A high quality forage for the subtropics and tropics. Circular S-356, Agricultural Experiment Station, Institute of Food and Agricultural Sciences, University of Florida, Gainsville, Florida, 1-18.

Sollenberger, L. E., Prine, G. M., Ocumpaugh, W. R., Schank, S. C., Kalmbacher, R. S. & Jones Jr., C. S. (1987). Dwarf napiergrass: A high quality forage with potential in Florida and the tropics. *Proceedings - Soil & Crop Science Society of Florida*, 46, 42–46.

Sunusi, A. A., Ito, K., Ishii, Y. & Ueno, M. (1999). Effect of the level of fertilizer input on dry matter productivity of two varieties of napiergrass (*Pennisetum purpureum* Schumach). *Grassland Science*, 45, 35–41.

Tudsri, S. & Ishii, Y. (2007). Improvement of pasture production and sustainability under smallholder condition in Thailand. *Bulletin of the Faculty of Agriculture, Miyazaki University*, 53, 1–10.

Utamy, R. F., Ishii, Y., Idota, S., Harada, N. & Fukuyama, K. (2011). Adaptability of dwarf napiergrass under cut-and carry and grazing systems for smallholder beef farmers in southern Kyushu, Japan. *Journal of Warm Regional Society of Animal Science, Japan*, 54, 87 − 98.

Utamy, R. F., Ishii, Y., Idota, S. & Khairani, L. (2012). Effect of weed control management on herbage yield, quality and wintering ability in the established dwarf napiergrass (*Pennisetum purpureum* Schumach). *Journal of Warm Regional Society of Animal Science, Japan*, 55, 17–26.

Utamy, R. F., Ishii, Y., Iwamura, K. & Idota, S. (2014). Effect of weed control on etablishment and herbage production in dwarf napiergrass. *Journal of Life Sciences*, 8, 46–50.

Wadi, A., Ishii, Y., Hasyim, H., Utamy, R. F., Idota, S. & Fukuyama, K. (2008). Rotational grazing effects of dairy cows on yield and consumption in dwarf napiergrass at a paddok scale for 3 years from establishment in southern Kyushu, Japan. *In:* Organizing Committee of 2008 IGC/IRC Congress (ed.), *Multifunctional Grasslands in a Changing World II*. p. 135, Guangdong People's Publishing House, China.

Wadi, A., Ishii, Y. & Idota, S. (2004). Effects of cutting interval and cutting height on dry matter yield and overwintering ability at the established year in Pennisetum species. *Plant Production Science*, 7, 88–96. doi:10.1626/pps.7.88.

Yokoyama, M. (1996). Development of machine-assisted transplantation technique utilizing seedlings that are grown in a special-arranged cell system. *Bulletin of the Shizuoka Agricultural Experiment Station*, 41, 13–23.

12

Effects of drying and storage management on fungi (Aflatoxin B1) accumulation and rice quality in Cambodia

Vichet Sorn [a,*], Pyseth Meas [b], Tara Pin [c], Martin Gummert [d]

[a]*General Directorate of Agriculture, Ministry of Agriculture, Forestry and Fisheries, Phnom Penh, Cambodia*
[b]*Department of International Cooperation, Ministry of Agriculture, Forestry and Fisheries, Phnom Penh, Cambodia*
[c]*Chea Sim University of Kamchaymear, Prey Veng Province, Cambodia*
[d]*International Rice Research Institute, Metro Manila, Philippines*

Abstract

Rice postharvest practices of farmers incur losses that limit supply and affect global production. Aside from physical losses, quality can be affected, leading to a possible accumulation of aflatoxin B1 (AFB1) that is harmful to humans when ingested. This is particularly important for countries like Cambodia that aim for both food security and rice exports. The objective of the research was to determine the effects of different field drying and storage practices on AFB1 accumulation and milled rice quality in Cambodia. The study had four drying treatments and four storage treatments, in a randomized complete block (RCB) design. Tests were done for moisture content (MC), milling quality, germination rate, and AFB1 accumulation. High-performance liquid chromatography (HPLC) method was used to determine AFB1 contamination and one-way analysis of variance (ANOVA) was performed using CropStat 7.2. No significant AFB1 content was detected. Different field drying treatments used, as well as duration and type of storage also had no significant effect on the accumulation of AFB1 in rice. Milled rice quality was higher with limited or no field drying ($P < 0.01$). Storing in IRRI-Superbag at 14 % MC resulted in higher germination ($P < 0.01$) than in other treatments. Storing in IRRI-Superbag at 16 % MC, however, resulted in lower head rice recovery than in the other three treatments. Reducing field drying and storing hermetically at 14 % MC could therefore potentially reduce rice postharvest losses. Field drying practices of 12 days or less can keep AFB1 contamination at bay.

Keywords: field drying, germination, hermetic storage, milling quality, moisture content, postharvest, rice

1 Introduction

The growing population of the world demands a high and stable supply of rice, with about 900 million poor people being dependent on rice as producers and consumers (Pandey *et al.*, 2010). This dependence on rice puts increasing pressure on the agricultural sector to meet supply and quality requirements. Global per capita consumption has consistently increased since the 1960s, with 88 % of the total amount consumed in Asia (Timmer, 2010). In Cambodia, per capita consumption is 143 kg per year while the population is 14.9 million and growing at a rate of 1.7 % (MAFF, 2010; IndexMundi, 2013).

The Cambodian government, which seeks to re-establish the country as a major exporter, cannot stop by focusing on supply alone. Like that of other crops, rice quality is affected by variety and conditions in pre- and postharvest handling (Gummert *et al.*, 2010). Aside

---
*Corresponding author
General Directorate of Agriculture,
Ministry of Agriculture, Forestry and Fisheries,
#54B/49F, Street 395-656, Sangkat Toeuk Laak3,
Khan Tuol Kok, Phnom Penh, Cambodia
Email: sornvi70@gmail.com

from spills during handling, rice is easily contaminated by fungi and insects, in particular if the un-threshed rice panicles and grains are left in the field for several days (field drying). This is a common practice in Cambodia which is to pre-dry the grains, as farmers wait for available threshers. Potentially, the fungi contaminate the rice more easily when it has cracks or broken grains. Relative humidity, particularly in the rainy season, rises to levels that cause the moisture content of rice in storage to increase, resulting in cracks in the milling process. Physical losses have been estimated to be 15–25 % throughout the different postharvest activities (ibid.).

Moreover, climate conditions in Cambodia, as in other countries in Asia with temperatures of 26–39 °C and relative humidity of 67–98 %, are conducive to fungi growth and contamination in rice (Sales & Yoshizawa, 2005). Previous studies have detected fungi of the genera *Aspergillus*, *Penicillium*, and *Rhizopus* in stored rice (Phillips *et al.*, 1988). Mycotoxins are found to result from the secondary metabolites of fungi, and can cause illness and death to humans (Bennett & Klich, 2003). Fungi such as *Aspergillus*, *Fusarium*, and *Penicillium* genera produce aflatoxins, of which *Aspergillus flavus* is one that can seriously harm humans (Moss 1991). Of the four main types of aflatoxin, aflatoxin B1 (AFB1), AFB2, AFG1, and AFG2, it is AFB1 that has been strongly linked to causing cancer in people (Coker, 1994). Recent studies from different countries have detected aflatoxin B1, B2, G1, and G2 levels in rice that were above the regulatory limits in samples tested (Rahmani *et al.*, 2011; Firdous *et al.*, 2012). The worldwide regulations for mycotoxin content in foodstuffs have been set to 4 ppb of AFB1 and 8 ppb of total aflatoxins (B1+B2+G1+G2) as the maximum tolerance level (Van Egmond, 1999; Dohlman, 2003). The European Union (EU) standard is 2 ppb for AFB1 and 4 ppb for total aflatoxins (EC, 1999). The United States, however, have set 20 ppb for total aflatoxins as tolerance level (FAO, 2004).

The 5–10 % physical losses in storage (Gummert *et al.*, 2010) can be reduced by using hermetic storage. This is a new technology in which the grains are enclosed in a hermetically sealed container that prevents them from absorbing water from the ambient air, kills insects, and prevents new infestation. A study comparing open and hermetic storage systems and their effects on the quality of milled rice for 8 months of storage have been done in Vietnam (Diep *et al.*, 2006). Milled rice recovery from paddy stored under open conditions dropped 2.9 % compared with initial conditions while in hermetic storage (such as Super Bags developed by

the International Rice Research Institute, IRRI) milled rice recovery dropped by only 0.76 % (ibid.). However, such studies have documented losses under laboratory conditions, but not quantified the effects of practices of farmers, such as long field drying periods and different storage techniques, on physical and quality losses, as well as on the accumulation of mycotoxins.

This study aimed at establishing whether AFB1 is present in rice grains from sub-optimal postharvest systems such as those in Cambodia. It also aimed at quantifying the amount of mycotoxins accumulated in different treatments. We hypothesized that field drying will affect AFB1 contamination in rice. For drying, we examined the effects of different field drying periods on AFB1 accumulation and milled rice quality. For storage, we tested whether different methods, moisture contents, and duration of storage have an effect on AFB1 accumulation and milled rice quality.

2   Materials and methods

2.1   Drying experiments

The drying experiments were executed on farmers' rice fields in Thnong Khang Kerth Village, Smorng Cherng Commune, Kamchaymear District, Prey Veng Province. This allowed similar conditions as in the farmers' practice of field drying, with some control against cows and birds. An improved traditional rice variety, Phka Romduol, was selected and three rice fields were used. In each rice field, 12 plots were used for four treatments with three replications per treatment, providing a total of 36 plots. A randomized complete block (RCB) design was used in the experiments. Experiments were carried out in November-December 2011 (end of the rainy season), with ambient temperatures of 25–36 °C, and relative humidity of 45–85 %.

The field drying study used four treatments: (1) No FD (no field drying, the manually cut crop was immediately threshed); (2) 4-day FD, the manually cut crop was left for field drying for 4 days; (3) 8-day FD, with field drying for 8 days; and (4) 12-day FD, with field drying for 12 days. For each treatment, the cut crop was threshed using a mechanical thresher and 30 kg of the obtained paddy was sun-dried to 13–14 % moisture content (MC) in case the MC was higher than 14 %.

From the primary sample, a secondary sample of 5 kg per replication was obtained for AFB1 analysis. Another secondary sample of 1 kg per replication was obtained for milled rice quality assessment.

2.2 Storage experiments

The storage experiments were done at the Ministry of Agriculture, Forestry and Fisheries (MAFF) in an area where birds and rats can be controlled. It was implemented in December 2011 to May 2012 (dry season). This storage experiment included two different storage methods, and two initial moisture contents: (1) IRRI-Super Bag at 14% MC (SB-14%), (2) polypropylene bag at 14% MC (PPB-14%), (3) IRRI-Super Bag at 16% MC (SB-16%), and (4) polypropylene bag at 16% MC (PPB-16%). The four treatments were stored either for 2, 4 or 6 months. All treatments had three replications. These were compared with initial samples that had either 14% MC or 16% MC. The IRRI-Super Bag is a type of hermetic or airtight storage that minimizes gas and moisture transfer from the ambient air (Gummert *et al.*, 2010). The polypropylene bag is the woven plastic material commonly used by farmers for grain storage. It is permeable to relative humidity, water, and insects.

The rice used for this experiment was bought from a farmer (total of 1,400 kg; Phka Romduol variety). The farmer's rice field was harvested using a combine harvester and the paddy was sun-dried to reduce the MC from 22% to either 16% or 14% for the different treatments. After sun-drying, samples for each treatment were re-cleaned using a mechanical cleaner, and then mixed. After mixing, nine bags (30 kg per bag) were obtained for each of the treatments (three bags for each treatment and storage period).

From each 30-kg bag per replication, a sample of 5 kg was obtained and sent for AFB1 analysis. Furthermore, another sample of 1 kg per bag was obtained for milled rice quality assessment after 2, 4 and 6-month period and for assessment of the germination rate after 6 month storage period.

2.3 Aflatoxin analysis

To test for AFB1 contamination from each 5-kg sample, 1-kg sub-samples were obtained and crushed. The AFB1 concentrations for calibration curve were eluted by methanol and quantified using high-performance liquid chromatography (HPLC). Then we used Methanol:Acetonitrile:Water = 1:1:3 (Merck), and LC–10AVP HPLC (Shimadzu, Japan) plus LC-10ATVP Pump. All analyses (36 samples for the field drying experiment and 42 samples for the storage experiment) were executed at the Southern Sub-Institute of Agricultural Engineering and Postharvest Technology (SIAEP) laboratory in Vietnam.

2.4 Rice quality analysis

A total of 36 samples for the field drying experiment and another 42 samples for the storage experiment (1 kg per sample) were used to test the germination rate, milled rice quality, and head rice recovery (percentage of whole milled rice plus broken milled rice that have retained >80% of the whole) by using a laboratory rice mill.

2.5 Statistical analysis

One-way analysis of variance (ANOVA) was performed on experimental data collected using CropStat 7.2. Separation of treatment means was done using LSD at the 5% level of significance.

3 Results

3.1 Effects of drying practices on aflatoxin content and grain quality

3.1.1 Aflatoxin content

AFB1 was detected in all samples but no significant differences between the treatments were found and the mean AFB1 content (Table 1) was much lower than the current EU limit of 2 ppb for cereals.

Table 1: *Mean Aflatoxin B1 content (in parts per billion) for the field drying (FD) treatments.*

Treatment	AFB1 content (ppb)	Standard deviation
No FD	0.392	0.08
4-day FD	0.437	0.06
8-day FD	0.297	0.14
12-day FD	0.160	0.13

3.1.2 Moisture content (%)

The MC of paddy dropped from 22% to 15.3% after 4 days of field drying, 11.8% after 8 days, and 11.5% after 12 days. Hence, treatments 1 and 2 had to be dried further before milling to achieve 13–14% MC required. Moreover, there was a significant difference ($P < 0.05$) in MC at milling between treatments 2 and 3 as well as between 2 and 4 (Table 2). The reduction of moisture content during the first 4 days of drying was similar among treatments 2, 3 and 4. The rate of MC reduction, decreased however over time. Between days 5–8 (treatments 3 and 4), the average MC reduction was 0.87 per day, and between days 9–12 (treatment 4), it was only 0.07 per day.

Table 2: *Moisture content (MC) at milling.*

Treatment (days of field drying)	*MC at milling*	*SD*
No FD	13.1	1.1
4-day FD	13.2	0.4
8-day FD	11.8	0.7
12-day FD	11.5	0.4

3.1.3 Percentage of rice recovery

After milling, the No FD treatment obtained the highest milled rice recovery (65.6 %), with a decreasing, but non-significant, trend as the field drying period became longer (Fig. 1).

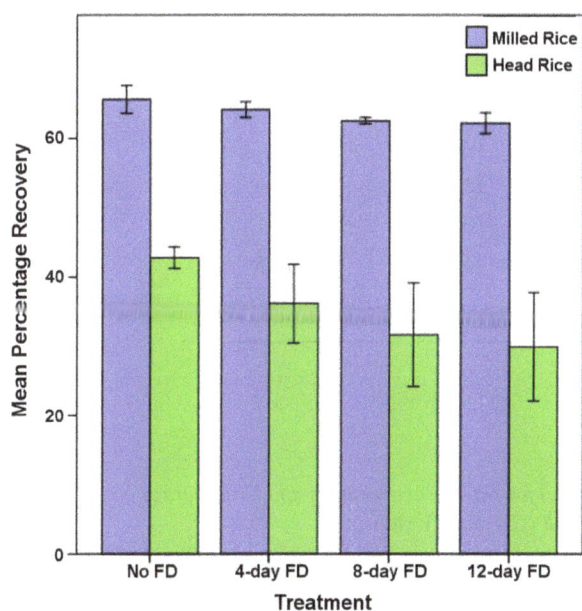

Fig. 1: *Mean milled rice and head rice recovery (%) for different field (FD) drying treatments.*

The No FD treatment obtained also the highest quality, with 43 % head rice recovery (Fig. 1). With increasing field drying period head rice recovery decreased significantly as compared to the control ($P < 0.01$). No FD had a 7 %, 11.1 % and 12.9 % higher head rice recovery ($P < 0.01$) as compared to 4-day FD, 8-day FD and 12-day FD, respectively.

3.2 Effects of storage practices on aflatoxin content and grain quality

3.2.1 Aflatoxin (AFB1) content

AFB1 was detected in 92 % of the paddy samples; however, the content in the samples stored for 2, 4, and 6 months was lower than the EU limit of 2 ppb (Table 3). No significant differences were found between the treatments.

3.2.2 Moisture content (%)

The changing moisture content of paddy during storage is given in Table 4. Paddy stored hermetically (IRRI Super Bag (SB) treatments) showed less fluctuation in MC than paddy stored in polypropylene bags where air and moisture could penetrate (PPB treatments). Paddy stored with 16 % initial MC fluctuated more than paddy stored with 14 % initial MC. Of these, the SB treatment fluctuated less, with a maximum of 0.5.

3.2.3 Germination rate after 6 month period

The SB-14 % treatment had a significantly higher germination rate ($P < 0.01$) than the other three treatments (Fig. 2). The germination rate of SB-16 % was even significantly lower than the two PPB treatments ($P < 0.01$). These two PPB treatments did not differ from each other. The hermetic storage (SB) results in higher germination rates than storage in polypropylene bags if grains are stored at an initial MC of around 14 %. Under airtight conditions, higher moisture contents reduce germination.

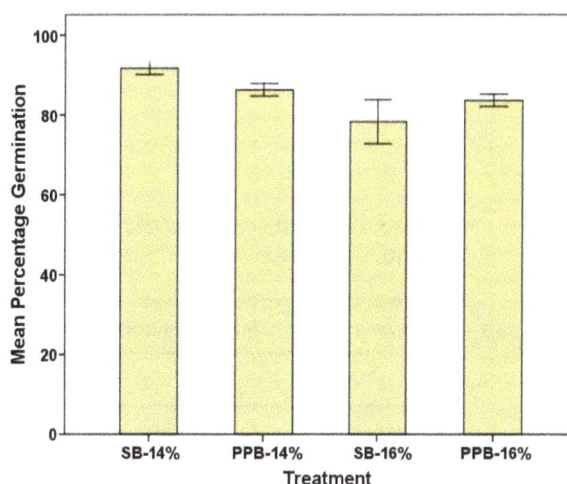

Fig. 2: *Comparison of germination percentage for two storage methods after 6 months of storage (SB: IRRI Super Bag; PPB: polypropylene bag; 14 % and 16 % initial paddy moisture content).*

3.2.4 Percentage rice recovery after 2 to 6 months of storage

Before storage, milled rice recovery was 64.5 % for the 14 % MC treatments, and 60 % for the 16 % treatments (Table 5). Milled rice recovery of PPB-16 %, rose to 62 %, 64 %, and 65 % after 2, 4, and 6 months of storage because its MC dropped to between 14.1–14.5 %, making it more suitable for milling. For SB-14 % and PPB-14 %, milled rice recovery was not significantly different.

Table 3: *Aflatoxin B1 (AFB1) content in parts per billion (ppb) found in paddy samples for two storage methods and three storage periods at two initial moisture contents (SD in brackets).*

Treatment	Initial	After 2 months	After 4 months	After 6 months
SB-14%	0.09 (0.02)	0.65 (0.48)	0.54 (0.02)	0.56 (0.10)
PPB-14%	0.09 (0.02)	0.44 (0.12)	0.64 (0.09)	0.72 (0.16)
SB-16%	0.68 (0.20)	0.60 (0.65)	0.74 (0.17)	0.55 (0.12)
PPB-16%	0.68 (0.20)	0.51 (0.09)	0.80 (0.20)	0.60 (0.17)

SB: IRRI Super Bag; PPB: polypropylene bag; 14% and 16% initial paddy moisture content

Table 4: *Comparison of moisture content (%) of paddy for two storage methods and three storage periods.*

Treatment	Initial	After 2 months	After 4 months	After 6 months
SB-14%	14.0	14.2	14.0	14.3
PPB-14%	14.0	13.9	13.3	14.4
SB-16%	15.9	16.5	16.2	16.1
PPB-16%	15.9	14.3	14.1	14.5

SB: IRRI Super Bag; PPB: polypropylene bag; 14% and 16% initial paddy moisture content

Table 5: *Comparison of milled and head rice recovery (%) for two storage methods and three storage periods, compared to the initial paddy before storage, with LSD and CV values.*

Treatment	Initial (control)	After 2 months	After 4 months	After 6 months
Milled rice recovery				
SB-14%	64.51	62.83	62.35	65.50
PPB-14%	64.51	61.84	59.63	64.28
SB-16%	60.18	56.81	54.65	61.07
PPB-16%	60.18	62.27	63.75	65.50
LSD	1.46	1.33	1.40	0.75
CV (%)	1.30	1.20	1.30	0.60
Head rice recovery				
SB-14%	44.53	43.53	42.80	42.08
PPB-14%	44.53	41.07	38.74	36.06
SB-16%	33.95	33.47	32.90	32.43
PPB-16%	33.95	40.96	39.95	37.11
LSD	1.62	1.0	0.76	0.92
CV (%)	2.20	1.3	1.10	1.30

SB: IRRI Super Bag; PPB: polypropylene bag; 14% and 16% initial paddy moisture content

Head rice recovery differed between treatments as well as between storage durations, especially compared with initial values representing optimal conditions for milling (Table 5). At the start of the experiment, head rice recovery for 14 % MC treatments was 44.5 % and for 16 % MC treatments 34.0 %.

The head rice recovery of paddy in the SB-14 % treatments dropped less (around 2 % after 6 months) compared to the PPB-14 % treatment (around 8 %) (Table 5). For PPB-16 %, head rice recovery went up to 41 % after 2 months because its MC dropped to 14.3 %, making it more suitable for milling. After 4 months, however, its head rice recovery started to decline. Head rice recovery for both SB treatments had the least fluctuation.

4 Discussion

This study explored whether there would be AFB1 contamination in rice grains from different field drying and storage treatments in Cambodia. The mean AFB1 contaminations found were not higher than 1 ppb and therefore much lower than the current standard EU limitations of 2 ppb (Tables 1 and 3). A study in India for example, showed that one out of 35 samples was contaminated with AFB1 and B2 at levels of 15–30 ppb (Siruguri *et al.*, 2012). In Vietnam, a study also found AFB1 contamination in rice (Nguyen *et al.*, 2007). These two studies however used samples that were either rain-damaged grains or milled rice samples from markets (stored as milled rice). Nguyen *et al.* (2007) showed that AFB1 contamination is higher when there is high ambient moisture content such as in the rainy season. In the current study the samples tested on AFB1 were fresh grains bought from a farmer and stored during dry season months. This study provided preliminary data for Cambodian conditions, wherein local field drying practices combined with non-rainy (dry) conditions did not result to AFB1 contamination in rice.

The different field drying periods as well as duration and type of storage had no significant effect on the presence of AFB1 in rice. With enough sunlight resulting in favourable high temperatures (mean of 35.6 °C) and a relative humidity of around 75 %, the moisture content of paddy dropped to about 15.3 % within 4 days of field drying. Furthermore, the AFB1 analysis in the current study was on samples that included the rice husk. A study in China by Liu *et al.* (2006) found that AFB1 content could be reduced by removing the husk. The experiment should therefore be repeated during the wet-season, and AFB1 content analysis to include samples without husk.

Although the field drying method did not affect AFB1 content, it affected the rice quality. Lengthening the field drying period reduced head rice recovery significantly even if the milled rice recovery was not affected. Meas (2012) documented that, already after 24 hours of storing wet rice grains, its quality started to deteriorate. Hence, quality can be optimised by reducing grain MC to 14 % immediately after harvest.

The low head rice recovery for 8- and 12-day FD is due to its low MC at threshing and milling (<12 % MC), where 13–14 % MC is seen as optimal (IRRI, 2009). Also, longer field drying periods increase paddy cracking due to the fact that grains absorb dew at night and dry again during daytime, resulting in lower head rice recovery.

Storing paddy at 14 % MC in Super Bags gave the best quality in terms of germination rate and head rice recovery compared to the other storage treatments. This is the case even with prolonged periods of storage (up till 6 months). Our findings concur with those of Sim (2010) that the drop in head rice recovery is lower (4 %) in hermetic storage than in open storage (15 %) after 8 months. The moisture content of paddy under hermetic storage fluctuated less than with PPB storage. These results concur with other studies looking at rice storage. In a study done in Vietnam, the MC in hermetic storage (comparable to the SB treatments) fluctuated less compared with storage where moisture exchange with surrounding air could happen (Diep *et al.*, 2006). Other studies found that the MC of paddy stored hermetically fluctuated up to 0.2 % while paddy stored in storage such as PPB fluctuated up to 1.2 % (Diep *et al.*, 2006; Sim, 2010; Ouk, 2011).

However, it is not advisable to store paddy hermetically at high MC of 16 % because it has a negative effect on germination rate, as well as on milled rice and head rice recovery. These findings support the current recommendation for storing paddy at 13–14 % MC using hermetic storage to maintain rice quality for 6–12 months (IRRI, 2009). The preservation of quality is indicated by germination rate. On this note, it was found previously that there is a 15 % difference in germination between paddy stored hermetically and in open storage (Sim, 2010; Ouk, 2011).

5 Conclusion

Studies on aflatoxin in rice commonly analysed damaged grains, parboiled rice, rice bran as well as brown rice; this study looked into Cambodian conditions where rice is stored as paddy and milled as white rice. Our findings demonstrate that the practice of Cambodian farmers of field drying for 12 days or less after crop harvest, as well as different duration and type of storage during dry season had no significant effect on the presence of AFB1 in rice. AFB1 content detected was much lower than current standard EU limitations. Field drying however, had significant effect on rice quality especially head rice recovery. Future studies relating to aflatoxins on rice storage could examine the effects of other common practices of farmers such as keeping wet grains over different periods, piling bundles of cut crop for different periods (not only over the dry-season harvest but also for the monsoon-crop harvest). Also, the effect of storage at higher MC such as 18 % or 20 %, for rice stored during rainy season, on AFB1 accumulation could be explored.

Acknowledgements

The authors would like to thank

- The Asian Development Bank and the International Rice Research Institute through the Postharvest Project for funding of this research
- Ms. Rica Joy Flor for editing help and comments on the paper,
- Ms. Jennifer Niem from IRRI, Dr. Pham Van Tan, and Ms. Tram Anh San from the Southern Sub-Institute of Agricultural Engineering and Postharvest Technology (SIAEP) laboratory for insights on mycotoxin study and analysis.

References

Bennett, J. W. & Klich, M. (2003). Mycotoxins. *Clinical Microbiology Review*, 16, 497–516. doi: 10.1128/CMR.16.3.497-516.2003.

Coker, R. D. (1994). The biodeterioration of grain and the risk of mycotoxins. *In:* Proctor, D. L. (ed.), *Grain storage techniques: Evolution and trends in developing countries.* FAO Agricultural Services Bulletin, No. 109, pp. 25–39, Food and Agriculture Organization of the United Nations (FAO), Rome.

Diep, B. C., Phan, V. L., Nguyen, T. D., Gummert, M. & Rickman, J. (2006). Effects of hermetic storage in super bag on seed quality and milled rice quality of different varieties in Bac Lieu, Vietnam. Proceedings of the 2nd International Rice Congress 2006. New Delhi, India. p. 567.

Dohlman, E. (2003). Mycotoxin Hazards and Regulations: Impacts on Food and Animal Feed Crop Trade.

In: Buzby, J. C. (ed.), *International Trade and Food Safety. Economic Theory and Case Studies. Agricultural Economics Report 828.* United States Department of Agriculture, Economic Research Service.

EC (1999). Council Directive 1999/29/EC on the undesirable substances and products in animal nutrition. Official Journal of the European Community (EC), April 1999. Available at: http://europa.eu.int/eur-lex/pri/en/oj/dat/1999/l_115/l_11519990504en00320046.pdf (last accessed: 06.13.2002).

FAO (2004). *Worldwide Regulations for Mycotoxins in Food and Feed in 2003.* Food and Agriculture Organization (FAO), Rome. p. 180

Firdous, S., Ejaz, N., Aman, T. & Khan, N. (2012). Occurrence of aflatoxins in export-quality Pakistani rice. *Food Additives and Contaminants*, 5 (2), 121–125. doi:10.1080/19393210.2012.675360.

Gummert, M., Hien, P. H., Meas, P., Rickman, J., Schmidley, A. & Pandey, S. (2010). Emerging technological and institutional opportunities for efficient postproduction operations. *In:* Pandey, S., Byerlee, D., Dawe, D., Dobermann, A., Mohanty, S., Rozelle, S. & Hardy, B. (eds.), *Rice in the Global Economy: Strategic Research and Policy Issues for Food Security.* Ch. 2.6, pp. 333–355, International Rice Research Institute (IRRI), Los Baños, Philippines.

IndexMundi (2013). Cambodia Demographics Profile 2013. Available at: http://www.indexmundi.com/cambodia/demographics_profile.html

IRRI (2009). Storage. Rice Knowledge Bank, International Rice Research Institute (IRRI). Available at: http://www.knowledgebank.irri.org/storage/

Liu, Z., Gao, J. & Yu, J. (2006). Aflatoxins in stored maize and rice grains in Liaoning Province, China. *Journal of Stored Products Research*, 42 (4), 468–479.

MAFF (2010). Annual report (2010). The Ministry of Agriculture, Forestry and Fisheries (MAFF), Cambodia.

Meas, P. (2012). Technical Implementation for Rice Storage. Booklet produced for IRRI Postharvest Project.

Moss, M. O. (1991). Mycology of cereal grain and cereal products. *In:* Chelkowski, J. (ed.), *Cereal grain: mycotoxin fungi and quality in drying and storage.* Elsevier Science Publishers B.V., Amsterdam.

Nguyen, M., Tozlovanu, M., Tran, T. & Pfohl-Leszkowicz, A. (2007). Occurrence of Aflatoxin B1, citrinin and ochratoxin A in rice in five provinces of the central region of Vietnam. *Journal of Food Chemistry*, 105 (1), 42–47.

Ouk, M. (2011). *Integrated technology for improving rice crop production*. Cambodian Agricultural Development and Research Institute (CARDI). Phnom Penh, Cambodia. pp. 8–9.

Pandey, S., Byerlee, D., Dawe, D., Dobermann, A., Mohanty, S., Rozelle, S. & Hardy, B. (eds.) (2010). *Rice in the Global Economy: Strategic Research and Policy Issues for Food Security*. International Rice Research Institute (IRRI), Los Baños, Philippines. pp. 1–10.

Phillips, S., Widjaja, S., Wallbridge, A. & Cooke, R. (1988). Rice yellowing during post-harvest drying by aeration and during storage. *Journal of Stored Products Research*, 24, 173–181.

Rahmani, A., Soleimanya, F., Hosseinib, H. & Nateghic, L. (2011). Survey on the occurrence of aflatoxins in rice from different provinces of Iran. *Food Additives and Contaminants*, 5, 185–190. doi: 10.1080/19393210.2011.599865.

Sales, A. C. & Yoshizawa, T. (2005). Updated profile of aflatoxin and *Aspergillus* section *Flavi* contamination in rice and its byproducts from the Philippines. *Food Additives & Contaminants*, 22 (5), 429–436. doi:10.1080/02652030500058387.

Sim, C. (2010). Reduction loss of paddy rice during storage in different methods. Unpublished thesis. The Royal University of Agriculture (RUA), Phnom Penh, Cambodia.

Siruguri, V., Kumar, P. U., Rahgu, P., Rao, M., Sesikeran, B., Totega, G. S., Gupta, P., Rao, S., Satyanarayana, K., Katoch, V. M., Bharaj, T. S., Mangat, G. S., Sharma, N., Sandhu, J., Bhargav, V. K. & Rani, S. (2012). Aflatoxin contamination in stored rice variety PAU 201 collected from Punjab, India. *Indian Journal of Medical Research*, 136 (1), 89–97.

Timmer, C. P. (2010). Rice and structural transformation. *In:* Pandey, S., Byerlee, D., Dawe, D., Dobermann, A., Mohanty, S., Rozelle, S. & Hardy, B. (eds.), *Rice in the Global Economy: Strategic Research and Policy Issues for Food Security*. Ch. 1.2, pp. 37–59, International Rice Research Institute (IRRI), Los Baños, Philippines.

Van Egmond, H. (1999). Worldwide Regulations for Mycotoxins. Third Joint FAO/WHO/UNEP International Conference on Mycotoxins, Tunis, Tunisia, 3–6 March 1999. MYCCONF/99/8a.

13

Throughfall and soil properties in shaded and unshaded coffee plantations and a secondary forest:a case study from Southern Colombia

Lucía Gaitán [a,c,*], Inge Armbrecht [b], Sophie Graefe [a]

[a]*Georg-August-Universität Göttingen, Tropical Silviculture and Forest Ecology, Göttingen, Germany*
[b]*Universidad del Valle, Departamento de Biología, Cali, Colombia*
[c]*International Center for Tropical Agriculture (CIAT), Pham Van Dong, Tu Liem, Hanoi, Vietnam*

Abstract

In Colombia coffee production is facing risks due to an increase in the variability and amount of rainfall, which may alter hydrological cycles and negatively influence yield quality and quantity. Shade trees in coffee plantations, however, are known to produce ecological benefits, such as intercepting rainfall and lowering its velocity, resulting in a reduced net-rainfall and higher water infiltration. In this case study, we measured throughfall and soil hydrological properties in four land use systems in Cauca, Colombia, that differed in stand structural parameters: shaded coffee, unshaded coffee, secondary forest and pasture. We found that throughfall was rather influenced by stand structural characteristics than by rainfall intensity. Lower throughfall was recorded in the shaded coffee compared to the other systems when rain gauges were placed at a distance of 1.0 m to the shade tree. The variability of throughfall was high in the shaded coffee, which was due to different canopy characteristics and irregular arrangements of shade tree species. Shaded coffee and secondary forest resembled each other in soil structural parameters, with an increase in saturated hydraulic conductivity and microporosity, whereas bulk density and macroporosity decreased, compared to the unshaded coffee and pasture. In this context tree-covered systems indicate a stronger resilience towards changing rainfall patterns, especially in mountainous areas where coffee is cultivated.

Keywords: coffee, saturated hydraulic conductivity, precipitation, shade trees, throughfall

1 Introduction

Coffee production systems represent an example of how land-use change produced distinct transformations in the Colombian landscape, a tendency that can be also observed for other coffee producing areas in Latin-America (Perfecto *et al.*, 1997; Armbrecht *et al.*, 2005). Since 1970, coffee farming in Colombia has continuously moved from shade-grown cultivation to low-shade or no-shade cultivation (Cárdenas, 1993; Guhl, 2004), for instance, from 2007 to 2013 the area of unshaded coffee plantations increased from 4200 to 5530 km^2 (FNC, 2014).

Due to its floral complexity coffee that is grown under a shade canopy has a forest-like structure, providing different, albeit neglected benefits to the agroecosystem, including climate regulation, protection from pathogens and insects, improvement of soil fertility, biodiversity conservation and carbon sequestration (Perfecto *et al.*, 1996; Beer *et al.*, 1998; Lin, 2007). Furthermore, it has allowed some farmers to access premium eco-certification programs, to gain additional revenues and to contribute to environmental welfare (Castro *et al.*, 2004; Vaast *et al.*, 2005).

*Corresponding author
Email: lucia.gaitan.sanchez@gmail.com

Climate projections for Colombia show that temperatures will most likely increase by 2.5 °C and precipitation by 2.5 % by the year 2050. The country is also affected through climate variations related to El Niño and La Niña (Lau *et al.*, 2010), i.e. in the season 2011–2012 La Niña reduced the number of dry months in the Colombian coffee region from four to one (FNC, 2012). There is a strong tendency towards an increase in the variability of precipitation, with the wettest periods becoming wetter and the driest periods becoming less dry (Lau *et al.*, 2010). Such variations in rainfall will alter flowering dates of coffee and the hydrological cycles in the plantations, and will most likely result in yield reductions (Porter & Semenov, 2005; Läderach *et al.*, 2011). Through simulation models Van Oijen *et al.* (2010) found that doubling the amount of rainfall would negatively affect coffee production systems in 28 % of the cases in Central America, whereas at the same time, 50 % higher nitrogen losses and 336 % higher soil losses would occur. Additionally, a 5 °C increase in temperature resulted in a 19 % decrease of coffee yield in unshaded coffee compared to shaded coffee.

Preventing shaded coffee plantations from being converted to monocultures or conversely, introducing shade trees in open plantations may be important strategies to mitigate the effects of extreme rainfall and temperature events, since trees modulate the micro-environment by buffering humidity and soil moisture availability and by enhancing soil organism activity (Martius *et al.*, 2004; Siles *et al.*, 2010b). Furthermore, the Colombian Coffee Growers Federation has a strong interest in developing compensation schemes for hydrological services provided by trees in river basins of the coffee growing region (Sosa & Moreno, 2014).

Depending on the canopy characteristics of vegetation, different amounts of water are intercepted by the canopy and evaporated directly into the atmosphere (Hodnett *et al.*, 1995; Grip *et al.*, 2004). This is also true for shaded vs. unshaded coffee, which differ significantly in the amount of vegetation. Measuring throughfall is an efficient way to determine the amount and distribution of water that reaches the soil (net rainfall) in this context. Throughfall is mainly influenced by the vegetation structure, the intensity and duration of precipitation (Crockford & Richardson, 2000). For instance, light rain with small raindrops will be rather intercepted by the canopy, whereas heavy rains will saturate the upper canopy storage capacity, transmitting most of it to the lower strata (Salas, 1987). Also, the presence of shade trees and other vegetation creates a second layer above the coffee plants, which increases the variability

of vertical water distribution and reduces the amount of water reaching the soil (Herwitz, 1985; Siles *et al.*, 2010a). Once rainfall reaches the soil, many factors influence water storage and movement, such as infiltration rate, porosity and texture (Zimmermann *et al.*, 2006; Scheffler *et al.*, 2011). These properties depend on land use type and management practices. Forest cover promotes water infiltration and groundwater recharge during the rainy season, and is able to supply water during the dry season. However, those soil hydrological dynamics of forests are often lost by land conversion (Bruijnzeel, 1989; Sandström, 1998; Malmer *et al.*, 2009). In general, land use practices involving any sort of farming (crop cultivation, grazing) affect the soil physical properties by the decrease of water storage in the soil and a reduction in the movement through the soil profile (Jaramillo, 2002b).

The present study aimed at measuring throughfall and soil hydrological properties in four different land use types in a coffee producing region of South-Western Colombia: shaded coffee, unshaded coffee, secondary forest (positive control) and pasture (negative control). It was assumed that the presence of shade trees in coffee plantations reduces the net rainfall compared to unshaded coffee due to a higher interception loss. Furthermore, shade trees might improve soil hydrodynamics in a way that the net rainfall input in the tree-covered systems will infiltrate and conduct a greater amount of water into the soil profile with a possible runoff reduction.

2 Materials and methods

2.1 Study sites

The field study was conducted during the months of highest rainfall, April and May 2013, at "Vereda El Rosal" (02° 51′ 856″ N, 0.76° 33′ 916″ O), which belongs to the district of Mondomo in the municipality of Caldono, department of Cauca. The study sites were located on the eastern slopes of the western Andean mountain range (Cordillera Occidental) at elevations of 1340 to 1430 m a.s.l. The climate is mild and humid, with a mean annual temperature of 22 °C and annual rainfall of 1600 mm (Jaramillo *et al.*, 2011). The rainfall patterns are bimodal with rainfall peaks recorded during the months of April–May and October-November, and a dry period lasting from May to October, which is interrupted by a smaller rainy season in July–August (Pardo-Locarno *et al.*, 2005). The soils are derived from deposits of volcanic ashes over igneous rocks, and are deep and well drained (IGAC, 2009).

Average rainfall registered at the El Madrigal Meteorological Station from 1997 to 2010 was 108 mm for the month of April and 103 mm for the month of May (FNC, 2013). However, during the period of 2006–2013 above average rainfall values were already recorded except for the year 2009, and the year-to-year variability was high (Fig. 1a). The number of rainfall events during the months of April and May was between 9 to 18 days month^{-1}, except for the rather dry year of 2009 (Fig. 1b).

The study was conducted on four plots representing the land use systems shaded coffee, unshaded coffee, secondary forest and pasture. The unshaded cof-

fee was located at a distance of 68 m from the pasture and 258 m from the forest, whereas the shaded coffee was 120 m from the pasture. The experimental plots had a size of approximately 40×30 m, and were located within each patch of land use system. The shaded coffee was planted in 2010 beneath approximately 20 years old shade trees, which were located within the coffee row. It comprised a total area of 2 ha, with spacing of 1.4 m between the coffee rows and 1.2 m within the coffee row, resulting in 5674 coffee plants ha^{-1} (Fig. 2b). Shade trees were mainly composed of leguminous *Inga* species and non-leguminous mango (*Mangifera indica*) trees, which were planted within the coffee rows at spa-

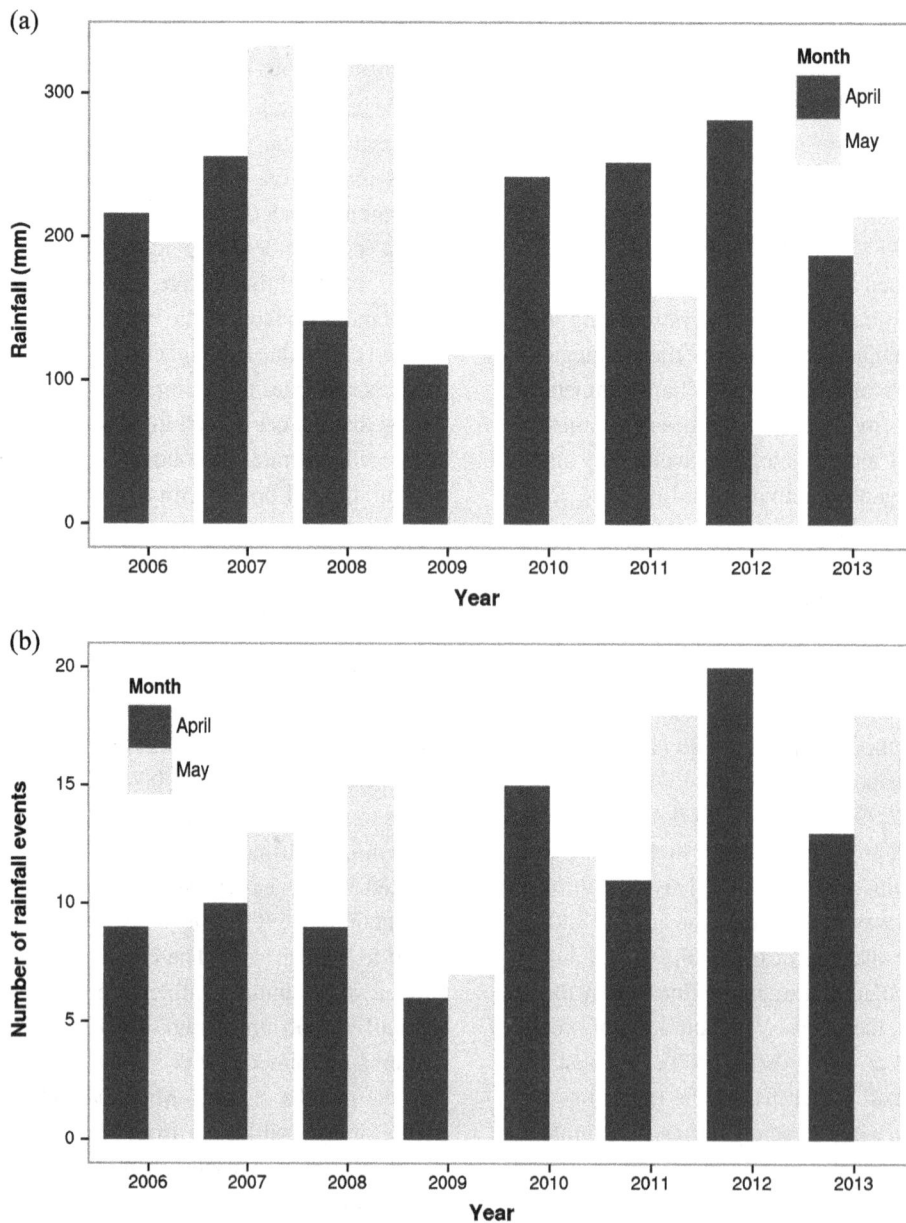

Fig. 1: *(a) Gross precipitation and (b) frequency of rainfall events during the months of April and May. Data from 2006–2012 was taken from El Madrigal Meteorological Station (FNC, 2013), whereas the values of 2013 were obtained in the present study.*

cing of 6×6 to 7×7 m. The unshaded coffee was planted in 2009 and comprised an area of 1 ha, with spacing of 1.5 m between the coffee rows and 1 m within the coffee rows, yielding in a density of 6666 coffee plants ha^{-1} (Fig. 2c; personal communication with farmers 2013). Previously to coffee establishment this plot was a cattle pasture. The area under secondary forest was 4 ha and it existed for at least 14 years. It consisted of a fragment nearby a stream and selective logging was practiced for large diameter trees dominated by the Lauraceae family (personal communication with farmer 2013). The pasture comprised an area of 3 ha. Each hectare was separated and used a rotational grazing system based on 60 days.

2.2 Throughfall

In order to determine throughfall, the approach of Siles *et al.* (2010a) was slightly modified (Fig. 2). In each plot nine sets of rain gauges were placed randomly with a minimum distance of 4 m from each other, and at 20 cm height from the ground floor. The rain gauges were made of cylindrical plastic bottles with a diameter of 5 cm, and the volume was standardized through a graduated cylinder. To obtain a temporal comparability throughfall was collected every morning between 7–10AM during two months. Each set consisted of two rain gauges that were placed at distances of 0.1 and 0.5 m from the target plant (Fig. 2a). In the shaded coffee, each of the nine sets comprised two distances from the shade tree (1 m and 2.3 m, Fig. 2b). In this system gauges were placed below four different shade tree species, which were *Inga densiflora* (n=1), *Inga edulis* (n=3), *Mangifera indica* (n=4) and *Oreopanax floribundus* (n=1). In the unshaded coffee and the secondary forest nine sets of rain gauges were placed following the same spacing (Fig. 2c, 2d). In the secondary forest saplings with the same DBH (diameter at breast height) of coffee plants (i.e. 1.9–2.2 cm, in the following called target plant) were chosen, in order to place the rain gauges at the corresponding distances. The distance between the rain gauges and the trees was on average 6.6 m. The same arrangement was followed for the pasture to measure gross precipitation (P_g), i.e. nine sets were placed randomly. To ensure preciseness of the field data P_g was compared to recordings of the Madrigal Meteorological Station located 1.1 km from the study area.

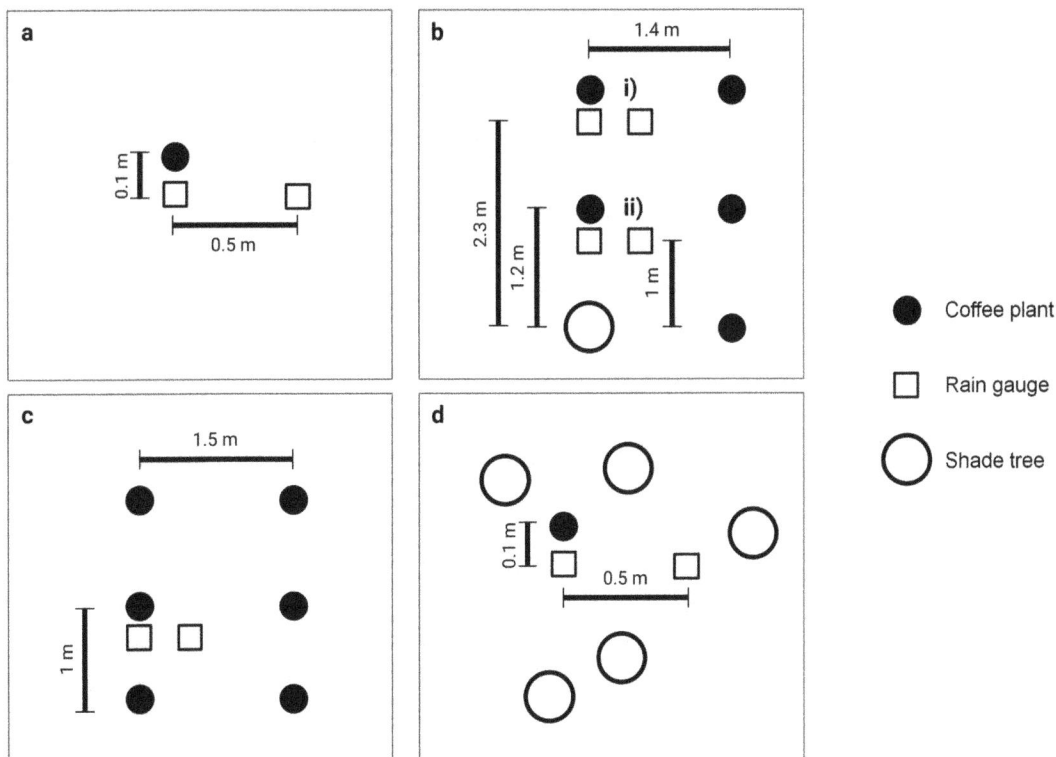

Fig. 2: *Sketch of the experimental design for measuring throughfall in each land use type, based on Siles et al. (2010a). (a) represents one set of rain gauges placed at a defined distance to the coffee plant, (b) set-up in the shaded coffee located at distances of i) 1 m and ii) 2.3 m from the shade tree, (c) set-up in the unshaded coffee, and (d) set-up in the secondary forest, where the sets of rain gauges were placed next to trees with a similar diameter as coffee plants. Since there were no coffee plants or shade trees in the forest, in this case, the black dot represents the target plant referred as to in the text.*

2.3   Stand inventory

In the shaded coffee system, the DBH of the shade trees and the distance to each coffee plant were recorded. In the secondary forest DBH and distance to the rain gauge were recorded for all trees that influenced the set with their canopy. However, all the trees on the plot were counted to calculate stem density. Additionally, the percentage canopy cover was measured for each set with a GRS densitometer in both tree-covered systems. In all four land use types the slope was calculated by measuring the ground level between two points through the steepest part of the slope (FAO, 2013).

2.4   Soil hydrological properties

In order to evaluate major properties determining soil water movement and storage the parameters texture, porosity, bulk density and saturated hydraulic conductivity (K_{sat}) were measured in each land use system. Soil samples were taken to a depth of 30 cm with four replicates from each system. Analyses were performed with established methods (IGAC, 2006) at the Laboratory of Agricultural Soils and Water at the Universidad del Valle in Cali. Soil texture was determined with the Bouyoucos Hydrometer Method. For porosity undisturbed soil samples were taken using a cylinder with a height of 5 cm and a diameter of 5 cm. The samples were weighted at a saturated (M_s), drained (M_d) and dried stage (M_{dry}). Porosity (P_t) was calculated according to the following equation (IGAC, 2006), where V is the volume of the cylinder:

$$\%P_t = \left(\frac{M_s - (M_d + M_{dry})}{V} \right) * 100$$

Bulk density was determined by the core method with the same cylinder. The Porchet method was applied to measure hydraulic conductivity.

2.5   Statistical analysis

R (2.15.3) for Statistical Computing was used to perform statistical analyses. To evaluate the influence of canopy cover and rainfall intensity on the distribution of throughfall a two way analysis of variance was conducted. Given that the assumptions of normality or variance homogeneity were not met for the response variable, a Wilcoxon signed-rank test was conducted. All systems were tested against each other in matched samples. The same was done to assess the effect of land use system on soil parameters.

3   Results

3.1   Stand inventory

All four sites were located on moderately steep (15–30%) to steep (30–60%) slopes (Table 1). Compared to the secondary forest, which had 358 trees ha^{-1} with diameters greater than 10 cm, the tree density in the shaded coffee was lower (Table 1). The mean DBH of trees in the secondary forest was significantly lower than the DBH of the shade trees in the coffee system. The percentage canopy cover in the shaded coffee system was high (76%), but significantly lower than in the secondary forest (92%, Table 1).

Table 1: *Stand inventory with main characteristics of the four land use systems.*

Land use	Tree species, genus, family	Mean DBH (cm) *	Canopy cover (%)	Stem density (trees ha^{-1})	Average distance of trees to rain gauge (m)	Slope (%)
Shaded coffee	*Inga edulis* *Inga densiflora* *Mangifera indica* *Oreopanax floribundus*	20.9 ± 2.5^a	75.7 ± 7.1^a	278	1.7 ± 0.2^a	20.1
Secondary forest	*Cupania* sp. *Cinnamomum* sp. Melastomateceae Mimosaseae Lauraceae Myrtaceae	16.0 ± 1.8^b	92.2 ± 0.9^b	358	6.6 ± 2.4^a	43.2
Unshaded coffee	–	–	0	0	–	37.7
Pasture	–	–	0	0	–	31.3

* Only trees with DBH larger than 10 cm were considered

3.2 Rainfall patterns and throughfall

During the two months of the study period rainfall was quite high, with 188 mm in April and 215 mm in May 2013. The amount of water collected as throughfall was strongly influenced by the placement of the rain gauges in all three systems, excluding the pasture (Fig. 1). Significantly lower throughfall was recorded when the rain gauges were placed in the coffee row compared to the coffee inter-row. The same was true for the secondary forest, where the rain gauges were placed at similar distance to the target plants as in the coffee systems (Table 2).

Significantly lower throughfall was measured in the shaded coffee, when rain gauges were placed at a distance of 1 m to the shade tree compared to a distance of 2.3 m to the shade tree, which was due to the higher canopy cover above the coffee at a shorter distance Throughfall and canopy cover, in contrast, had more similar values in the secondary forest at both distances (Tables 1 and 2). In the unshaded coffee system, throughfall was more than twice as high in the coffee inter-row compared to the coffee row (Table 2).

The highest throughfall was recorded in coffee inter-rows in the unshaded coffee system, whereas lowest throughfall was recorded in coffee rows in the shaded coffee at a distance of 1 m to the coffee plant. The variability of throughfall in both coffee systems was higher in the coffee inter-row compared to the coffee row (Table 2).

In the shaded coffee system, significantly higher throughfall was collected below mango trees (*Mangifera indica*) compared to *Inga* spp. For this latter species it was found that throughfall sometimes exceeded P_g (Table 3).

Throughfall was obviously higher in the coffee inter-row than in the coffee row in both the shaded and unshed coffee system for nearly all precipitation ranges, but it did not increase with increasing rainfall intensity. However the spatial variability of throughfall increased with increasing rainfall intensity and increasing distance from the rain gauge to shade trees and coffee plants (Table 4). The secondary forest showed a more similar throughfall distribution between the two rain gauge placements.

3.3 Soil hydrological properties

Soil texture was classified as clay in all four systems. A higher amount of sand minerals was recorded in the shaded coffee and secondary forest, whereas unshaded coffee and pasture had more clay minerals (Table 5). In the shaded coffee and the secondary forest, soil porosity was higher compared to the unshaded coffee and pasture (Table 5). Microporosity dominated all four land use types, indicating higher water storage potential albeit potential problems with drainage and aeration, but pasture and secondary forest showed higher values than the coffee systems. The highest macroporosity was recorded in the shaded coffee and secondary forest. Shaded coffee and secondary forest had also significantly lower bulk densities than the other two systems (Table 5). Highest K_{sat} was found in the secondary forest, followed by shaded coffee and unshaded coffee with significant differences between the tree-covered systems and the pasture. In all four systems K_{sat} was classified as moderately high, except for the secondary forest, which was classified as high (Table 5).

Table 2: *Cumulative throughfall (expressed in % P_g) measured in the coffee row (0.1 m) and coffee inter-row (0.5 m) in the four land use systems. In the shaded coffee throughfall was measured at a distance of 1.0 and 2.3 m from the shade tree.*

Land use		Coffee row (%)	Coffee inter-row (%)
Shaded coffee	1 m	$17.8 \pm 4.3^{\,a}$	$39.9 \pm 9.7^{\,a}$
	2.3 m	$60.3 \pm 3.9^{\,b}$	$80.5 \pm 7.2^{\,b}$
Unshaded coffee		$40.3 \pm 3.3^{\,b}$	$92.9 \pm 3.7^{\,b}$
Secondary forest*		$64.2 \pm 3.2^{\,bc}$	$73.2 \pm 3.2^{\,b}$
Pasture (control)		$100 \pm 2.6^{\,c}$	$100 \pm 2.9^{\,b}$

Mean values in a column with same letter are not significantly different ($p < 0.05$)
(Mean ± standard error, n = 32)
* The target plants that simulated the coffee plants in the secondary forest were chosen according to the coffee plant DBH

Table 3: *Cumulative throughfall (expressed in % P_g) below the two most common shade tree species. Throughfall was measured in the coffee row (0.1 m) and coffee inter-row (0.5 m) and at two distances from the shade tree (1.0 and 2.3 m).*

Shade tree species	Distance	Coffee row (%)	Coffee inter-row (%)	Average throughfall (%)
Inga sp.	1 m	10.6 ± 3.2^a	56.2 ± 5.7^a	39.1 ± 5.0^a
	2.3 m	39.2 ± 18.6^{bc}	50.8 ± 4.7^a	
Mangifera indica	1 m	27 ± 9.1^{ac}	72.2 ± 4.7^{ab}	63.1 ± 4.6^b
	2.3 m	41.3 ± 11.5^b	111.7 ± 15^b	

Mean values in a column with same letter are not significantly different ($p < 0.05$) (Mean ± standard error, n = 32)

Table 4: *Cumulative throughfall (expressed in % of P_g) grouped into rainfall intensities in the three land use systems. In the shaded coffee throughfall was measured at distances of 1.0 and 2.3 m from the shade tree.*

Rainfall range	<5 mm	5–10 mm	10–20 mm	20–30 mm	>30 mm
Frequency of rainfall event	11	8	6	4	3
Coffee row (0.1 m)					
Shaded coffee (1 m)	24.8 ± 0.4	7.3 ± 1.1	25.4 ± 1.7	12.6 ± 1.1	20.5 ± 4.0
Shaded coffee (2.3 m)	54.6 ± 0.3	62.5 ± 0.4	66.7 ± 1.8	55.9 ± 2.6	71.6 ± 1.0
Unshaded coffee	46.3 ± 0.3	44.9 ± 0.6	45.0 ± 1.2	51.4 ± 2.1	21.3 ± 2.0
Secondary forest*	59.6 ± 0.3	68.6 ± 0.5	71.8 ± 1.6	60.7 ± 0.1	72.6 ± 3.7
Coffee inter-row (0.5 m)					
Shaded coffee (1 m)	30.4 ± 0.2	41.4 ± 1.6	35.0 ± 2.1	39.5 ± 4.1	14.8 ± 1.4
Shaded coffee (2.3 m)	67.4 ± 0.4	89.6 ± 1.4	96.3 ± 1.6	97.5 ± 6.8	76.9 ± 8.1
Unshaded coffee	94.1 ± 0.5	96.4 ± 1.0	97.3 ± 1.6	87.3 ± 2.8	93.8 ± 5.3
Secondary forest*	64.0 ± 0.3	70.3 ± 0.6	85.8 ± 1.3	75.8 ± 2.3	83.0 ± 2.9

*In the secondary forest rain gauges were placed next to trees with similar DBH as coffee plants (see Fig. 2)

Table 5: *Evaluated soil properties in the four land use systems.*

Soil parameter	Shaded coffee	Unshaded coffee	Secondary forest	Pasture
Sand (%)	23.7 ± 0.5^a	21.8 ± 5.0^a	25.1 ± 2.4^a	16.4 ± 2.5^a
Clay (%)	66.3 ± 0.5^a	61.3 ± 3.9^{ab}	55.8 ± 1.3^b	68.3 ± 2.2^a
Silt (%)	10 ± 0.8^a	17.5 ± 1.3^a	19 ± 1.6^a	15.2 ± 2.7^b
Texture	Clay	Clay	Clay	Clay
Microporosity (%)	46.0 ± 2.54^{ab}	43.4 ± 0.93^a	49.1 ± 3.26^b	50.0 ± 1.36^b
Macroporosity (%)	9.0 ± 4.09^{ab}	6.7 ± 1.95^a	7.7 ± 2.89^a	2.5 ± 1.08^b
Total porosity (%)	55.0 ± 2.5^a	50.1 ± 1.3^b	56.8 ± 0.78^a	52.5 ± 1.08^b
Classification of porosity*	Excellent	Satisfactory	Excellent	Satisfactory
Bulk density (g cm^{-3})	0.97 ± 0.08^a	1.12 ± 0.07^b	0.99 ± 0.05^a	1.16 ± 0.08^b
K_{sat} (cm h^{-1})	2.1 ± 0.27^a	1.8 ± 0.3^{ab}	3.6 ± 1.5^a	1.1 ± 0.27^b
Classification of K_{sat} [†]	Moderately high	Moderately high	High	Moderately high

Mean values in a row with same letter are not significantly different ($p < 0.05$) (Mean ± standard error, n = 32)
* According to Kaurichev (1984); [†] According Soil Survey Division Staff (1993)

4   Discussion

4.1   Impact of stand structure and rainfall intensity on throughfall

Throughfall is expected to decrease with increasing leaf area index (LAI) and tree height (Dietz *et al.*, 2006; Siles *et al.*, 2010a). This was reflected in the shaded coffee, which displayed lower throughfall under conditions where rain gauges were placed at a shorter distance to the shade tree. In the shaded coffee a high variability of throughfall was observed, which was not the case for the secondary forest, which can be explained by its natural and randomly distributed stand structure resembling an old-growth forest.

The highest throughfall was recorded in the unshaded coffee when rain gauges were placed in the coffee inter-row. The canopy openness of the coffee plants in the unshaded system was higher and displayed larger gaps, which allowed the rain to pass through freely. On the contrary, the presence of trees and the relatively closed canopy in the tree-covered systems intercepted more water, which was also proven by other authors (e.g. Tobón Marín *et al.*, 2000; Park & Cameron, 2008; Siles *et al.*, 2010a). Through the presence of a canopy the velocity of water is reduced, which decreases its kinetic energy and lowers the removal capacity of upper soil layers (Blanco & Lal, 2008). This buffering is of great importance considering the increase of rainfall in this mountainous region. Taking into account the spatial distribution of shade trees in the coffee plantations (on average 6–7 m from each other), coffee plants placed at a closer distance to the shade tree will rather benefit from a higher canopy cover during events of high rainfall. However, a close spacing can also result in competition for water and nutrients between shade trees and coffee plants (Beer *et al.*, 1998). Cannavo *et al.* (2011) found that soil water content was similar between shaded and unshaded coffee in the upper layers, suggesting that shade trees rather use water from deeper soil layers. Their study was conducted in an area where annual rainfall exceeds evapotranspiration, same as in the present study. However, during the dry season evaporation may exceed rainfall for a certain time period, which can limit water availability in the system. Yet better soil properties and a slower drying of the soil due to shaded conditions may compensate the lower net rainfall in shaded coffee systems.

Throughfall values obtained in the present study were comparable to results of Siles *et al.* (2010a) from Costa Rica, who recorded throughfall of 54 % and 94 % in rain gauges placed at distances of 0.1 and 0.5 m to the coffee plant in shaded coffee, as well as 64 % and 97.5 % in unshaded coffee. Dietz *et al.* (2006) reported throughfall of 79 % in a forest with small timber extraction, which could be compared to the secondary forest of the present study, and 82 % in a cacao agroforestry system in Sulawesi, Indonesia. Tobón Marín *et al.* (2000) recorded throughfall of 82 to 87 % in different forest types in the western Amazon. Extremely high throughfall recorded in the shaded coffee can be explained by the concentration of water at a single point. In this case leaves may act as a funnel, and in some rain gauges throughfall was twice as high as P_g. This pattern was also observed by Jaramillo (2003) in coffee shaded with eucalypt, and to a lesser extend in a forest.

It was not possible to measure stemflow in the present study; thus net-rainfall might be underestimated to a certain extent. Trees may benefit from stemflow, which causes a concentration of rainfall near the stems (Schroth *et al.*, 1999). Several authors found stemflow of only 1–3 % of P_g in tropical forest (e.g. Hölscher *et al.*, 2004; Dietz *et al.*, 2006; Macinnis-Ng *et al.*, 2013), while a study conducted by Siles *et al.* (2010a) found 7 % stemflow in a coffee monoculture and 10 % in a shaded coffee system in Costa Rica.

In our study a trend towards a higher throughfall with increasing rainfall intensity was only observed for the secondary forest, and the variation across land use types was high. E.g. rainfall events in the range of 20–30 mm day^{-1} contributed to 12.7 % of throughfall in shaded coffee, but 60.7 % of throughfall in secondary forest. The significantly higher throughfall below *Mangifera indica*, compared to *Inga* spp. can be attributed to its lower extension of branches and a rather smooth surface of the leaves. This points to the fact that structural parameters such as stem density and plant architecture (e.g. structure and arrangement of branches and leaves) have a stronger influence on throughfall than the intensity of rainfall events, which was also found by other authors (e.g. Tobón Marín *et al.*, 2000; Siles *et al.*, 2010a; Bäse *et al.*, 2012).

4.2   Influence of vegetation cover on soil hydrological properties

In general, soil hydrological properties followed a similar pattern in the two tree-covered systems and the unshaded systems, respectively. Tree-covered systems allow the movement of water into deeper soil profiles, reducing the possibility of runoff and soil erosion, which is of relevance when extreme rainfall events occur. Both tree-covered systems existed for more than 14 years, and could have a similar influence on soil properties. Similarity might also be attributed by soil composition, since soil texture influences aeration, drainage and water retention (Jaramillo, 2002a). In general, clay content and particle distribution due to its colloidal constitution does not obstruct water movement completely, yet it creates a longer and difficult way for water to run freely (Jiménez & Rodríguez, 2008). This could explain the significantly higher hydraulic conductivity in the secondary forest and shaded coffee compared to pasture, since pasture showed the highest clay and lowest sand content. On the contrary, the higher percentage of sand and lower percentage of clay in the tree-covered systems could account for the faster water movement through the soil. In addition, it has been reported that higher clay content can lead to compaction processes (Jaramillo, 2002a).

Hassler *et al.* (2011) analysed the effect of secondary forest succession on the hydraulic conductivity of a site that was previously covered by pasture in Central Panama. The study reported the recovery of K_{sat} after 12 years of secondary succession and suggested that these forests could provide similar hydrological services as undisturbed forests. The secondary forest site of our study could have reached a similar stage considering its age and the fact that timber extraction is not high.

The unshaded coffee and the pasture showed higher bulk densities and lower porosity compared to the two tree-covered systems, which can be related to soil disturbance and compaction through animals or human activity (Ziegler *et al.*, 2004; Zimmermann *et al.*, 2006), and may limit water infiltration and increase rain splashing (Cannavo *et al.*, 2011). The low macroporosity of the pasture may be an indicator that the soil is packed, and extremely fine pore sizes may hinder roots to penetrate (Benegas *et al.*, 2014). De Moraes *et al.* (2006) reported a reduction of macroporosity and hydraulic conductivity when forest was replaced by pasture over a time span of 30 years in the eastern Amazon, whereas soil moisture increased, promoting frequent saturation. In fact, several authors suggested that lower infiltration rates and K_{sat}, together with increased bulk density, runoff and soil erosion in pastures are consequences of a lack of woody vegetation together with inappropriate tillage methods (Freebairn & Gupta, 1990; Beer, 1987; Cannavo *et al.*, 2011).

Runoff and soil erosion can be even more relevant considering the steep slopes on which the analysed land use systems were located. Sharma *et al.* (2001) reported that agroforestry interventions such as planting trees reduced soil losses by 22 % on slopes in the Himalayas of Nepal. Verbist *et al.* (2010) found significantly lower runoff in forests and shaded coffee plantations than in a coffee monoculture in Sumatra, Indonesia on slopes of 20–30 %.

Furthermore, compensation schemes for hydrological services might also encourage farmers to adopt the use of shade trees. This approach is currently being promoted by the Colombian Coffee Growers Federation, which is part of a strategy for environmental hydric services in river basins through economic incentives, such as tax waivers, to those coffee growers that conserve natural forests and increase incorporation of trees in their coffee crops (Sosa & Moreno, 2014). The vision is that increasing natural vegetation and trees, especially in the higher mountain belts where water springs are born, will protect both soil and water availability for towns and cities located downward in the Andean mountains.

4.3   Stand structural characteristics of shaded coffee and secondary forest

The secondary forest canopy cover in the present study was quite similar or even higher to canopy covers obtained from other studies that were conducted in undisturbed tropical forests, e.g. 90 % in Sulawesi, Indonesia (Dietz *et al.*, 2007) and 83–89 % in the western Amazon (Tobón Marin *et al.*, 2000). Different studies indicate that during secondary succession many forest functions such as canopy closure (but not canopy species composition) approach values of primary forests in a time scale of 5 to 20 years (e.g. Finegan, 1996; Guariguata & Ostertag, 2001) and thus resemble a near natural vegetation. Usually agroforests have a lower canopy cover compared to natural forests or forests with a certain amount of timber extraction, which was also confirmed by Dietz *et al.* (2007). The high variability of canopy cover in the shaded coffee can be related to its stand structural characteristics, i.e. different DBH classes of shade trees, a lower density and variable distances between the shade trees and coffee plants.

The lower DBH of trees in the secondary forest compared to shade trees in the shaded coffee could indicate timber extraction activities in the area. Even though the site seemed quite undisturbed, some farmers stated that timber is used sporadically for furniture and firewood purposes (personal communication with farmer, 2013). In contrast, shade trees in the coffee system, such as mango, citrics, avocado and guamo (*Inga edulis*, *Inga densiflora*) remained for over 20 years since their fruits are used for local consumption.

5 Conclusion

Our case study showed that throughfall was rather influenced by stand structural characteristics than by rainfall intensity. On average throughfall was lower in shaded coffee compared to unshaded coffee and secondary forest, despite of the latter displaying a higher stem density and canopy cover. Tree-covered systems resembled in soil hydrology properties, and showed a better porosity and saturated hydraulic conductivity than unshaded coffee and pasture. In the face of climate change, coffee systems without shade trees could become severely affected by an increase in volume and frequency of rainfall, which can trigger erosive processes and landslides. The use of shade trees will help to reduce runoff and probably also erosion when heavy rainfall occurs. Furthermore, where unshaded coffee exists, a proper conversion and arrangement to shaded coffee could lead to a sustainable production system, with the provision of ecosystem services, and economic incentives to coffee growers through compensation schemes for hydrological services.

Acknowledgements

The Konrad Krieger-Stiftung and the Internationalization program of the Faculty of Forest Sciences and Forest Ecology at the Georg-August-Universität Göttingen provided funding for this study. We are grateful for the support of Dr. James Montoya Lerma, Aldemar Reyes and the students of the Entomology department at the Universidad del Valle in Cali, who helped during project development. Meldy Morales and Catherine Gomez from the Laboratory of Agricultural Soils and Water at Universidad del Valle performed the analyses of soil hydrological properties. We further thank the people from the village El Rosal, especially Doña Martha Paz, for their hospitality and support, as well as the field assistants Maria Ximena Urrutía, Joan Manuel Gaitán and Viviana Alarcón, without whom it would not have been possible to carry out this study.

References

Armbrecht, I., Rivera, L. & Perfecto, I. (2005). Reduced diversity and complexity in the leaf litter ant assemblage of Colombian coffee plantations. *Conservation Biology*, 19(3), 897–907.

Bäse, F., Elsenbeer, H., Neill, C. & Krusche, A. V. (2012). Differences in throughfall and net precipitation between soybean and transitional tropical forest in the southern Amazon, Brazil. *Agriculture, Ecosystems & Environment*, 159, 19–28.

Beer, J. (1987). Advantages, disadvantages and desirable characteristics of shade trees for coffee, cacao and tea. *Agroforestry Systems*, 5(1), 3–13.

Beer, J., Muschler, R., Kass, D. & Somarriba, E. (1998). Shade management in coffee and cacao plantations. *In:* Nair, P. K. R. & Latt, C. R. (eds.), *Directions in Tropical Agroforestry Research*. pp. 139–164, Springer, The Netherlands.

Benegas, L., Ilstedt, U., Roupsard, O., Jones, J. & Malmer, A. (2014). Effects of trees on infiltrability and preferential flow in two contrasting agroecosystems in Central America. *Agriculture Ecosystems & Environment*, 183, 185–196.

Blanco, H. & Lal, R. (2008). Agroforestry. *In:* Blanco, H. & Lal, R. (eds.), *Principles of soil conservation and management*. pp. 259–283, Springer, The Netherlands.

Bruijnzeel, L. A. (1989). (De)forestation and dry season flow in the tropics: a closer look. *Journal of Tropical Forest Science*, 1(3), 145–161.

Cannavo, P., Sansoulet, J., Harmand, J. M., Siles, P., Dreyer, E. & Vaast, P. (2011). Agroforestry associating coffee and Inga densiflora results in complementarity for water uptake and decreases deep drainage in Costa Rica. *Agriculture Ecosystems & Environment*, 140(1), 1–13.

Cárdenas, J. (1993). La Industria del café en Colombia. *Ensayos sobre Economía Cafetera*, 6(9), 3–15.

Castro, F., Montes, E. & Raine, M. (2004). Centroamérica La Crisis Cafetalera: Efectos y Estragias para Hacerle Frente. The World Bank, Latin America and Caribbean Region. Environmentally and Socially Sustainable Development Department (LCSES). Sustainable Development Working Paper 23.

Crockford, R. H. & Richardson, D. P. (2000). Partitioning of rainfall into throughfall, stemflow and interception: effect of forest type, ground cover and climate. *Hydrological Processes*, 14(16–17), 2903–2920.

De Moraes, J., Schuler, A., Dunne, T., Figueiredo, R. & Victoria, R. (2006). Water storage and runoff processes in plinthic soils under forest and pasture in Eastern Amazonia. *Hydrological Processes*, 20(12), 2509–2526.

Dietz, J., Hölscher, D. & Leuschner, C. (2006). Rainfall partitioning in relation to forest structure in differently managed montane forest stands in Central Sulawesi, Indonesia. *Forest Ecology and Managament*, 237(1), 170–178.

Dietz, J., Hölscher, D., Leuschner, C., Malik, A. & Amir, M. A. (2007). Forest structure as influenced by different types of community forestry in a lower montane rainforest of Central Sulawesi, Indonesia. *In:* Tscharnkte, T., Leuschner, C., Zeller, M., Guhardja, E. & Bidin, A. (eds.), *Stability of Tropical Rainforest Margins.* pp. 131–146, Springer, Heidelberg.

FAO (2013). Chapter 4: Measuring vertical angles and slopes. FAO Training Series: Simple methods for aquaculture, Volume 16, 2: Topographical surveys. URL `ftp://ftp.fao.org/fi/cdrom/fao_training/ FAO_Training/General/x6707e/x6707e04.htm` (last accessed: 23.09.2015).

Finegan, B. (1996). Pattern and process in neotropical secondary rain forests: the first 100 years of succession. *Trends in Ecology & Evolution*, 11 (3), 119–124.

FNC (2012). Comportamiento de la Industria Cafetera Colombiana 2012. Informes de Gestión. Federación Nacional de Cafeteros de Colombia (FNC). URL `http://www.federaciondecafeteros. org/particulares/es/quienes_somos/ publicaciones/` (last accessed: 23.09.2015).

FNC (2013). Anuario Metereológico Cafetero 2006–2012. Federación Nacional de Cafeteros de Colombia (FNC), Cenicafe, Colombia.

FNC (2014). Área cultivada según exposición solar por departamento. Cultivos de café en Colombia. Federación Nacional de Cafeteros de Colombia (FNC). URL `http://www.federaciondecafeteros. org/particulares/es/quienes_somos/119_ estadisticas_historicas/` (last accessed: 23.09.2015).

Freebairn, D. M. & Gupta, S. C. (1990). Microrelief, rainfall and cover effects on infiltration. *Soil & Tillage Research*, 16 (3), 307–327.

Grip, H., Fritsch, J. M. & Bruijnzeel, L. A. (2004). Soil and water impacts during forest conversion and stabilization of new land use. *In:* Bonnell, M. & Bruijnzeel, L. A. (eds.), *Forests, Water and People in the Humid Tropics.* pp. 561−589, Cambridge University Press, Cambridge.

Guariguata, M. R. & Ostertag, R. (2001). Neotropical secondary forest succession: changes in structural and functional characteristics. *Forest Ecology and Managament*, 148 (1), 185–206.

Guhl, A. (2004). Café y cambio de paisaje en la zona cafetera colombiana, 1970–1997. *Cenicafé*, 55 (1), 29–44.

Hassler, S. K., Zimmermann, B., Van Breugel, M., Hall, J. S. & Elsenbeer, H. (2011). Recovery of saturated hydraulic conductivity under secondary succession on former pasture in the humid tropics. *Forest Ecology and Managament*, 261 (10), 1634–1642.

Herwitz, S. R. (1985). Interception storage of tropical rainforest canopy trees.

Hodnett, M. G., Pimentel da Silva, L., da Rocha, H. R. & Cruz Senna, R. (1995). Seasonal soil water storage changes beneath central Amazonian rainforest and pasture. *Journal of Hydrology*, 170, 233–254.

Hölscher, D., Köhler, L., van Dijk, A. I. J. M. & Bruijnzeel, L. A. (2004). The importance of epiphytes to total rainfall interception by a tropical montane rain forest in Costa Rica. *Journal of Hydrology*, 292 (1), 308–322.

IGAC (2006). Métodos analíticos del Laboratorio de Suelos. Instituto Geográfico Agustín Codazzi (IGAC), Bogotá.

IGAC (2009). Estudio general de suelos y zonificación de tierras del departamento del Cauca a escala 1:100.000. Instituto Geográfico Agustín Codazzi (IGAC), Bogotá.

Jaramillo, Á. (2003). La lluvia y el transporte de nutrimentos dentro de ecosistemas de bosque y cafetales. *Cenicafé*, 54 (2), 134−144.

Jaramillo, Á., Ramírez, V. H. & Pulgarín, J. A. (2011). Patrones de distribución de la lluvia en la zona cafetera. *Avances Técnicos Cenicafé*, 410, 1–12.

Jaramillo, D. F. (2002a). La textura del suelo. *In:* Jaramillo, D. F. (ed.), *Introducción a la ciencia del suelo.* pp. 163–179, Universidad Nacional de Colombia Facultad de Ciencias, Medellín, Colombia.

Jaramillo, D. F. (2002b). Manejo del medio físico del suelo. *In:* Jaramillo, D. F. (ed.), *Introducción a la ciencia del suelo.* pp. 255–294, Universidad Nacional de Colombia Facultad de Ciencias, Medellín, Colombia.

Jiménez, I. & Rodríguez, L. (2008). Diagnóstico de la infiltración y permeabailidad en los suelos de la zona de recarga del acuifero Morroa en el area Sincelejo, Corozal y Morroa. Facultad de Ingenieria, Universidad de Sucre, Sincelejo (Sucre), Colombia. pp. 151.

Kaurichev, I. S. (1984). *Prácticas de edafología.* Mir, Moscow.

Läderach, P., Lundy, M., Jarvis, A., Ramírez, J., Pérez, P., Schepp, E. & Eitzinger, K. A. (2011). Predicted impact of climate change on coffee supply chains. *In:* Leal Filho, W. (ed.), *The economic, social and political elements of climate change.* ch. 42, pp. 703–724, Springer, Berlin.

Lau, C., Jarvis, A. & Ramírez, J. (2010). Colombian agriculture: Adapting to climate change. CIAT Policy Brief no. 1. Centro Internacional de Agricultura Tropical (CIAT), Cali, Colombia.

Lin, B. B. (2007). Agroforestry management as an adaptive strategy against potential microclimate extremes in coffee agriculture. *Agricultural and Forest Meteorology*, 144 (1), 85–94.

Macinnis-Ng, C. M. O., Flores, E. E., Müller, H. & Schwendenmann, L. (2013). Throughfall and stemflow vary seasonally in different land-use types in a lower montane tropical region of Panama. *Hydrological Processes*, 28 (4), 2174–2184.

Malmer, A., Murdiyarso, D., Bruijnzeel, L. A. & Ilstedt, U. (2009). Carbon sequestration in tropical forests and water: a critical look at the basis for commonly used generalizations. *Global Change Biology*, 16 (2), 599–604.

Martius, C., Höfer, H., Garcia, M., Römbke, J., Förster, B. & Hanagarth, W. (2004). Microclimate in agroforestry systems in central Amazonia: does canopy closure matter to soil organisms? *Agroforestry Systems*, 60 (3), 291–304.

Pardo-Locarno, L. C., Montoya-Lerma, J., Bellotti, A. C. & Van Schoonhoven, A. (2005). Structure and composition of the white grub complex (Coleoptera: Scarabaeidae) in agroecological systems of northern Cauca, Colombia. *Florida Entomologist*, 88 (4), 355–363.

Park, A. & Cameron, J. (2008). The influence of canopy traits on throughfall and stemflow in five tropical trees growing in a Panamanian plantation. *Forest Ecology and Managament*, 255 (5), 1915–1925.

Perfecto, I., Rice, A. R., Greenberg, R. & Van der Voort, M. E. (1996). Shade coffee: A disappearing refuge for biodiversity. *BioScience*, 46 (8), 598–608.

Perfecto, I., Vandermeer, J., Hanson, P. & Cartin, V. (1997). Arthropod diversity loss and the transformation of a tropical agroecosystem. *Biodiversity and Conservation*, 6 (7), 935–945.

Porter, J. R. & Semenov, M. A. (2005). Crop responses to climatic variation. *Philosophical Transactions of the Royal Society B: Biological Sciences*, 360 (1463), 2021–2035.

Salas, G. (1987). Suelos y ecosistemas forestales; con énfasis en América Tropical. Turrialba, IICA.450.

Sandström, K. (1998). Can forests 'provide' water: widespread myth or scientific reality? *Ambio*, 27 (2), 132–138.

Scheffler, R., Neill, C., Krusche, A. V. & H. Elsenbeer, H. (2011). Soil hydraulic response to land-use change associated with the recent soybean expansion at the Amazon agricultural frontier. *Agriculture Ecosystems & Environment*, 144 (1), 281–289.

Schroth, G., da Silva, L. F., Wolf, M.-A., Teixeira, W. G. & Zech, W. (1999). Distribution of throughfall and stemflow in multi-strata agroforestry, perennial monoculture, fallow and primary forest in central Amazonia, Brazil. *Hydrological Processes*, 13, 1423–1436.

Sharma, E., Rai, S. C. & Sharma, R. (2001). Soil, water and nutrient conservation in mountain farming systems: Case-study from the Sikkim Himalaya. *Journal of Environmental Management*, 61 (2), 123–135.

Siles, P., Vaast, P., Dreyer, E. & Hermand, J. M. (2010a). Rainfall partitioning into throughfall, stemflow and interception loss in a coffee (*Coffea arabica* L.) monoculture compared to an agroforestry system with *Inga densiflora*. *Journal of Hydrology*, 395 (1), 39–48.

Siles, P., Vaast, P. & Harmand, J. M. (2010b). Effects of Inga densiflora on the microclimate of coffee (*Coffea arabica* L.) and overall biomass under optimal growing conditions in Costa Rica. *Agroforestry Systems*, 78 (3), 269–286.

Soil Survey Division Staff (1993). Soil survey manual. Soil Conservation Service. U.S. Department of Agriculture. Handbook 18.

Sosa, C. & Moreno, C. (2014). Esquema de compensación por servicios ambientales hídricos en el sector productivo cafetero. Caso microcuenca Toro, Valle del Cauca. FNC, Cali.

Tobón Marin, C., Bouten, W. & Sevink, J. (2000). Gross rainfall and its partitioning into throughfall, stemflow and evaporation of intercepted water in four forest ecosystems in western Amazonia. *Journal of Hydrology*, 237 (1), 40–57.

Vaast, P., Beer, J., Harvey, C. & Harmand, J. M. (2005). Environmental services of coffee agroforestry systems in Central America: a promising potential to improve the livelihoods of coffee farmers' communities. *In:* Wallace, H. (ed.), *Intregrated management of environmental services in human–dominated tropical landscapes: Inter-American Scientific 4th Conference Series.* pp. 35 – 39, CATIE, Turrialba.

Van Oijen, M., Dauzat, J., Harmand, J. M., Lawson, G. & Vaast, P. (2010). Coffee agroforestry systems in Central America: II. Development of a simple process-based model and preliminary results.

Verbist, B., Poesen, J., van Noordwijk, M., Widianto, Suprayogo, D., Agus, F. & Deckers, J. (2010). Factors affecting soil loss at plot scale and sediment yield at catchment scale in a tropical volcanic agroforestry landscape. *CATENA*, 80 (1), 34–46.

Ziegler, A. D., Giambelluca, T. W., Tran, L. T., Vana, T. T., Nullet, M. A., Fox, J., Viene, T. D., Pinthongf, J., Maxwell, J. F. & Evetth, S. (2004). Hydrological consequences of landscape fragmentation in mountainous northern Vietnam: evidence of accelerated overland flow generation. *Journal of Hydrology*, 287 (1), 124–146.

Zimmermann, B., Elsenbeer, H. & De Moraes, J. M. (2006). The influence of land-use changes on soil hydraulic properties: implications for runoff generation. *Forest Ecology and Managament*, 222 (1), 29 – 38.

14

Effects of time-controlled and continuous grazing on total herbage mass and ground cover

Gholamreza Sanjari [a,*], Hossein Ghadiri [b], Bofu Yu [b]

[a] *Research Institute of Forests and Rangelands, Tehran, Iran*
[b] *Griffith School of Environment; Griffith University, Australia*

Abstract

Grazing practices in rangelands are increasingly recognized as a management tool for environmental protection in addition to livestock production. Long term continuous grazing has been largely documented to reduce pasture productivity and decline the protective layer of soil surface affecting environmental protection. Time-controlled rotational grazing (TC grazing) as an alternative to continuous grazing is considered to reduce such negative effects and provides pasture with a higher amount of vegetation securing food for animals and conserving environment. To research on how the grazing system affects herbage and above ground organic materials compared with continuous grazing, the study was conducted in a sub-tropical region of Australia from 2001 to 2006.

The overall results showed that herbage mass under TC grazing increased to 140 % in 2006 compared with the first records taken in 2001. The outcomes were even higher (150 %) when the soil is deeper and the slope is gentle. In line with the results of herbage mass, ground cover under TC grazing achieved significant higher percentages than continuous grazing in all the years of the study. Ground cover under TC grazing increased from 54 % in 2003 to 73 %, 82 %, and 89 % in 2004, 2005, and 2006, respectively, despite the fact that after the high yielding year of 2004 herbage mass declined to around 2.5 ton ha^{-1} in 2005 and 2006. Under continuous grazing however there was no significant increase over time comparable to TC grazing neither in herbage mass nor in ground cover. The successful outcome is largely attributed to the flexible nature of the management in which grazing frequency, durations and the rest periods were efficiently controlled. Such flexibility of animal presence on pastures could result in higher water retention and soil moisture condition promoting above ground organic material.

Keywords: Currajong, herbage mass, ground cover, time-controlled grazing, continuous grazing

1 Introduction

It is well accepted that continuous grazing largely exposes desirable species to repeated defoliations of differing intensities, compared with the less desirable and undesirable (Dyksterhuis, 1949; Holechek *et al.*, 1998). For this reason, even when stocking rates are low in relation to the pasture carrying capacity, the most palatable and nutritious plants are subject to higher grazing pressures (Wilson & Harrington, 1984). Depending on the pastoral ecosystem characteristics such a continual selective grazing over time results in a decrease in the quantity of desirable species and in turn reduces pasture productivity.

Rotational grazing which includes some periods of grazing exclusion, helps to minimize this repeated defoliation of the species. To include rest periods, at least two paddocks are needed; however a higher number of paddocks, as many as 50 or more, could be involved. The more paddocks in the rotation, the greater is the flexibility in management such as the option of skipping

* Corresponding author
Research Institute of Forests and Rangelands, Tehran, Iran
Email: g.sanjari@gmail.com

one paddock during a rotation cycle without imposing significant stress on the remaining paddocks (Norton, 2003). Time-controlled (TC) grazing as a new variant of rotational grazing involves short periods of intensive grazing followed by long rest durations. Grazing intensity is much higher than that of the normal carrying capacity usually estimated for continues grazing, therefore selective grazing is greatly reduced. The grazing system was put forward by Savory & Parsons (1980). The terms "The Savory Grazing System", "Short Duration Grazing" and "Cell Grazing" are the various names given to the grazing practice.

Field trials by graziers under commercial livestock production in the USA, South Africa and Australia have shown quite significant improvements in some pasture attributes under TC grazing (Alsemgeest & Alchin, 2002; Detterling, 1999; Johnson, 1998; Joyce, 2000; Sayre, 2001; Sparke, 2000; Suther, 1991), however, uncertainty in relation to the ability of TC grazing to increase desirable species, livestock production, labour costs and herbage mass still exist (Gammon, 1978; Holechek et al., 1998; Valentine, 2000; Wilson, 1986).

The successful outcomes from commercial trials in grazing lands reported earlier have resulted in an increasing adoption of the new grazing practice in Australia as well as throughout the world. However, due to the periods of intensive grazing, balance between pasture utilisation and environmental conservation is of great concern. Such a balance is more emphasised in areas where soil depth and available moisture are limiting factors of herbage growth.

To investigate how the grazing system of TC grazing affects above ground organic matter and land surface protection, the current study was conducted using research paddocks of the commercial property of Currajong in southeast Queensland Australia over the period of 2001 to 2006. In this study, the focus is only on pasture attributes that affect above ground organic materials and land surface protection. The paper is then aimed to report on the general impacts of the grazing treatments (TC, Continuous grazing) on total herbage mass and ground cover.

2 Materials and methods

2.1 Study sites

The study area is located 40 km West of Stanthorpe in the semi-arid region of southeast Queensland Australia (28°33′ S, 151°33′ E, altitude 675 m). Long term average rainfall (119 years) is 645 mm of which 70 %

is falling in summer within the months of October to March. The rain in summer (grass growth season) is characterised by relatively high frequency of medium to large events of short (thunderstorms) and long (cyclonic depressions) duration. Mean temperature is 17.3°C with July being the coldest (10°C) and January the warmest months (24°C).

Natural vegetation in the study area is an Eucalypt open woodland that has been extensively cleared over the past century for agricultural and grazing activities. Understory vegetation is dominated mostly by Queensland blue grass [Dichanthium sericeum (R. Br.) A. Camus]. The remaining desirable species include but are not limited to Silky browntop [Eulalia aurea (Bory) Kunth], Wallaby grass [Danthonia tenuior (Steud.) Conert] and Hairy Panic [Panicum effusum R.Br.]. The next grouping of plants, in the order of dominance, comprises native species of Wiregrass (Aristida sp.) known as less desirable. There is also another group of species that has a medium palatability in the area such as Pitted blue grass [Bothriochloa decipiens (Hack.) C.E. Hubb.] and Digitaria [Digitaria breviglumis (Domin) Henrard]. Coolati grass [Hyparrhenia hirta (L.) Stapf] along with African lovegrass [Eragrostis curvula (Schrad.) Nees] are less desirable to sheep than to cattle, while both species are invasive to the area.

The study area has shallow to moderately deep soils with a brown to dark clay loam underlaid by a bleached A2 horizon. The soil analysis showed for the surface soil (0–10 cm) a pH and EC of 5.6 and 0.06 mS respectively; soil organic carbon of 26 ton ha^{-1}; NO$_3$ of 0.6 kg ha^{-1} and extractable P of 17 kg ha^{-1} (Sanjari et al., 2008). The area is the headwater for a number of streams and visibly eroded by sheet erosion due to the lack of vegetative cover, channel incisions and re-incision of alluvial deposits in valley floors (Sanjari, 2008).

2.2 Treatments

This research was conducted on a commercial grazing property, which was in the process of converting from long term continuous grazing into TC grazing. The application of TC grazing required the existing large paddocks to be sub-divided into 21 smaller paddocks using electric fences. One of the paddocks under TC grazing was assigned to this research for data collection. There was also another research paddock with similar geomorphology and soils as the TC paddock to represent the continuous grazing system. Following the assignment of the paddocks to the grazing treatments, they were each subdivided into two sub-treatments that based on the differences in slope and soil depth are henceforth called "deep flat" and "shallow sloppy" (Table 1). Under this arrangement, the sub-treatments T.deep flat and T.shallow sloppy belong to time-controlled and the sub-treatments C.deep flat and C.shallow sloppy belong to continuous grazing. The combined results of the sub-treatments are reported as the grazing treatment effects. The similarities (in terms of slope and soil depth) between sub-treatments T.deep flat and C.deep flat on the one hand and between T.shallow sloppy and C.shallow sloppy on the other hand, provided the experiment with a chance of reducing *between treatment errors* when comparing the two grazing treatments.

Under TC grazing, a large herd of livestock is moved between a number of paddocks for short periods of time. These periods of grazing are considerably shorter than the rest durations. A general recommendation suggests 30–90 days for the rest durations, which shortens during rapid plant growth and lengthens as plant growth slows (Gillen *et al.*, 1991). Such flexibility in rest duration is also the case for grazing periods and stocking rates, therefore a different numbers of stocks could be moved between paddocks at any time depending on grass growth rate and feed on offer (Fig 1). In our study a sheep herd of merinos with different sizes in different grazing events (1750 – 4577 DSE) were moved between the 21 paddocks over the study period. The grazing details for our research paddocks (one per grazing system) are summarized in Figure 1 and Table 2.

The stocking rate (SR) for the two grazing treatments is expressed as dry sheep equivalent (dse) per hectare. Dry sheep equivalent is defined as the nutritional or metabolisable energy needed to maintain a 50 kg dry sheep (non-lactating). A 50 kg wether has a dry sheep equivalent of 1, animals requiring more feed have a

higher rating and animals with less feed requirements have a lower rating. The history of stocking rate summarised in Table 2 shows that the paddock with TC grazing was heavily stocked with an average number of 12.3 dse ha^{-1} over a mean grazing period of 14 days and then rested for various time (101 ± 60 days). In the continuous grazing, the pasture was stocked with a constant stocking rate of around 1.6 dse/ha throughout the years of the study that is considered normal in the region and exerts a light to moderate pressure on the pasture.

Maximum efforts were made to keep the same overall grazing pressure in both grazing treatments. This was achieved by the similar total number of dse.day ha^{-1} (DDH) reported in Table 2 despite the fact that the grazing systems had major differences in stocking rates, grazing durations and rest periods. The similar DDH between the paddocks indicates that the stocking management by the grazier kept the overall stocking pressure equal between the treatments.

2.3   Sampling

To undertake field data collection, 44 permanent sampling location sites were selected across the research paddocks that were distributed equally between the two grazing treatments and sub-treatments in catenary sequences to include all landform components. Sampling with 10 replications was performed randomly at each permanent location using a quadrate of 0.25 m^2 to measure main components of vegetation and residue cover (i.e. herbage mass and ground cover). The area in which quadrates were placed was a circle with 25 meter radius centred at each permanent location. The quadrate was thrown to fall at random within the site area. Cautious was taken to avoid any overlap between the new and the previous sampling locations.

Table 1: *Summary of soil characteristics assigned to the grazing treatments*

Sub treatments	Depth (cm)	Area (ha)	Slope %	Size fraction (%)			pH	EC (mS)
				sand	silt	clay		
TC grazing								
T.deep flat	40	50	10.2	34.6	28.7	36.7	5.9	0.07
T.shallow sloppy	28	42	15.3	28.1	34.0	38.0	5.4	0.03
Continuous grazing								
C.deep flat	42	128	10.0	31.6	31.0	37.4	5.9	0.08
C.shallow sloppy	27	110	14.8	45.6	25.3	29.1	5.1	0.06

T: time-controlled grazing; C: continuous grazing

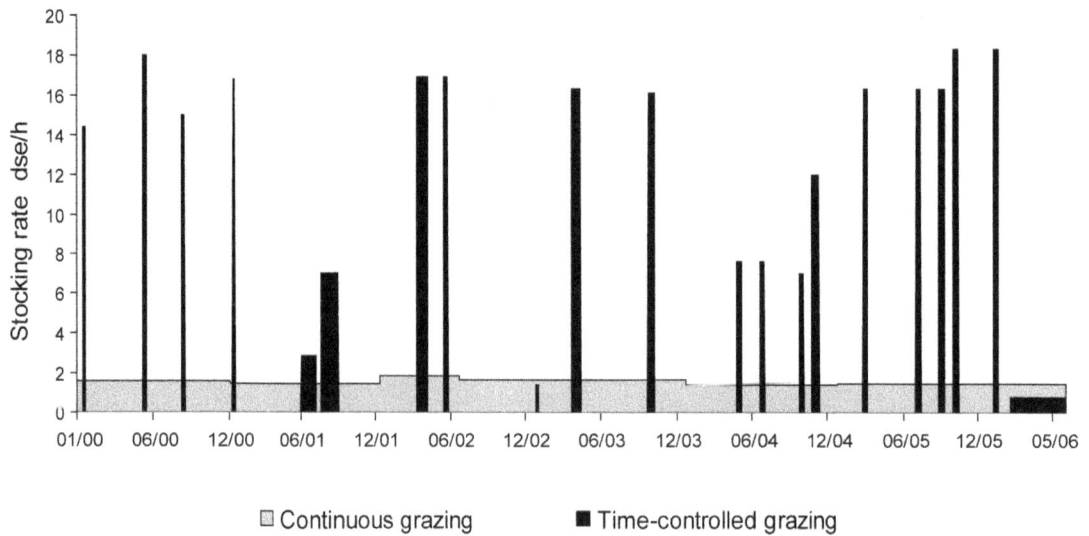

Fig. 1: *Stock density, grazing durations and rest periods under time-controlled (dark bars) and continuous grazing (grey area) systems*

Table 2: *Mean stocking details for the two grazing treatments (2000–2006)*

Grazing treatments	Grazing periods (days)	Rest periods (days)	SR (dse/ha)	DDH dse.day/ha
Time-controlled	14 ± 9 [‡]	(101 ± 60) [‡]	12.6 ± 6 [‡]	3608
Continuous	365	0	1.6 ± 0.2	3529

DDH- Number of dse days per hectare over the whole study period; dse = dry sheep equivalent
[‡] Means ± SD; *SR*: Stocking rate

Herbage was sampled each year at the end of growth season (Mid February to first May) in both grazing treatments. These times were set based on the coincidence of rest periods with end of grass growth season so that there have been always 10 to 30 days before the commencement of the next grazing period in TC grazing. Herbage harvested at ground level comprised both green and dead materials of all existing plant species. The harvested material refers only to the total (green + dead) mass of standing plant materials and excludes individual measurements of the species in the quadrat. The samples were oven dried at 40°C and reported as unit weight of dry matter per hectare ($kg\,DM\,ha^{-1}$).

Ground cover in this paper refers to any non-soil materials remained on or near ground surface that protect the soil surface against erosive forces of raindrops and overland flow (McIvor *et al.*, 1995). It includes any form of living and dead plant material as well as dung and stones. Ground cover in soil erosion studies has some advantages over canopy cover. While both the ground

and canopy covers protect the soil against raindrop impact, only ground cover effectively interrupts overland flow and bears a fraction of its flow shear stress thus reducing soil erosion (Proffitt & Rose, 1991).

The definition of ground cover is originally based on the commonly used method of aerial plant cover (Greig-Smith, 1983) measuring the proportion of the ground occupied by perpendicular projection of the aerial parts of plants plus other non-soil components. To estimate ground cover from the randomly laid quadrates, the results of two methods of Visual estimation (Zhou *et al.*, 1998) and digital image analysis (Abramoff *et al.*, 2004) were averaged. For ground cover estimation at any permanent site, 5 out of 10 replications were assessed by visual estimation and the remaining 5 replications by digital image analysis. Visual estimation gives a relatively quick and reliable estimate of ground cover compared with those obtained by more objective and time consuming methods (Murphy & Lodge, 2002; Vanha-Majamaa *et al.*, 2000).

For image analysis, a digital camera with focal length of 35 mm was used. The photographs were taken vertically from 160 cm above the centre of the quadrates and analysed by ImageJ, a Java image processing program. The process is based on grayscale image where white pixels correspond mostly to bare ground and as the colour turns to grey and black, it includes stone, litter and standing vegetation. For ground cover to be measured by digital ImageJ analysis, the most appropriate cut-off value distinguishing the bare and non-bare ground areas needs to be determined. This threshold could be obtained by crosschecking the binary images produced under a range of different cut-off values of grey colour intensity. While ImageJ has been widely used for digital image analysis in biology, no records of such application on rangeland monitoring were found in literature.

2.4 Data analysis

A two tails T-test analysis of variance was used to compare the paired values of herbage mass and ground cover taken at the beginning and at the end of the study period. This analysis simply examines the differences between the means of the two groups of samples. The second test employs regression lines analysis of variance (Tsutakawa & Hewett, 1978) that compares the overall changes in herbage mass over time using the slopes and intercepts of two lines corresponding to the grazing treatments. All the data analyses were carried out using Statistix9 and SPSS 15.

3 Results

3.1 Herbage mass

The records of herbage mass sampled at the beginning of the study in 2001 (Fig 2a) showed an almost equal herbage mass (1.9–2.0 ton DM ha^{-1}) for the two grazing treatments. This was expected as before the start of this research both paddocks had been grazed continuously for a very long period of time. Herbage yield is in perfect relation with rain received between any two consecutive harvest times in both grazing treatments but with higher values achieved with TC grazing than continuous grazing. It should be noted that the total rainfall in 2004 was 23 % above the long term average and lead to a production of 3.25 ton herbage DM ha^{-1} under TC grazing but only 2.2 ton herbage DM ha^{-1} under continuous grazing.

Herbage mass under TC grazing (Fig 2a) fluctuated over the years, peaked on March 2004 and then sustained at 2.7 ton DM ha^{-1} in 2006 which was significantly higher (p $\leq$ 0.01, Table 3) than the initial mass in

May 2001. Under continuous grazing the same pattern of herbage fluctuations as under TC grazing was found. However, under continuous grazing the herbage mass in May 2006 only reached to 2.2 ton DM ha^{-1} and was not significantly different from the initial herbage mass in May 2001. When the soil condition is favourable, as it is relatively the case under deep flat, the response to the grazing treatments is more pronounced than under shallow sloppy soil conditions. As it is shown in Fig 2b and Table 3, the increased herbage mass in T.deep flat from 2001 to 2006 accounts for 966 kg DM ha^{-1} while it reached to 314 kg DM ha^{-1} in C.deep flat. Under less favourable soil conditions (shallow sloppy), TC grazing again displayed a higher gain in herbage mass (560 kg DM ha^{-1}) than continuous grazing (106 kg DM ha^{-1}) by the final year.

Table 3 also shows the mean herbage mass and the trends of herbage accumulation along with the regression line parameters for the grazing treatments. The table illustrates the mean herbage accumulation of 2.45 ton DM ha^{-1} with gradient 0.43 for TC grazing and 2.1 ton DM ha^{-1} with gradient 0.17.

The results of regression line analysis (table 4) is in line with the outcomes of the T test ANOVA presented earlier and show that TC grazing produced a higher level of herbage mass than continuous grazing in the study area.

3.2 Ground cover

Ground cover showed a general decrease from 2001 to the end of 2003 (period 1) followed by an increase from 2004 to 2006 (period 2) under both grazing systems. In the first period, ground cover declined 10 % under TC grazing and 17 % under continuous grazing (Fig 4a). During the second period however, the ground cover under TC grazing increased to 75 % in 2004 to reach 90 % by 2006. Under continuous grazing it reached 62 % in 2004 ending up at 68 % in 2006. Ground cover in the sub-treatments (Fig 4b) relatively demonstrated the same decrease trends during period 1 with a lower rate of decrease in T.deep flat than the others. However over the second period a significant higher increase in ground cover was observed under TC grazing than under continuous grazing.

The results of a T test analysis of variance displayed in table 5 show details of the comparisons made on the ground cover achieved by the grazing treatments over the first and the last halves of the study period. For any comparison in the matrix, two values are given with the associated p values underneath. The table shows that ground cover in T.deep flat was significantly

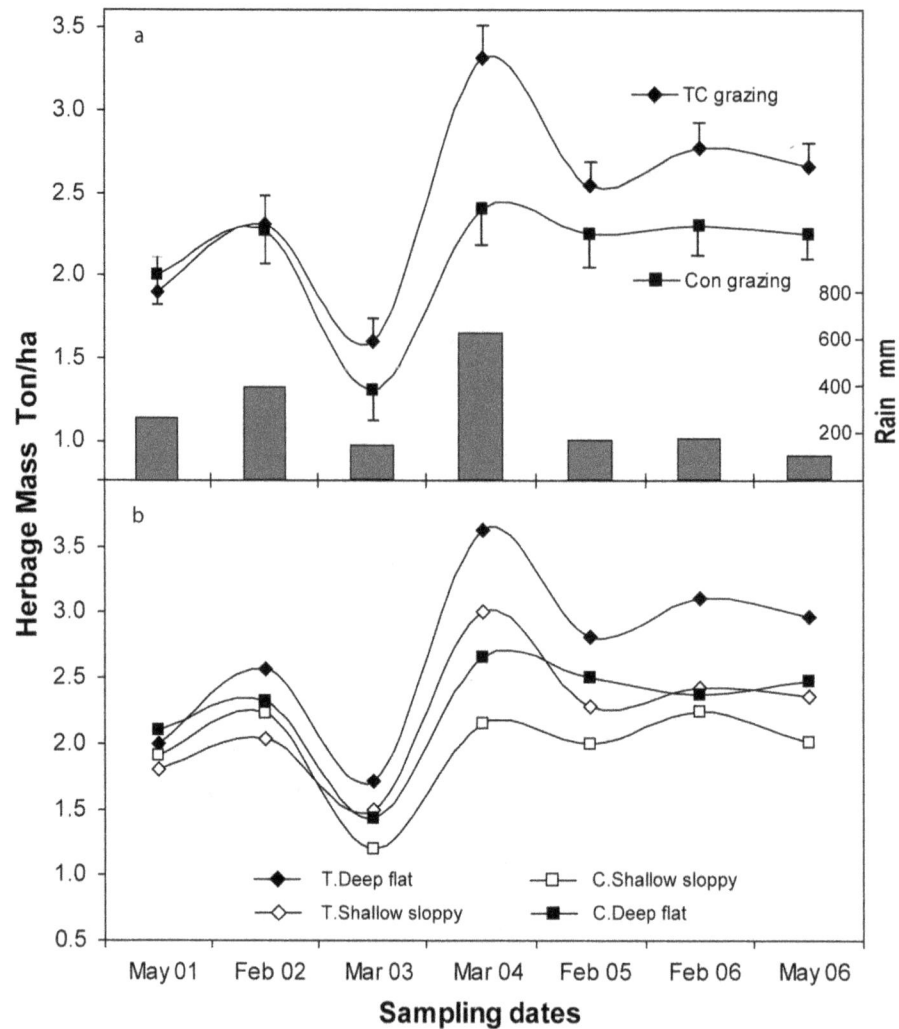

Fig. 2: *Herbage mass and the total rain received from the previous sampling date for the grazing treatments (a) and the sub-treatments (b); error bars show SDs*

Table 3: *Summary of herbage mass analysis for the period of May 2001 to May 2006*

Treatments	Herbage $(\mathrm{kg\,DM\,ha^{-1}})$ [†]	Herbage $(\mathrm{kg\,DM\,ha^{-1}})$ [‡]	r	slope [§]	intercept [§]
T.deep flat	+966 ***	2671 ± 615	0.75 **	0.51	2165
T.shallow sloppy	+560 **	2231 ± 438	0.63 *	0.35	1885
TC grazing (overall)	+764 ***	2451 ± 525	0.69 **	0.43	2025
C.deep flat	+314 ns	2279 ± 371	0.45 ns	0.24	2043
C.shallow sloppy	+106 ns	1918 ± 325	0.23 ns	0.11	1812
Con. grazing (overall)	+240 ns	2098 ± 318	0.38 ns	0.17	1928

[†] Results of T test ANOVA and the values are the products of the mean herbage harvested in May 2006 minus the mean herbage in May 2001;
[‡] Mean ± SD, is the mean of herbage harvested per sampling ± SD
r: Correlation coefficient; [§] Regression line identities; *: $p \leq 0.10$; **: $p \leq 0.05$; ***: $p \leq 0.01$; ns: not significant
T: time-controlled grazing; C: continuous grazing

Table 4: *The results of regression line analysis between treatments and sub-treatments*

Grazing Treatments	T.deep flat	T.shallow sloppy	C.shallow sloppy	TC grazing (overall)
T.shallow sloppy	6.22* ; 4.32*			
C.shallow sloppy	16.32*** ; 10.35**	5.44* ; 3.22*		
C.deep flat	7.42* ; 3.62*	1.31ns ; 1.06ns	0.47ns ; 3.92*	
Continuous grazing (overall)	—	—	—	7.69* ; 3.79*

The paired values in the table represent F_{slope}; $F_{intercept}$, respectively. T: time-controlled grazing; C: continuous grazing; * $p \leq 0.10$; ** $p \leq 0.05$; *** $p \leq 0.01$; ns: not significant

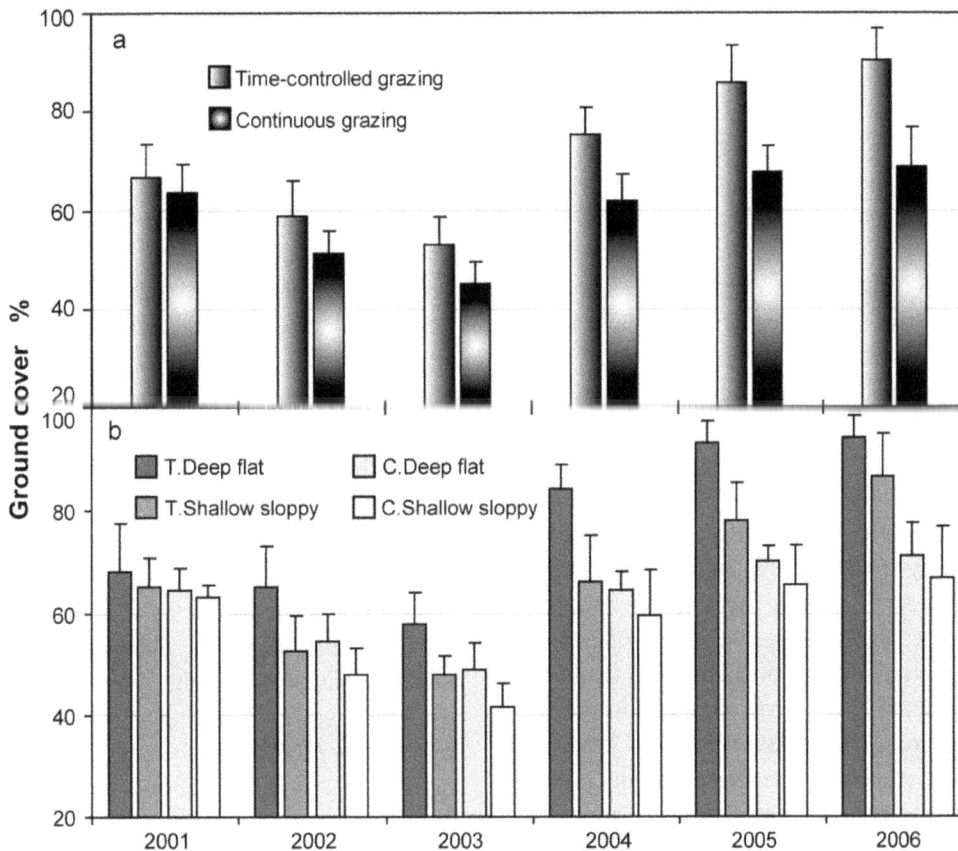

Fig. 3: *The percentage of ground cover achieved under time-controlled and continuous grazing in the study area*

Table 5: *Ground cover analysis of variance between sub-treatments for the first (2001–2003) and the second (2004–2006) periods of the study*

Sub-treatments	T.deep flat		T.shallow sloppy		C.deep flat	
	1st	2nd	1st	2nd	1st	2nd
T.shallow sloppy	%9 ***	%16 ***				
C.deep flat	%8 ***	%22 ***	%1 ns	%6 ***		
C.shallow sloppy	%14 ***	%27 ***	%5 ns	%11 ***	%6 *	%5 **

Values in the table indicate differences between the sub-treatments;
$p < 0.1$; ** $p \leq 0.05$; *** $p < 0.01$; ns: non significant; T: time-controlled grazing; C: continuous grazing

higher than the other sub-treatments including T.shallow sloppy, both in the first and in the second periods of the study. The T.shallow sloppy received higher significant cover than the two continuous grazing sub-treatments only over the second period.

The significant increase in ground cover in 2004 as compared with 2003 for both treatments (Fig 3a) is most likely associated with the high rainfall in 2004. Although the annual rainfall declined to somewhat below the long term average in 2005 and 2006 (Fig 2a), the ground cover under TC grazing continued to increase leading to significant different levels compared to continuous grazing over the last two years of the study (Fig 3).

4 Discussion

Herbage mass, as the primary source of above ground organic materials, provides grazing animals and soil micro-organisms with nutrient and energy and protects soil by increasing litter and ground cover. The improvement in herbage mass and ground cover under TC grazing could be mainly attributed to the adequate long rest periods, appropriate grazing durations and stocking rates. The periods of animal exclusion from paddocks, are necessary for maintaining plant vigour, seeding, seedling establishment and plant recovery after defoliations (Wilson et al., 1984; Lodge, 1995). In order for plant recovery to be fully implemented, the length of rest periods under TC grazing should be long enough to decrease and control the negative effects of intensive defoliations on soil conservation. The length of rest period or defoliation interval has a direct positive effect on herbage mass (Hill, 1989; Binnie & Chestnutt, 1991). In our experiment, the periods of grazing exclusion differed with time and on average lasted around 3 months. While the applied rest periods in the research paddocks seem to be adequate, a scientifically based threshold has yet to be developed for this region.

Stocking rates and grazing durations are also critical factors to manage a successful outcome by TC grazing. The physiological research on plants under intensive single defoliation (common under TC grazing) of about 50% of shoot volume, showed that the root growth was retarded for 6–18 days in 7 out of the 8 perennial grasses investigated (Crider, 1955). Richards (1993), who reviewed a large number of studies in relation to plant physiological responses to defoliation, believes that following an intensive defoliation, the plants go through a temporal phase of rapid changes in available carbon and nitrogen ultimately resulting in partial root mortality. This mechanism restores the shoot:root ratio of the plant required to begin a recovery process with a fast photosynthesis rate.

Apart from the physiological processes involved in plant recovery, the roots pruned and decayed this way are added to soil profile as the main source of organic matter. Depending on the intensity of defoliations under TC grazing, the amounts of root pruned could be substantial, leaving a large amount of soil pores after dying off facilitating infiltration.

Over the processes of fast photosynthesis rates, if the defoliated plants have access to adequate water and nutrients, the recovery would be fully implemented. In the southeast Queensland including the study area, there is a high chance of having a number of consecutive rainfall events in summer during the long rest periods. This provides the pasture with the maximum transpiration and nutrient uptake during the fast recovery period. Over the growth season of 2003–2004, our paddock of TC grazing was rested for the longest duration of 156 days, during which it received a total rainfall of 480 mm. Although such an excessively long duration of grazing exclusion is somewhat exceptional, the coincidence of the rest period with the favourable wet condition resulted in the highest level of herbage mass in 2004 (Fig 2a). Following this period, the grazier applied more frequent but shorter grazing durations which sustained herbage mass and ground cover until the end of the study period resulting in a far better outcome than under continuous grazing.

The increase in ground cover over 2005–2006 under TC grazing shows the incremental additive effect of herbage mass on ground cover implemented either by physical action of grazing animals or by natural decay. McIvor (2002) who found a logarithmic relationship between ground cover and herbage mass showed that herbage growth can occur even when ground cover is 100%. Such cumulative effect of herbage mass on ground cover under TC grazing practice can build up an adequate layer of surface organic materials during periods of favourable growth conditions providing soil protection in subsequent years. The results on ground cover in this study are supported by the earlier results on ground litters (Sanjari et al., 2008) and soil erosion control (Sanjari et al., 2009) under TC grazing.

In contrast to TC, under continuous grazing a lower herbage mass was found as well as a lower ground cover leaving the soil surface with less organic protective layer against water erosion. Continuous grazing even at light stocking rate exert high pressure on desirable species (Beattie, 1993; Tainton & Walker, 1993; Lodge, 1995; Parsons, 1995) and leave the rest of the plants un-grazed.

The results of this study have shown that under prevailing condition of the study area, time-controlled grazing has a potential to significantly increase total herbage mass compared with yearlong continuous grazing despite the fact that the total DDH (dse.days/ha) numbers for the two grazing practices were similar. Time-controlled grazing also achieved a high significant level of ground cover, providing a reliable layer of organic material to protect the soil surface against water erosion. The improvement in total herbage mass and ground cover under TC grazing is largely attributed to the proper inclusion of grazing frequency and duration, as well as the provision of adequate rest periods in the study area.

Acknowledgements

The work is dedicated to Mr. Cyril Cisiolka who diagnosed with leukaemia and died in 2013. He was passionate about time-controlled grazing and had major contribution to this paper. The authors acknowledge the Queensland Inglewood Landcare and Natural Heritage Trust for their funds and support. They also thank Rick & Louise Goodrich, the owners of the property and Mr. Eugene Creek for his assistance with field work.

References

Abramoff, M. D., Magelhaes, P. J. & Ram, S. J. (2004). Image processing with ImageJ. *Biophotonics International*, 11, 36–42.

Alsemgeest, V. & Alchin, B. (2002). Comparison of continuous and cell grazing on brigalow country in central Queensland. *In:* Nicholson, S. & Wilcox, D. (eds.), *The 12th Biennial Conference of the Australian Rangeland Society, 2002, Kalgoorlie, Western Australia.* pp. 265–266, Australian Rangeland Society.

Beattie, A. S. (1993). Grazing for grassland management in the high rainfall, perennial grassland zone of Australia. *In:* Kemp, D. R. & Michalk, D. L. (eds.), *Grassland management: technology for the 21st century.* pp. 62–69, CSIRO, Melbourne.

Binnie, R. C. & Chestnutt, D. M. B. (1991). Effect of regrowth interval on the productivity of swards defoliated by cutting and grazing. *Grass and Forage Science*, 46, 343–350.

Crider, F. J. (1955). Root growth stoppage resulting from defoliation of grass. USDA, No. Technology Bulletin 1102.

Detterling, D. (1999). Daily rotation raises carrying capacity 450%. *In:* The Progressive Farmer. Vol. March 1999. pp. 65–68.

Dyksterhuis, E. J. (1949). Condition and management of rangeland based on quantitative ecology. *Journal of Range Management*, 2, 104–115.

Gammon, D. M. (1978). A review of experiments comparing systems of grazing management on natural pastures. *Procedings of Grassland Society of Southern Africa*, 13, 75–82.

Gillen, R. L., Mccollum, F. T., Hodges, M. E., Brummer, J. E. & Tate, K. W. (1991). Plant community responses to short duration grazing in tallgrass prairie. *Journal of Range Management*, 44, 124–128.

Greig-Smith, P. (1983). Quantitative plant ecology, 3rd edn. *WileyBlackwell, Oxford.*

Hill, M. J. (1989). The effect of differences in intensity and frequency of defoliation on the growth of *Phalaris aquatica* L. and *Dactylis glomerata* L. *Australian Journal of Agricultural Research*, 40 (2), 333–343.

Holechek, J. L., Pieper, R. D. & Herbel, C. L. (1998). *Range Management: Principles and Practices. 3rd edn.* Prentice Hall College Div, New Jersey.

Johnson, G. D. (1998). Controlled grazing: it's an attitude. *Rangelands*, 20, 24–25.

Joyce, S. (2000). Change the management and what happens - a producer's perspective. *Tropical Grasslands*, 34, 223–229.

Lodge, G. M. (1995). The role of grazing management in sustaining the pasture community. *In:* Ayres, J. F., Michalk, D. L. & Davies, H. L. (eds.), *Proceedings of the 10th Annual Conference of the Grassland Society of NSW, 1995, Armidale.* pp. 34–42, Grassland Society of NSW.

McIvor, J. G. (2002). Pasture management in semi-arid tropical woodlands: effects on ground cover levels. *Tropical Grasslands*, 36, 218–226.

McIvor, J. G., Williams, J. & Gardener, C. J. (1995). Pasture Management Influences Runoff and Soil Movement in the Semiarid Tropics. *Australian Journal of Experimental Agriculture*, 35, 55–65.

Murphy, S. R. & Lodge, G. M. (2002). Ground cover in temperate native perennial grass pastures. I. A comparison of four estimation methods. *Rangeland Journal*, 24 (2), 288–300.

Norton, B. E. (2003). Spatial management of grazing to enhance both livestock production and resource condition: A scientific argument. *In:* Allsopp, N., Palmer, A. R., Milton, S. J., Kirkman, K. P., Kerley, G. I. H., Hurt, C. R. & Brown, C. J. (eds.), *VIIth International Rangeland Congress, 2003, Durban, South Africa.* pp. 810–820, International Rangeland Congress.

Parsons, S. D. (1995). *Putting profit into grazing.* Resource Consulting Services, Yeppoon, Qld.

Proffitt, A. P. B. & Rose, C. W. (1991). Soil-Erosion Processes. 1. The Relative Importance of Rainfall Detachment and Runoff Entrainment. *Australian Journal of Soil Research,* 29, 671–683.

Richards, J. H. (1993). Physiology of plants recovering from defoliation. *In:* Baker, M. J. (ed.), *The 17th Internatinal Grasslands Congress, 1993, Wellington, New Zealand.* pp. 85–94.

Sanjari, G. (2008). *Evaluating the Impacts of Time Controlled and Continuous Grazing Systems on Soil Properties, Surface Hydrology and Pasture Production in South-East Queensland.* Ph.D. thesis, Griffith University, Australia.

Sanjari, G., Ghadiri, H., Ciesiolka, C. A. A. & Yu, B. (2008). Comparing the effects of continuous and time-controlled grazing systems on soil characteristics in Southeast Queensland. *Australian Journal of Soil Research,* 46, 348–358.

Sanjari, G., Yu, B., Ghadiri, H., Ciesiolka, C. A. A. & Rose, C. W. (2009). Effects of Time-Controlled Grazing on Runoff and Sediment Loss. *Australian Journal of Soil Research,* 47, 796–808.

Savory, A. & Parsons, S. D. (1980). The Savory grazing method. *Rangelands,* 2, 234–237.

Sayre, N. F. (2001). *The New Ranch handbook: A guide to restoring western rangelands.* The Quivira Coalition, Sante Fe, New Mexico.

Sparke, R. (2000). Cell Grazing - a producer's perspective. *Tropical Grasslands,* 34, 219–222.

Suther, S. (1991). Rethinking the ranch. *Beef Today,* Vol. February 1991, 44–49.

Tainton, N. M. & Walker, B. H. (1993). Grasslands of southern Africa. *In:* Coupland, R. T. (ed.), *Ecosystems of the world 8B: Natural Grasslands. 1st edn.* pp. 265–290, Elsevier, Amsterdam.

Tsutakawa, R. K. & Hewett, J. E. (1978). Comparison of Two Regression Lines Over a Finite Interval. *Biometrics,* 34, 391–398.

Valentine, J. F. (2000). *Grazing Management. 2nd edn.* Academic Press, San Diego, California.

Vanha-Majamaa, I., Salemaa, M., Tuominen, S. & Mikkola, K. (2000). Digitized Photographs in Vegetation Analysis: A Comparison of Cover Estimates. *Applied Vegetation Science,* 3, 89–94.

Wilson, A. D. (1986). Principles of grazing management systems. *In:* Joss, P. J., Lynch, P. W. & Williams, O. B. (eds.), *Rangelands: A Resource Under Siege. Proceedings of the Second International Rangeland Congress, 1986, Canberra.* pp. 221–225.

Wilson, A. D. & Harrington, G. N. (1984). Grazing ecology and animal production. *In:* Harrington, G. N., Wilson, A. D. & Young, M. D. (eds.), *Management of Australia's rangelands.* pp. 63–77, CSIRO, Melbourne.

Wilson, A. D., Harrington, G. N. & Beale, I. F. (1984). Grazing management. *In:* Harrington, G. N., Wilson, A. D. & Young, M. D. (eds.), *Management of Australia's rangelands.* pp. 129–139, CSIRO, Melbourne.

Zhou, Q., Robson, M. & Pilesjo, P. (1998). On the ground estimation of vegetation cover in Australian rangelands. *International Journal of Remote Sensing,* 19, 1815–1820.

15

Technical efficiency and production potential of selected cereal crops in Senegal

Yodai Okuyama [a], Atsushi Maruyama [b,*],
Michiko Takagaki [b], Masao Kikuchi [b]

[a]*RECS International Inc., Chiyoda-ku, Tokyo, 102-0075, Japan*
[b]*Graduate School of Horticulture, Chiba University, Matsudo, Chiba, 271-8510, Japan*

Abstract

This study focused on the production outcomes for five crops cultivated in Senegal: upland rice, lowland rice, groundnut, maize, and pearl millet. Technical efficiency (TE) of the production of each crop was estimated using data envelopment analysis, and the determinants of TEs were assessed using generalised linear regression analyses. Data were collected in face-to-face interviews with 66 farmers in the Kaolack region of Central Senegal during November 2011–February 2012. Average TEs for upland rice, lowland rice, groundnut, maize, and pearl millet were estimated as 0.76, 0.88, 0.89, 0.94, and 0.90, respectively. The identified factors that had a positive impact on TE were years of cultivation experience, amount of nitrogen fertiliser applied, and participation in a farmers' association. Weeding hours, seeding rate, size of the cultivated area, and delays in sowing time were negatively associated with TE. The factors that significantly affected TE differed among the crops. Optimising these factors could enable potential yield increase of upland rice, lowland rice, groundnut, maize, and pearl millet by 24, 12, 11, 6, and 10 %, respectively.

Keywords: production function, agricultural extension, data envelopment analysis (DEA), rice, West Africa

1 Introduction

More than 54 % of Senegalese people live in rural areas (ANSD, 2014), and almost 25 % of these are estimated to suffer from malnutrition (FAO *et al.*, 2015). Based on the estimation that the national population will reach 23 million in 2030 (UN, 2015), the number of malnourished people is expected to increase to more than 3 million by 2030. Therefore, the Senegal government put high priority to increase agricultural productivity (FAO, 2015; IMF, 2013).

The main crops cultivated in Senegal are rice, groundnut, maize, and pearl millet. Rice is one of the most important crops with an annual consumption per capita of 91 kg in 2014/2015 (USDA, 2015). However,

Senegal produces only 20–30 % of the total consumed rice (USDA, 2015). To improve the production, the Senegal government has set a national goal to produce 1.6 million t in 2017. Although this is a challenging target as production in 2015 was only 623,000 t (USDA, 2016), it is important for Senegal to achieve this goal because the price of imported rice is very volatile and large price variations make the livelihoods of Senegalese people unstable.

Crop yield can be increased in two ways: expanding the cultivated area and/or increasing the yield per unit of harvested area. Expanding the area under cultivation is almost not feasible in Senegal as the available uplands are being cropped already. Competition for uplands is very intense, and according to our interviews, almost all the surveyed farmers cultivate their uplands continuously, without a sufficient fallow period (Grosenick *et al.*, 1990; Diop, 1999). Furthermore, labour shortages

* Corresponding author
Address: Matsudo, Chiba, 271-8510, Japan
Email: a.maruyama@faculty.chiba-u.jp

restrict expansion of the lowlands. This labour shortage is related to climatic conditions, as the demand for farming labourers is largely restricted to the four months rainy period. As there are few employment opportunities in rural areas during dry season, people tend to migrate to urban areas or overseas to look for jobs. Once people have moved away from rural areas, they are unlikely to return home and tend to stay in their new location to maintain employment. Consequently, only 37 % of the potentially cultivable lowland area is in use for crop production in Senegal (Frenken, 2005).

Per-area yield can be improved by optimising farming (i.e. sowing, fertiliser application, or weeding), which requires little additional cost. Increasing the amount and quality of production inputs will increase the per-area yield as well. However, the purchase of inputs as chemical fertiliser or high quality seeds is restricted as Senegalese farmers do not have adequate financial resources and their access to financial services is limited (AFAP & IFDC, 2014). Therefore, an effective option to increase yield may be to adjust and optimise current farming practices.

The concept of technical efficiency (TE) is very useful for comparing levels of production efficiency, and TE can be used to identify the factors to improve the productivity of a decision-making unit (e.g. farm). Moreover, TE is useful to evaluate the disparity of technical level between individual farmers for each crop. This can be helpful to identify higher priority crops or farmers requiring technical support. Therefore, many studies have investigated the TE of farms in West Africa, but only three studies have been conducted in Senegal so far (see Table 4).

In the present study, we estimated TEs for five selected crops that are cultivated widely in Senegal: upland rice, lowland rice, groundnut, maize, and pearl millet. We examined the main factors influencing TE using a regression analysis with a generalised linear model. Our analyses indicated the change in farming practices required to improve crop yield. These findings will be helpful for extension programmes targeting resource poor farmers to increase their crop productivity.

2 Materials and methods

2.1 Study area and data collection

The study area was the Médina Sabakh community, located at 13°36′ N 15°35′ W in the Kaolack region, Central Senegal. This community had 34,263 persons living in 4,104 households in 2014–2015 (data from the local authority). There are two seasons: a rainy season from the end of June to the middle of October; and a dry season from the end of October to the middle of June. Rainfall occurs only during the rainy season, and the average annual precipitation is 766 mm (1988–2014, data from the Ministry of Agriculture in Senegal). A preliminary survey revealed that farmers usually cultivate rice, groundnut, maize, or pearl millet during the rainy season. Groundnut is cultivated as the main cash crop, and the other crops are mainly grown for personal consumption.

The survey was conducted from 27 November 2011 to 12 February 2012 through face-to-face interviews using structured questionnaires. The farmers included in the survey were selected randomly from all farms that cultivated rice in 2010 in the Médina Sabakh community, according to our preliminary survey. In total, 66 farmers were interviewed about the crops cultivated during the rainy seasons from 2009 to 2011 (upland rice, lowland rice, groundnut, maize, or pearl millet). The average number of crops cultivated was 2.3 per farm. The number of valid responses was 33 for upland rice, 19 for lowland rice, 38 for groundnut, 29 for maize, and 36 for pearl millet.

The items identified for the survey were age of the head of the household, number of family members working on the farm, cultivated area of the respective crops (ha), yield of each individual crop (t ha^{-1}), amount of nitrogen fertiliser applied (kg-N ha^{-1}), seeding rate (kg ha^{-1}), number of hours spent weeding (person h ha^{-1}), years of experience cultivating the particular crops (years), seeding sequence of crops on the farm, and experience of participating in a farmers' association. Data on yield and production inputs were based on the largest field of each crop cultivated by each farmer.

The amount of nitrogen applied was estimated by multiplying the weight of chemical fertiliser and animal manure by the amounts of nitrogen in these substances. The farmers provided the figure for the amount of nitrogen in the chemical fertiliser (6–15 %) and the amount of nitrogen in animal manure was set at 2.6 % as reported by Pratt & Castellanos (1981).

The time spent weeding (total weeding hours per ha) was used as a measure of total labour input. The weeding hours per ha were calculated from the time spent on inter-tillage weeding with animal traction hoes plus the time spent on manual weeding with hand hoes. Labour inputs from other farming practices, such as scaring away birds or applying fertilisers, were not included in the analysis for the following reasons. First, although scaring birds is a common and labour-intensive activity in rice and pearl millet farming in Africa, the farmers in the sample area did not follow this practise. Instead, they try to minimise bird damage by synchronising the maturation and harvest periods of rice and pearl millet. When synchronisation fails, the maturation and harvesting periods extend and damages from bird increase. Second, the amount of labour involved in fertiliser application is relatively low. Le Moigne (1980) estimated that labour input for fertiliser application was the equivalent of only 2–6 % of the total labour input in Senegal. In addition, our preliminary survey indicated that the labour input for fertiliser application was perfectly correlated with the amount of the fertiliser applied (unpublished).

The years of cultivation-experience possessed by the head of the farm household was included only in the analysis of upland and lowland rice production because we could not obtain reliable data for the other crops. The head of the household had cultivated crops other than rice from their youth on; the years of cultivation experience were closely related to their ages.

2.2　Analytical framework

2.2.1　Technical background

TE is an indicator that is defined as the ratio of a measured production level to the potential production level with given level of inputs and production technology (Farrell, 1957; Coelli et al., 2005). The potential production level is on the frontier production function estimated as an envelopment surface of observed production data (Coelli et al., 2005). Thus, a perfectly efficient farm has TE = 1, whereas an inefficient farm has $0 \leq TE < 1$. The value $1 - TE$ indicates the inefficiency level of a farm.

The frontier production function and TE can be estimated by two different approaches: a non-parametric approach, such as data envelopment analysis (DEA), or a parametric approach, such as stochastic frontier analysis (SFA). TE estimation by DEA is appropriate when the distribution of TE is not known a priori, and has an advantage in terms of identifying efficient farmers, who can act as role models for inefficient farmers. Therefore, DEA is applied in this analysis.

2.2.2　Data envelopment analysis

DEA has two calculation models: the variable returns to scale (VRS) model and the constant returns to scale (CRS) with an input-orientation or an output-orientation. In the survey area, as scale merits in cereal productions do not exist (Kelly et al., 1996),

CRS model is employed. Although an input-oriented or an output-oriented assumption is selected based on a purpose to measure TE by a proportional reduction in input usage or a proportional increase in output production, the two measuring methods provide the same value of TE under CRS model (Coelli, 1996). The input-oriented CRS DEA model applied in this study is specified as:

$$\min_{\theta_{im}, \lambda_{i1} \cdots \lambda_{in_i}} \theta,$$

subject to

$$-y_{im} + \sum_{j=1}^{n_i} y_{ij}\lambda_{ij} \geq 0,$$

$$\theta x_{kim} - \sum_{j=1}^{n_i} x_{kij}\lambda_{ij} \geq 0,$$

$$\lambda_{ij} \geq 0, i = 1, 2, \ldots, 5; j = 1, \ldots, m, \ldots, n_i; k = 1, 2, 3$$

where θ is TE of the ith crop of the jth farmer, y_{im} is the output of the ith crop of an observed farmer, λ_{ij} denotes weights which define the linear combination of the peer of the ith crop of the jth farmer, x_{kim} is the kth input of the ith crop of an observed farmer, and x_{kij} is the kth input of the ith crop of the jth farmer. Estimation is carried out by using the program DEAP version 2.1.

2.2.3　TE distribution

A crop having low TE has large potential of increasing production and therefore high priority in terms of agricultural extension. In order to decide the priority of crops, an average TE of each crop was estimated and the mean difference was tested by the Tukey-Kramer method. Furthermore, the TE distribution of each crop was compared using the kernel density distribution, which was calculated using Analytical Methods Committee software with a Gaussian kernel and a bandwidth h where,

$$h = 0.9 \times \min(\text{sample standard deviation}, \text{IQR}/1.34) \times n^{-1/5}$$

where IQR denotes the inter-quartile range of the data.

2.2.4　Determinant analysis of TE

We identified TE determinants through regression analyses. Since the dependent variable, TE, is bound within the range 0–1, it was not appropriate to use an ordinary least square method. Therefore, we applied a generalised linear model with logit link function:

$$\ln\left(\frac{\text{TE}_i}{1 - \text{TE}_i}\right) = \beta_{i0} + \sum_{l=1}^{7} \beta_{il}X_{il} + w_i$$

where TE_i is estimated technical efficiency of i th crop, X_{il} represents l th explanatory variable of the i th crop, β_{il} is unknown parameters to be estimated, and w_i is the error term. $t\,ha^{-1}$ To estimate the TE, amount of nitrogen applied (kg-$N\,ha^{-1}$), seeding rate ($kg\,ha^{-1}$), and weeding hours ($person\,h\,ha^{-1}$) were used as inputs; the yield per cultivated area ($t\,ha^{-1}$) was used as the output variable. Possible crop variety effects were not accounted for.

3 Results

3.1 Descriptive statistics

The basic data collected are presented in Table 1. Lowland rice had the highest yield ($1.5\,t\,ha^{-1}$), followed by maize ($1.0\,t\,ha^{-1}$) and the other crops (0.59–$0.73\,t\,ha^{-1}$). Nitrogen was intensively applied for upland rice and maize (72–$74\,kg\,ha^{-1}$), and it was mainly derived from chemical fertiliser. The seeding rate for groundnut ($71\,kg\,ha^{-1}$) was the highest, followed by upland rice, lowland rice, and maize (18–$28\,kg\,ha^{-1}$). The seeds of all plants were directly sown with a seeder pulled by draft animal. More hours were spent weeding in upland rice and lowland rice than in the other crops. The cultivated areas of groundnut and pearl millet were the highest among the crops. Years of cultivation experience were 2.51 and 3.81 for upland rice and lowland rice, respectively.

3.2 TEs and their distributions

The TEs obtained for the five crops are shown in Table 2. The average TEs of upland rice, lowland rice, groundnut, maize, and pearl millet were 0.76, 0.88, 0.89, 0.94, and 0.90, respectively. From these results, crop yields may be potentially increased in the range of 6 % (maize) to 24 % (upland rice). The TE for upland rice was significantly lower than those for the other crops.

The TE distributions for upland and lowland rice have two peaks (Fig. 1a, b). The peaks in upland rice appear at 0.5 and 0.8 and the distribution has a long tail. In contrast, the peaks in lowland rice appear between 0.8 and 1.0 and the variation is small. The distributions of the other crops have a single peak at 0.9 (Fig. 1c, d, e). The pooled TEs of all five crops had a single peak at 0.9 (Fig. 1f).

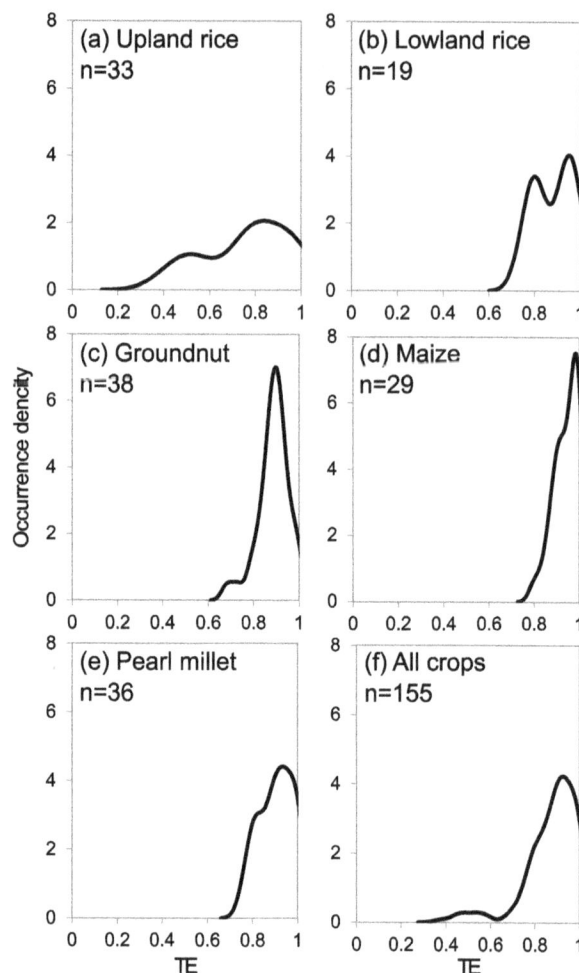

Fig. 1: *Distribution of technical efficiency (TE) estimated by kernel density in selected crops (a–e), and for all crops (f).*

3.3 Determinants of TE

Variables that might contribute to TE were examined using generalised linear regression analyses (Table 3). A significant positive factor affecting the TE of upland rice was year of cultivation experience ($P < 0.01$), whereas a significant negative factor was delay in sowing time ($P < 0.01$). For lowland rice, year of cultivation experience ($P < 0.01$) and participation in a farmers' association ($P < 0.05$) were positive factors, while weeding hours ($P < 0.01$), seeding rate ($P < 0.01$), cultivated land area ($P < 0.01$) and delay in sowing time ($P < 0.05$) were significant negative factors. For groundnut, the amount of nitrogen applied ($P < 0.05$) was a significant positive factor, while weeding hours ($P < 0.01$) and delay in sowing time ($P < 0.05$) were significant negative factors. For pearl millet, a significant positive factor was the amount of nitrogen applied ($P < 0.05$), while significant negative factors were weeding hours ($P < 0.05$) and seeding rate ($P < 0.01$). No significant variables were identified for maize production.

Table 1: *Basic statistics of the sample (average over the years 2009–2011).*

Categories	Variables	Upland rice	Lowland rice	Groundnut	Maize	Pearl millet
Output	Yield (t ha^{-1})	0.64 [0.55] [a]	1.5 [0.81] [c]	0.73 [0.31] [ab]	1.0 [0.54] [b]	0.59 [0.33] [a]
Inputs	Fertiliser (kg-N ha^{-1})	74 [49] [b]	16 [24] [a]	1.5 [2.9] [a]	72 [32] [b]	19 [17] [a]
	⟨% of N derived from chemical fertiliser⟩	⟨86⟩	⟨80⟩	⟨57⟩	⟨91⟩	⟨79⟩
	Seed (kg ha^{-1})	26 [16] [b]	28 [19] [b]	71 [26] [c]	18 [6.6] [b]	4.9 [2.8] [a]
	Weeding (h ha^{-1})	470 [230] [b]	430 [130] [b]	150 [57] [a]	150 [72] [a]	120 [39] [a]
Characteristics	Plot area (ha)	0.53 [0.33] [a]	0.61 [0.57] [a]	2.8 [2.1] [b]	1.3 [0.80] [a]	2.8 [1.5] [b]
	Cultivation experience (years)	2.51 [1.15] [a]	3.84 [2.14] [b]	n.a.	n.a.	n.a.
	Age (years)	54 [9.7] [a]	52 [10] [a]	52 [11] [a]	54 [11] [a]	52 [11] [a]
	No. family labourers (persons)	12 [7.2] [a]	14 [21] [a]	15 [17] [a]	15 [17] [a]	14 [16] [a]
Sample size		33	19	38	29	36
Share of the farmers (%)		50	29	58	44	55

[a,b,c] Means with a different superscript are significantly different (*P* < 0.05), Standard deviation in brackets; n.a. not applicable.

Table 2: *Summary statistics of technical efficiency.*

	Upland rice	Lowland rice	Groundnut	Maize	Pearl millet	Total
Mean	0.76 [a]	0.88 [b]	0.89 [b]	0.94 [b]	0.90 [b]	0.87
Std Dev.	0.19	0.09	0.07	0.05	0.07	0.12
Min	0.38	0.73	0.68	0.80	0.76	0.38
Max	1.00	1.00	1.00	1.00	1.00	1.00
Sample size	33	19	38	29	36	155

[a,b,c] Means with a different superscript are significantly different (*P* < 0.05)

Table 3: *Determinants of technical efficiency.*

	Upland rice	Lowland rice	Groundnut	Pearl millet
ln (Weeding hours)	n.s.	−1.65 ** [−5.04]	−1.10 ** [−3.26]	−0.73 * [−2.47]
ln (Amount nitrogen applied)	n.s.	n.s.	0.42 * [2.44]	0.24 * [2.35]
ln (Seeding rate)	n.s.	−0.93 ** [−2.57]	n.s.	−0.70 ** [−2.73]
ln (Years of cultivation experience)	1.04 ** [3.72]	1.25 ** [4.67]	n.a.	n.a.
ln (Cultivated area)	n.s.	−1.10 ** [−5.17]	n.s.	n.s.
Delayed sowing time (Dummy)	−0.74 ** [−2.57]	−0.50 * [−2.09]	−0.37 * [−2.37]	n.s.
Participation in a farmers' association (Dummy)	n.s.	0.85 * [2.36]	n.s.	n.s.
Constant	0.70 * [2.49]	12.6 ** [6.20]	7.46 ** [4.46]	6.16 ** [4.66]
n	33	19	38	36
AIC	−36	−56	−104	−93
BIC	−32	−49	−97	−87

* *P* < 0.05; ** *P* < 0.01; *t*-values in parenthesis;
AIC: Akaike's Information Criterion; BIC: Bayesian Information Criterion.

4 Discussion

4.1 Production output and inputs

The average yield of upland rice and lowland rice were 0.64 and 1.5 t ha^{-1}, respectively. These yields are considerably lower than the national averages of 3.92 t ha^{-1} (FAO Stat, 2009–2014), perhaps because rice crops in the study area were cultivated rainfed, whereas irrigation is widely used elsewhere. However, the estimated yield of 1.5 t ha^{-1} for lowland rice is equivalent to that reported for rainfed rice in West and Central Africa (Nin-Pratt et al., 2011). The estimated yields for groundnut, maize, and pearl millet were comparable with the national yield averages, 0.85, 1.41, and 0.70 t ha^{-1}, respectively (FAO Stat, 2009–2014), as well as with the regional averages for West and Central Africa, 0.83, 1.24, and 0.72 t ha^{-1}, respectively (Nin-Pratt et al., 2011).

The recommended amounts of nitrogen application for upland rice, lowland rice, groundnut, maize, and pearl millet were 61–99 kg-N ha^{-1} (Akintayo et al., 2008; Ekeleme et al., 2008), 76–99 kg-N ha^{-1} (Ekeleme et al., 2008), 25 kg-N ha^{-1} (Ajeigbe et al., 2014), 40–120 kg-N ha^{-1} (Belfield & Brown, 2008; Sommer et al., 2013), and 20–60 kg-N ha^{-1} (Khairwal et al., 2007), respectively. Compared to these recommended amounts, the amounts applied to lowland rice (16 kg-N ha^{-1}) and groundnut (1.5 kg-N ha^{-1}) in this study were low. The comparatively small amount of nitrogen applied for lowland rice production is a consequence of the relatively high soil fertility. Fields used for lowland rice production are generally fertile because these are recently taken under cultivation and these receive nutrients from upland areas. In the case of groundnut, farmers are aware of the nitrogen-fixing capacity of groundnut so N application is low. For the other crops, access to subsidies for fertiliser (Druilhe & Barreiro-Hurlé, 2012) or to in-kind payment loans (farmers can receive chemical fertilisers at the beginning of the cultivation season without payment and then pay back the cost of the fertilisers in form of the harvested crop) provided by local NGOs contribute to appropriate application levels of nitrogen fertiliser.

Recommended seeding rates for upland rice, lowland rice, groundnut, maize, and pearl millet in Senegal or other West African countries are 40–60, 50–60, 60–66, 20, and 3.5–5.0 kg ha^{-1}, respectively (Havard, 1986; Freud et al., 1997; Akintayo et al., 2008; Ekeleme et al., 2008; Ragasa et al., 2013). The rates in the surveyed area for groundnut and maize were consistent with the recommended levels whereas that of pearl millet was slightly higher, and the rates of upland and lowland rice were at approximately 25–50 % of the recommended levels. These differences reflect variations in sowing methods in Senegal compared to other West African countries. Single row seeder (general type) is very in common use for Senegalese than Malian or Nigerian (Le Moigne, 1980; Schmitz et al., 1991). As a result, drilling has been adopted as the conventional sowing method in Senegal while in the countries where the availability of seeders is low, the crops are sown broadcast or by dibbling (making small holes and sowing seeds in the holes).

Weeding time spent on rice was 2.8–3.9 times longer than for the other crops. For upland rice the weeding time was 1.2–2.3 times longer than found in other West African countries (Dalton et al., 1998). This increased weeding effort in rice cultivation may be due to inefficient weed management arising from a relatively low cultivation experience (Linares, 2002), or due to a higher motivation for intensive management because of a higher market value or a higher consumption demand.

Each farmer devoted ca. 70 % of the cultivated land to groundnut and pearl millet (approximately 2.8 ha each; Table 1). This reflects the fact that groundnut is the main cash crop and pearl millet is the traditional staple crop. In contrast, the area devoted to rice and maize was approximately one-quarter to one-half of that devoted to groundnut and pearl millet. Rice and maize do not only require more inputs (e.g. fertiliser), these crops are also not as drought resistant as groundnut and pearl millet. Therefore, under the existing rainfed conditions it is a challenge for Senegalese farmers to expand the areas under rice and maize cultivation.

4.2 TEs and their distributions

The TE of upland rice was lower and more variable than that of other crops (Table 2). This reflects that farmers tend to easily start growing upland rice but the cultivation techniques are in fact difficult; the yield of upland rice is far more affected by farming practices or environmental conditions such as the amount and distribution of rainfall. Consequently, the TE in upland rice was relatively low and very variable (Fig. 1). The higher TE and lower variation found in lowland rice are the result of higher soil fertility and water availability. Furthermore, as the farmers have to invest in the development of lowland fields for planting rice, this act as a barrier for farmers with severe resource constraints. The initial investments are also an incentive for farmers to achieve a high yield to recover the cost as soon as possible. The lowland rice TE peaks at 0.8 and 0.9 (Fig. 1) suggest that two different environments, that accompany the distance from Gambia, coexist among the lowland rice farmers. The study area included land at the national border with Gambia (a country where people have long experience of rice cultivation), as well as areas far from this region. In the border area, farmers had opportunities to import suitable rice varieties as well as to adopt rice-farming techniques from Gambia. Thus, the TE level of farmers living in a village with easier access to Gambia (near a trading centre or national road) tended to be higher. Groundnut and pearl millet had high TEs with a peak at 0.9 (Fig. 1) underlining the long experience of the farmers growing these crops. Likewise, the very high TE of maize was the result of a well-established farming system and a strong incentive to recover the high costs of the necessary fertilisers.

TEs found in previous studies in West African countries were 0.35–0.90 for rice farming and 0.45–0.87 for other crops (Table 4). These are relatively lower than our results. Sherlund et al. (2002) showed that including natural environmental factors such as soil, topography, pests, weeds, and weather conditions lead to higher TE levels compared to no inclusion of these factors. This means that if environmental factors are controlled or identical among farmers, TE will be higher. Hence, the environmental conditions in our study area may have been comparable amongst farmers and therefore may not have (negatively) influenced TE levels.

4.3 Determinants of TEs

Weeding time was a significant negative factor affecting the TE of lowland rice, groundnut, and pearl millet (Table 3). In the study area, weeding is carried out by both inter-row weeding using an animal drawn hoe (a "houe sine" or "houe occidentale"; Starkey, 1989) using a donkey, cow, or horse, and within-row manual weeding using a hand hoe. This weeding method was almost the same among farmers. A previous study in Gambia demonstrated that the manual within-row weeding had little effect on final crop yield compared to mechanical inter-row weeding only (Remington & Posner, 2000). In our sample, 56–81% of weeding hours were manual weeding (data not shown). Therefore, a more selective weeding approach with a focus on mechanical inter-row weeding will decrease overall weeding time and increase TE levels.

The amount of nitrogen fertiliser applied in groundnut and pearl millet cultivation was a positive determinant factor of TE; however, it was not a significant factor for upland and lowland rice (Table 3). The significant and higher (compared to pearl millet) effect of nitrogen for groundnut is different from our expectation based

on the fact that groundnut root nodules provide nitrogen to the soil. If soil fertility is sufficiently high, the effect of applied nitrogen on TE will be none or small. Groundnut has been cultivated in Senegal without applying the necessary amount of nitrogen since the 1980s (Freud et al., 1997); the amount supplied in the study area is equivalent to only one-sixteenth of the recommended amount (Ajeigbe et al., 2014) and fields have been intensively cultivated without adequate fallowing, so the soils used for groundnut are deficient in nitrogen. Senegalese farmers may have overestimated the contribution of root nodules to the nitrogen supply in soil. Our results for pearl millet also show that inefficient pearl millet farmers need to increase the amount of nitrogen application in order to raise their TE. Senegalese farmers can obtain fertiliser through in-kind loan or subsidy program but the obtainable amount is limited in general. Thus, to increase the amount of nitrogen application, one solution is to make more use of animal manure because most farmers do not manage this resource properly.

The amount of nitrogen applied was not a significant factor for upland and lowland rice production. Water availability is a principal factor affecting the fertiliser effect on the yield. As upland rice easily suffers from drought and lowland rice is cultivated under more favourable water and fertility conditions, additional fertiliser application had little effect on increasing TE.

The seeding rate had a negative effect on the TE of lowland rice production; i.e. farmers need to decrease the seeding rate to increase TE. Surveyed farmers tend to sow more seeds than is efficient to avoid hazards such as bird attacks, competition with weeds, and losses from inappropriate management practices but the additional amount of the sown seeds did not contribute to increase the yield effectively. If there is no constraint in expanding inputs, the recommended seeding rate will produce a maximum yield, ceteris paribus. In case that, however, farmers' resource is restricted, it is not always appropriate to aim at the recommendation level. A feasible solution for inefficient farmers is to aim at the seeding level of efficient farmers who are under similar situation (i.e. production resource endowment). The seeding rate of pearl millet also had a negative effect on TE. However, the seeding situation of pearl millet is different from that of lowland rice. The average seeding rate of pearl millet is at the upper limit of the recommendation range, 3.5–5.0 kg ha^{-1} (Havard, 1986). Our result shows that the seeding rate of pearl millet, even in the range, can be reduced without decreasing the yield.

Table 4: *Overview of crop technical efficiency studies in some West African countries.*

Authors	Area	Crops	Estimated TE
Abdulai & Huffman (1998)	Ghana	Rice	0.72 (SFA)
Abdulai & Huffman (2000)	Ghana	Rice	0.73 (SFA)
Anang *et al.* (2016)	Ghana	Rice	0.61–0.63 (SFA)
Diagne *et al.* (2013)	Senegal	Rice	0.55–0.60 (SFA)
Okoruwa *et al.* (2006)	Nigeria	Rice	0.76–0.81 (SFA)
Seck (2016)	Senegal	Rice	0.27 (CRS), 0.32 (VRS)
Sherlund *et al.* (2002)	Côte d'Ivoire	Rice	0.56 (VRS), 0.35 (SFA) [†]
Sherlund *et al.* (2002)	Côte d'Ivoire	Rice	0.90 (VRS), 0.76 (SFA) [‡]
Abdulai *et al.* (2013)	Ghana	Maize	0.74 (SFA)
Aye & Mungatana (2010)	Nigeria	Maize	0.72 (CRS), 0.78 (VRS), 0.79 (SFA)
Aye & Mungatana (2013)	Nigeria	Maize	0.80 (CRS), 0.86 (VRS), 0.87 (SFA)
Binam *et al.* (2004)	Cameroon	Maize	0.75 (SFA)
Binam *et al.* (2004)	Cameroon	Groundnut	0.71 (SFA)
Kane *et al.* (2012)	Cameroon	Groundnut & Maize	0.44 (CRS), 0.67 (VRS)
Thiam & Bravo-Ureta (2003)	Senegal	Groundnut	0.70 (SFA)
This study	Senegal	Rice	0.76–0.88 (CRS)
		Maize	0.94 (CRS)
		Groundnut	0.89 (CRS)
		Pearl millet	0.90 (CRS)

Note: CRS: Data Envelop Analysis was conducted with Constant Return to Scale assumption, VRS: Data Envelopment Analysis was conducted with Variable Return to Scale assumption, SFA: Stochastic Frontier analysis was conducted.
[†]: Environmental factors were excluded in the TE measurement variables. [‡]: Environmental factors were included in the TE measurement variables.

Cultivated area was a negative factor for TE only for lowland rice. Land development for lowland rice production tends to occur first in the most favourable locations and then spreads to less favourable sites, which are located at the boundary between lowland and upland and therefore are at risk of soil degradation (e.g. soil erosion). As the proportion of less favourable land being cultivated increases, the area of cultivated land will have a negative effect on TE. This relationship between land area and TE was reported in previous studies also (Okoye *et al.*, 2009; Aye & Mungatana, 2010; Kane *et al.*, 2012). These authors focused on small-scale farmers (0.61–1.20 ha) using traditional farming practices (manual labour and crude implements) and they explained the negative relationship in terms of labour shortages (Kane *et al.*, 2012) or timing of input application (Aye & Mungatana, 2010). On the other hand, a positive relationship was found by Ogundele & Okuruwa (2006) in places where larger-scale farming (2.59–6.52 ha) was dominant. The technological difference between these small and larger farms typically appeared to be the degree of mechanisation. Mechanical equipment functioned more efficiently as the size of contiguous farmland increased.

A delay in sowing time is a significant negative factor for TE in upland rice, lowland rice, groundnut, and maize. Pearl millet is the exception, possibly because it is less sensitive to drought than the other crops. As the rainy season in the study area is very short (four months), drought-sensitive crops may be exposed to water shortages and drought-related damage when sowing is delayed. Therefore, a delay in sowing negatively affects the TE of crops that have low drought resistance.

Participation in a farmers' association positively affected the TE of lowland rice production especially, as reported also in earlier studies (Audibert, 1997; Kane *et al.*, 2012; Seck, 2016). This positive association might be the result of farmers sharing their experiences and exchanging ideas on cultivation techniques. The farming technique for lowland rice, especially for water management, is relatively new and difficult for farmers, which perhaps explains why the positive coefficient of participation was only significant for lowland rice in this study.

For upland rice and lowland rice, TE was positively related to the variable of years of cultivation experience as reported also by Seck (2016). To enhance the TE levels, therefore, upland and lowland rice farmers need to compensate for their short cultivation experience. Participating in a training program can offset the deficiencies in farming experience and knowledge.

4.4 Production potential and concluding remarks

The inefficiency rates of crop production $(1 - TE)$ were 24, 12, 11, 6, and 10 % in upland rice, lowland rice, groundnut, maize, and pearl millet, respectively. In case that the current efficiency levels are base setting and 100 % efficiency level is a target, or potential setting, the potential yields of the five crops will increase at 0.84, 1.70, 0.82, 1.06, and $0.66\,t\,ha^{-1}$, respectively. These yield levels can be achieved by optimising the existing farming practices. The fact that TE distribution and its influencing factors are different among crops, could be very useful for preparing an effective program of agricultural extension for the farmers in the study area. Although upland rice has the highest improvement potential, an expansion of upland rice may not be feasible for Senegalese farmers owing to its vulnerability to droughts. Therefore, a focus should be put on lowland rice, groundnut, and pearl millet.

It should be noticed, moreover, that even if the TE levels can be increased, there will still be a yield gap if compared to actual yield levels of some West African countries (Binam *et al.*, 2004; Aye & Mungatana, 2013; Anang *et al.*, 2016). Thus, in order to enhance the production levels, additional crop improvement strategies will be necessary. Amelioration of farmers' accessibility to improved agricultural technologies (e.g. hybrid seeds, chemical fertiliser, irrigation infrastructures, or agricultural mechanisation) should be the preferred approach.

References

Abdulai, A. & Huffman, W. E. (1998). An examination of profit inefficiency of rice farmers in Northern Ghana. Iowa State University, Economics department staff paper 296. Available at: http://ageconsearch.umn.edu/bitstream/18271/1/isu296.pdf (last accessed: 06.02.2016).

Abdulai, A. & Huffman, W. E. (2000). Structural adjustment and economic efficiency of rice farmers in Northern Ghana. *Economic Development and Cultural Change*, 48 (3), 503–520.

Abdulai, S., Nkegbe, P. K. & Donkoh, S. A. (2013). Technical efficiency of maize production in Northern Ghana. *African Journal of Agricultural Research*, 8 (43), 5251–5259.

AFAP & IFDC (2014). Senegal fertilizer assessment – In support of the African fertilizer and agribusiness partnership. Available at: https://ifdcorg.files.wordpress.com/2016/05/senegal-fertilizer-assessment.pdf (last accessed: 20.12.2016).

Ajeigbe, H. A., Waliyar, F., Echekwu, C. A., Ayuba, K., Motagi, B. N., Eniayeju, D. & Inuwa, A. (2014). A farmer's guide to profitable groundnuts production in Nigeria. International Crops Research Institute for the Semi-Arid Tropics (ICRISAT), Patancheru, India.

Akintayo, I., Cissé, B. & Zadji, L. D. (2008). Guide pratique de la culture des NERICA de plateau. Africa Rice Center, Cotonou, Bénin.

Anang, B. T., Bäckman, S. & Sipiläinen, T. (2016). Agricultural microcredit and technical efficiency: The case of smallholder rice farmers in Northern Ghana. *Journal of Agriculture and Rural Development in the Tropics and Subtropics*, 117 (2), 189–202.

ANSD (2014). Rapport définitif RGPHAE (Recensement Général de la Population et de l'Habitat, de l'Agricupture et de l'Elevage) 2013. Available at: http://www.fao.org/fileadmin/templates/ess/ess_test_folder/World_Census_Agriculture/Country_info_2010/Reports/Reports_5/SEN_FR_REP_2013pdf.pdf (last accessed: 28.12.2016).

Audibert, M. (1997). Technical inefficiency effects among paddy farmers in the villages of the 'Office du Niger', Mali, West Africa. *Journal of Productivity Analysis*, 8 (4), 379–394.

Aye, G. C. & Mungatana, E. D. (2010). Technical efficiency of traditional and hybrid maize farmers in Nigeria: comparison of alternative approaches. *African Journal of Agricultural Research*, 5 (21), 2909–2917.

Aye, G. C. & Mungatana, E. D. (2013). Evaluating the performance of small scale maize producers in Nigeria: an integrated distance function approach. *Review of Urban & Regional Development Studies*, 25 (2), 79–92.

Belfield, S. & Brown, C. (2008). Field crop manual: maize – A guide to upland production in Cambodia. NSW department of primary industries, New South Wales, Australia.

Binam, J. N., Tonyè, J., Wandji, N., Nyambi, G. & Akoa, M. (2004). Factors affecting the technical efficiency among smallholder farmers in the slash and burn agriculture zone of Cameroon. *Food Policy*, 29 (5), 531–545.

Coelli, T. J. (1996). A Guide to DEAP Version 2.1: A Data Envelopment Analysis (Computer) Program. CEPA Working Papers No. 8/96, Department of Econometrics, University of New England, Armidale NSW Australia.

Coelli, T. J., Rao, D. S. P., O'Donnell, C. J. & Battese, G. E. (2005). *An introduction to efficiency and productivity analysis.* Springer Sience+Business Media, Inc., New York.

Dalton, T. J., Becker, M. & Johnson, D. E. (1998). Stabilization of upland rice production under shortened fallow in West Africa: research priority setting in a dynamic environmental and economic climate. Paper prepared for the American Agricultural Economic Association Annual Meeting, Salt Lake City, Uta. Available at: http://ageconsearch.umn.edu/bitstream/20944/1/spdalt01.pdf (last accessed: 06.02.2016).

Diagne, M., Demont, M., Seck, P. A. & Diaw, A. (2013). Self-sufficiency policy and irrigated rice productivity in the Senegal River Valley. *Food Security,* 5 (1), 55–68.

Diop, A. M. (1999). Sustainable agriculture: new paradigms and old-practices? Increased production with management of organic inputs in Senegal. *Environment, Development and Sustainability,* 1, 285–296.

Druilhe, Z. & Barreiro-Hurlé, J. (2012). Fertilizer subsidies in sub-Saharan Africa. ESA Working paper No. 12-04.

Ekeleme, F., Kamara, A. Y., Omoigui, L. O., Tegbaru, A., Mshelia, J. & Onyibe, J. E. (2008). Guide to rice production in Borno state, Nigeria. International Institute of Tropical Agriculture (IITA), Ibadan, Nigeria.

FAO (2015). Country fact sheet on food and agriculture policy trends–Senegal. Available at: http://www.fao.org/3/a-i4841e.pdf (last accessed: 28.12.2016).

FAO, IFAD & WFP (2015). Meeting the 2015 international hunger targets: taking stock of uneven progress. FAO, Rome.

FAO Stat (2009–2014). Available at: http://www.fao.org/faostat/en/#data/QC (last accessed: 29.12.2016).

Farrell, M. J. (1957). The measurement of productive efficiency. *Journal of the Royal Statistical Society. Series A (General),* 120 (3), 253–290.

Frenken, K. (2005). L'irrigation en Afrique en chiffres – Enquête AQUASTAT 2005. FAO rapports sur l'eau 29.

Freud, C., Freud, E. H., Richard, J. & Thénevin, P. (1997). L'arachide au Sénégal : un moteur en panne. Editions KARTHALA 22–24, boulevard Arago 75013 Paris.

Grosenick, G., Djegal, A., King, J. W., Karsh, E. & Warshall, P. (1990). Senegal natural resources management assessment. Available at: https://rmportal.net/library/content/tools/community-based-natural-forest-management-USAID-Lessons-Learned/cbnfm/USAID-BDB-cd-2-data/pnabk606-sencgal.pdf/view (last accessed: 07.12.2016).

Havard, M. (1986). Le semis du mil au Super Eco en culture attelée. *Machinisme Agricole Tropical Information,* 93, 15–20.

IMF (2013). Senegal: Poverty reduction strategy paper. Country Report No. 13/194. Available at: https://www.imf.org/external/pubs/cat/longres.aspx?sk=40739.0 (last accessed: 28.12.2016).

Kane, G. Q., Fondo, S. & Oyekale, A. S. (2012). Efficiency of groundnuts/maize intercropped farms in Zoetele, South Cameroon: a data envelopement approach. *Life Science Journal,* 9 (4), 3955–3962.

Kelly, V., Diagana, B., Reardon, T., Gaye, M. & Crawford, E. (1996). Cash crop and foodgrain productivity in Senegal: Historical view, new survey evidence, and policy implications. MSU International Development Paper No. 20.

Khairwal, I. S., Rai, K. N., Diwakar, B., Sharma, Y. K., Rajpurohit, B. S., Nirwan, B. & Bhattacharjee, R. (2007). *Pearl millet: crop management and seed production manual.* International Crops Research Institute for the Semi-Arid Tropics (ICRISAT), Patancheru, India.

Le Moigne, M. (1980). Animal-draft cultivation in Francophone Africa. *In:* Ryan, J. G. & Thompson, H. L. (eds.), *Proceedings of the International Workshop on Socio-economic Constraints to Development of Semi-Arid Tropical Agriculture, 19–23 February, 1979, Hyderabad, India.* pp. 213–220, ICRISAT.

Linares, O. F. (2002). African rice (*Oryza glaberrima*): History and future potential. *Proceedings of the National Academy of Sciences of the United States of America,* 99 (25), 16360–16365.

Nin-Pratt, A., Johnson, M., Magalhaes, E., You, L., Diao, X. & Chamberlin, J. (2011). Yield gaps and potential agricultural growth in West and Central Africa. Research monograph of International food policy research institute (IFPRI).

Ogundele, O. O. & Okuruwa, V. O. (2006). Technical efficiency differentials in rice production technologies in Nigeria. AERC research paper 154.

Okoruwa, V. O., Ogundele, O. O. & Oyewusi, B. O. (2006). Efficiency and productivity of farmers in Nigeria: a study of rice farmers in North Central Nigeria. Poster paper prepared for presentation at the International Association of Agricultural Economists Conference, Gold Coast, Australia, August, 12-18 2006. Available at: https://www.researchgate.net/publication/23511325_Efficiency_and_Productivity_of_Farmers_in_Nigeria_A_Study_of_Rice_Farmers_in_North_Central_Nigeria (last accessed: 06.02.2016).

Okoye, B. C., Agbaeze, C. C., Asumugha, G. N., Aniedu, O. C. & Mbanaso, E. N. A. (2009). Small is Beautiful: empirical evidence of an inverse relationship between farm size and productive efficiency in small-holder cassava production in Ideato North LGA of Imo state. MPRA paper No. 17418. Available at: https://mpra.ub.uni-muenchen.de/17418/1/Small_is_Beautiful_1_.pdf (last accessed: 06.02.2016).

Pratt, P. F. & Castellanos, J. Z. (1981). Available nitrogen from animal manures. *California Agriculture*, 35 (7), 24.

Ragasa, C., Dankyi, A., Acheampong, P., Wiredu, A. N., Chapoto, A., Asamoah, M. & Tripp, R. (2013). Patterns of adoption of improved maize technologies in Ghana. Ghana strategy support programme. International Food Policy Research Institute Working Paper 36.

Remington, T. R. & Posner, J. L. (2000). On-farm evaluation of weed control technologies in direct-seeded rice in the Gambia. *In:* Starkey, P. & Simalenga, T. (eds.), *Animal power for weed control.* pp. 255–261, Animal Traction Network for Eastern and Southern Africa (ATNESA), Wageningen, Netherlands.

Schmitz, H., Sommer, M. & Walter, S. (1991). *Animal traction in rainfed agriculture in Africa and South America.* Vieweg, Berlin, Germany.

Seck, A. (2016). *Fertilizer subsidy and agricultural productivity in Senegal.* AGRODEP Working Paper 0024.

Sherlund, S. M., Barrett, C. B. & Adesina, A. A. (2002). Smallholder technical efficiency controlling for environmental production conditions. *Journal of Development Economics*, 69 (1), 85–101.

Sommer, R., Bossio, D., Desta, L., Dimes, J., Kihara, J., Koala, S., Mango, N., Rodriguez, D., Thierfelder, C. & Winowiecki, L. (2013). *Profitable and sustainable nutrient management systems for east and southern African smallholder farming systems: challenges and opportunities – A synthesis of the Eastern and Southern Africa situation in terms of past experiences, present and future opportunities in promoting nutrients use in Africa.* CIAT, Cali, Colombia; The University of Queensland, Australia; CIMMYT, El Batan, Mexico.

Starkey, P. (1989). *Harnessing and implements for animal traction – An animal traction resource book for Africa.* GATE Div./ Deutsche Gesellschaft für Technische Zusammenarbeit (GTZ) GmbH, Eschborn.

Thiam, A. & Bravo-Ureta, B. E. (2003). Technical efficiency measures for a sample of Senegalese groundnut producers using pooled cross-section time series data. *International Arachis Newsletter*, 23, 36–39.

UN (2015). World population prospects: The 2015 revision. Available at: https://esa.un.org/unpd/wpp/Download/Probabilistic/Population/ (last accessed: 28.12.2016).

USDA (2015). Senegal – Grain and feed annual – 2015 update West Africa rice annual. Global Agricultural Information Network Report. Available at: http://gain.fas.usda.gov/Recent%20GAIN%20Publications/Grain%20and%20Feed%20Annual_Dakar_Senegal_4-29-2015.pdf (last accessed: 07.12.2016).

USDA (2016). Senegal – Grain and feed annual – 2016 update West Africa rice annual. Global Agricultural Information Network Report. Available at: https://gain.fas.usda.gov/Recent%20GAIN%20Publications/Grain%20and%20Feed%20Annual_Dakar_Senegal_4-6-2016.pdf (last accessed: 19.03.2017).

16

Carcass and meat quality characteristics of Arsi-Bale goats supplemented with different levels of air-dried *Moringa stenopetala* leaf

Aberra Melesse [a,*], Sandip Banerjee [a],
Degnet H/Meskel [b], Aster Abebe [a], Amsalu Sisay [a]

[a]*School of Animal and Range Sciences, Hawassa University, Ethiopia*
[b]*Wachamo University, Hossana, Ethiopia*

Abstract

This study was conducted to assess the effect of air-dried *Moringa stenopetala* leaf (MSL) supplementation on carcass components and meat quality in Arsi-Bale goats. A total of 24 yearling goats with initial body weight of 13.6±0.25 kg were randomly divided into four treatments with six goats each. All goats received a basal diet of natural grass hay *ad libitum* and 340 g head^{-1} d^{-1} concentrate. The treatment diets contain a control diet without supplementation (T1) and diets supplemented with MSL at a rate of 120 g head^{-1} d^{-1} (T2), 170 g head^{-1} d^{-1} (T3) and 220 g head^{-1} d^{-1} (T4). The results indicated that the average slaughter weight of goats reared on T3 and T4 was 18.2 and 18.3 kg, respectively, being ($P < 0.05$) higher than those of T1 (15.8 kg) and T2 (16.5 kg). Goats fed on T3 and T4 diets had higher ($P < 0.05$) daily weight gain compared with those of T1 and T2. The hot carcass weight in goats reared on T3 and T4 diets was 6.40 and 7.30 kg, respectively, being ($P < 0.05$) higher than those of T1 (4.81 kg) and T2 (5.06 kg). Goats reared on T4 had higher ($P < 0.05$) dressing percentage than those reared in other treatment diets. The rib-eye area in goats reared on T2, T3 and T4 diets was higher ($P < 0.05$) than those of T1. The protein content of the meat in goats reared on T3 and T4 was 24.0 and 26.4 %, respectively being significantly higher than those of T1 (19.1 %) and T2 (20.1 %). In conclusion, the supplementation of MSL to natural grass hay improved the weight gain and carcass parts of Arsi-Bale goats indicating Moringa leaves as alternative protein supplements to poor quality forages.

Keywords: Arsi-Bale goats, carcass traits, meat quality, natural grass hay, *Moringa stenopetala* leaf, supplementation

1 Introduction

There are approximately 570 breeds and types of goats in the world of which 89 are found in Africa (Gala, 2005). The present estimated population of goats in Ethiopia is 29.3 million (CSA, 2014). The goats play an important role by improving the livelihood of resource challenged farmers by creating alternative employment opportunities, enhancing family income by sale of live animals, skin, manure, etc. Therefore, demand for goat's meat is on the rise throughout the world especially in Ethiopia. This is mainly due to increased human population coupled with income growth. However, regardless of their attributes, the productivity of goats remains low mainly due to diseases, genotype, management and nutrition (Simela & Merkel, 2008).

* Corresponding author
Aberra Melesse, P.O. Box 05, School of Animal and Range Sciences, Hawassa University, Hawassa, Ethiopia
Email: a_melesse@uni-hohenheim.de; a_melesse@yahoo.com

One of the most important nutritional constraints in goat production in the tropics is underfeeding mainly attributed to limitations of feed in both quantity and quality. The situation is aggravated during the dry season where natural pastures became over-matured being critically deficient in protein and energy contents. As a result, large flocks of productive livestock cannot be maintained on such feeds that hardly meet even the basic maintenance requirement of farm animals. It is thus imperative to supplement the available poor quality feed resources with some amount of concentrates for enhanced productivity of farm animals. However, the use of conventional feeds as a supplement is usually limited under smallholder livestock production systems due to inaccessibility and high cost of such feed ingredients.

In order to mitigate the problems associated with the lack of protein supplement, there is a need to look for alternative cheap protein sources from unconventional feed resources that are easily accessible by the smallholder farmers (Manaye et al., 2009). Leaves of Moringa spp. could be one possible source of protein both for humans and livestock. Among the Moringa species, Moringa stenopetala is endemic to southern Ethiopia, which is unique as it does not shed its leaves even during the dry season (Thurber & Fahey, 2009). Recent studies conducted by Negesse et al. (2009), Melesse et al. (2009, 2012) and Debela & Tolera (2013) have indicated that the leaves of M. stenopetala are rich in protein (28.2–36.1 %) and contain substantial amounts of essential amino acids as well as macro and trace minerals.

There are several studies indicating the use of the fresh foliage of Moringa spp. as a source of livestock feed. Studies by Sánchez et al. (2006) reported significant increase in milk yield of Creole dairy cows supplemented with Moringa oleifera leaves. Goats supplemented with M. oleifera leaves at 20 and 50 % levels of total daily forage allowance had higher live-weight gains, and higher digestibility of nutrients (Aregheore, 2002). Improvements in nutrient intake, growth performances and nutrient retention have been also obtained from sheep supplemented with air-dried M. stenopetala leaf to the basal diet of Rhodes grass hay (Gebregiorgis et al., 2011). Moreover, Melesse et al. (2011, 2013) reported enhanced improvements in the feed intake, growth performances and carcass characteristics of growing dual-purpose chickens fed with graded levels of M. stenopetala leaf meal by replacing roasted soybean seed. However, information on the feeding values of M. stenopetala leaf on the carcass components and meat quality of male goats in Ethiopia is not available. Thus, this research was carried out to assess the effect of feeding graded levels of air-dried M. stenopetala leaf supplemented to a basal diet of natural grass hay on the carcass characteristics and meat quality of Arsi-Bale goats.

2 Materials and methods

2.1 Experimental site

The experiment was carried out at the Animal Farm of the School of Animal and Range Sciences, Hawassa University (Ethiopia), which lies geographically between 7°55′N latitude and 38°N29′E longitude at an altitude of 1700 m asl. The average annual rainfall ranges from 800 mm to 1100 mm. The mean minimum and maximum temperatures in the study area are 13.5°C and 27.6°C, respectively (NMA, 2012).

2.2 Preparation of experimental rations

Fresh M. stenopetala leaves were collected from farmers of Mirab Abaya near to Arbaminch city which is located between 6°4′N latitude and 37°34′E longitude at an altitude of 1220 m asl. The fresh leaves were harvested from available trees regardless of tree age. The collected leaves were then trimmed from twigs on a plastic sheet and dried under the shade to prevent the loss of vitamins and volatile nutrients. During the drying process, regular turning of leaves was done to ensure uniform drying for safe storage. The air-dried Moringa leaves were finally transported to the experimental site and ground into coarse powder which hereafter is referred to as air-dried M. stenopetala leaf (MSL). The ground leaf was packed in bags of 100 kg and stored until used. Natural grass hay was bought from a nearby private farm and hand chopped into the size of 3 to 5 cm for ease of feeding. Samples of MSL and various ingredients of the experimental ration were subjected to chemical composition analysis before being used in the formulation of experimental diets.

2.3 Experimental animals and their management

In this experiment, about 1 year old (age determined by dentition) 24 Arsi-Bale male goats were purchased from local market and transported to the experimental site. They were then quarantined for a fortnight, during which they were treated with 250 mg of albendazole (Chengdu Qiankun veterinary pharmaceuticals Co. ltd. China), administered through drenching gun for deworming the animals. Moreover, 1.5 ml Oxytetracycline (ibid) was provided intravenous for three days to treat the animals from possible respiratory diseases and physical injuries incurred during the transportation. The barns were also sprayed with Diazinone (*ibid*) to take care of any ecto-parasite infestation.

At the end of the quarantine period, the goats were ear tagged and weighed (prior to being offered any feed) for two consecutive days and the body weight was averaged. Then they were housed in individual pens and each animal was provided with individual feeder and watering trough.

2.4 Experimental design and treatment diets

The feeding trial was a completely randomized design (CRD) consisting of one control and three supplemental treatment diets with six goats randomly assigned to each treatment. The trial was conducted for 75 days exclusive a 15 days adaptation period. All the experimental goats had *ad libitum* access to natural grass hay and water. They were also provided with a concentrate at a rate of 2.5 % of their body weight according to the recommendation of McDonald *et al.* (2010). Consequently, as the average initial body weight of the goats was 13.6 kg, the calculated concentrate supplied per goat and day was 340 g. The diets therefore contained 340 g concentrate only (T1), 340 g concentrate plus 35 % MSL supplementation (120 g) (T2), 340 g concentrate plus 50 % MSL supplementation (170 g) (T3) and 340 g concentrate plus 65 % MSL supplementation (220 g) (T4). Accordingly, the total concentrate/MSL mixture offered to T1, T2, T3 and T4 treatment groups were 340, 460, 510 and 560 g head^{-1} d^{-1}, respectively. The concentrate offered was a mixture of 50 % wheat bran, 35 % maize, 14 % Noug seed (*Guizotia abyssinica*) cake and 1 % salt. Natural grass hay and the concentrate/MSL mixture were offered separately.

2.5 Data collection procedures

2.5.1 Feed consumption

Samples of daily feed offered and refused were collected, measured and pooled over the experimental period for each animal and used for chemical analysis. The daily average feed intake was determined by the difference between the amounts of feed offered and refused. Body weight of each goat was recorded every fortnight after overnight fasting.

2.5.2 Parts of the carcass

For the evaluation of carcass components, all goats were fasted overnight, weighed (here after referred as slaughter weight) and slaughtered. The weights of the following carcass parts were then recorded: head, skin, neck, thorax, lumbar, rack, heart, lung, trachea, liver, gall bladder, spleen, testis, kidneys, trotters, rumen, small and large intestines. Hot carcass weight was determined after the removal of the head, hide, intestinal tract, and internal organs. Dressing percentage was then calculated as proportion of hot carcass weight to slaughter weight. The rib-eye muscle (*Longissimus dorsi*) area of each animal was determined according to standard procedure using a digital planimeter (portable area meter, model LI 3000A) as suggested by AOAC (1995). The cross-sectional area of rib-eye muscle between the 12th and 13th ribs were traced on transparency paper from the right and left side and measured by using a planimeter. The average of the right and left cross-sectional areas was then taken as a rib-eye muscle area.

2.5.3 Meat quality
Cooking loss

The cooking loss was determined by using 1.5 g of the meat sample from *L. dorsi* muscle in replication. The meat was placed in a test tube (for indirect boiling) and immersed in a water bath having a temperature of 85°C and cooked for 30 minutes after which the meat was cooled to room temperature and then weighed. The difference was then recorded as the cooking loss (AOAC, 1995).

Meat pH

The meat pH was measured at 50 minutes of post mortem in the *L. dorsi* muscle at the first lumbar using a digital pH meter (IS/ISO 3100), which was equipped with a penetrating electrode. The pH meter was calibrated with pH values of 4, 7 and 9 standard solutions before each measurement was undertaken.

Intramuscular fat

The determination of the intramuscular fat (ether extract) was carried out according to the procedures of AOAC (1995). The intramuscular fat in *L. dorsi* muscle was assessed by taking one gram of dried meat sample which was folded in a filter paper and placed in Soxhlet apparatus. It was then refluxed 30–40 times using petroleum benzene (boiling point 60–80°C) until all the fat from the sample was transferred to the petroleum ether. The defatted sample was then transferred to desiccators and allowed to cool overnight and reweighed using sensitive balance.

2.6 Chemical analysis of the feed

The analysis of dry matter (DM), ash and ether extract (EE) was performed according to AOAC (1995). Accordingly, the DM content of the feed was determined by drying the samples at 105°C overnight. Ash was determined by combusting the samples at 550°C for 5 h. Nitrogen (N) was extracted with Kjeldahl method and then the crude protein (CP) was calculated as $N \times 6.25$ (AOAC, 1995). The contents of acid detergent fibre (ADF) and neutral detergent fibre (NDF) were analysed using the method of Van Soest et al. (1991) in an ANKOM® 200 Fibre Analyser (ANKOM Technology Corp., Fairport, NY, USA). All samples were analysed in duplicates at the Animal Nutrition Laboratory of Animal and Range Sciences, Hawassa University.

2.7 Statistical analysis

The data were subjected to Analysis of Variance (ANOVA) using the GLM of SAS (SAS, 2010, ver. 9.3). When significant differences were observed among treatment means, they were separated by Duncan multiple range test. Comparisons with $P < 0.05$ were considered significant and all statements of statistical differences were based on this level unless noted otherwise.

The following linear model summarizes the statistics employed to analyse the data:

$$Y_{ij} = \mu + A_i + e_{ij} \text{ ; where:}$$

Y_{ij} = the observed j^{th} variable in the ith treatment diet (fixed factor)

μ = overall mean of the observed variable

A_i = effect due to i^{th} treatment diets (i = 0, 120, 170, 220 g head^{-1} d^{-1})

e_{ij} = random residual error

3 Results

3.1 Chemical composition of experimental feed

Chemical composition of the ingredients of the experimental ration is presented in Table 1. The DM content was more or less similar across the feed ingredients although it was slightly higher in grass hay. The ash content was lowest for the maize grain while it was highest in MSL. The CP content was highest for the MSL followed by that of the Noug seed cake. The results also indicated that the EE content was highest for Noug seed cake and lowest for grass hay. The NDF and ADF contents were highest for grass hay while they were lowest in MSL.

3.2 Nutrient contents of experimental diets

The results from Table 2 indicate that the DM content of the experimental diets was more or less similar while the highest ash and EE contents were observed with 170 and 220 g head^{-1} d^{-1} MSL supplemented diets. The CP content was highest in diets supplemented with 220 g MSL and was lowest in the control diet. The ADF and NDF contents were generally high for the control diet and decreased with increasing levels of MSL supplementation in the treatment diets.

3.3 Body weight

As presented in Table 3, the goats supplemented with 170 and 220 g MSL had significantly higher final body weight ($P < 0.01$) and weight gain ($P < 0.001$) than those of the control and 120 g supplemented group. There was no significant difference between goats supplemented with 170 and 220 g MSL as well as between those reared on the control and 120 g MSL supplemented diets.

Table 1: *Nutrient compositions (g kg^{-1} DM) of the ingredients of the concentrate, natural grass hay and air-dried Moringa stenopetala leaf*

Nutrients	Ingredients of the concentrate mix			Natural grass hay	Moringa stenopetala
	Maize grain	Wheat bran	Noug seed cake*		
Dry matter	933	936	939	956	944
Crude protein	65.0	160	249	51.0	295
Ether extract	96.0	54.2	105	28.6	59.0
NDF	412	365	325	612	178
ADF	49.0	175	215	257	165

* Guizotia abyssinica; DM = dry matter; NDF = neutral detergent fibre; ADF = acid detergent fibre

Table 2: *The average nutrient compositions (g kg⁻¹ DM) of the experimental diets supplemented with various levels air-dried Moringa stenopetala leaf*

Nutrients	Levels of Moringa leaf supplementation (g head⁻¹ d⁻¹)			
	0	120	170	220
Dry matter	923	927	936	946
Ash	62.0	71.0	80.1	120
Crude protein	160	167	172	187
Ether extract	57.0	61.0	87.0	95.0
Neutral detergent fibre	693	455	365	336
Acid detergent fibre	561	383	238	225

Table 3: *The effect of various levels of air-dried Moringa stenopetala leaf supplementation on the growth performances of Arsi-Bale goats*

Growth performances	Levels of Moringa leaf supplementation (g head⁻¹ d⁻¹)				SE	LS
	0	120	170	220		
Initial body weight (kg)	13.5	13.8	13.6	13.7	0.25	NS
Final body weight (kg)	15.8 ^{b}	16.5 ^{b}	18.2 ^{a}	18.5 ^{a}	1.07	**
Total weight gain (kg)	2.26 ^{b}	2.73 ^{b}	4.65 ^{a}	4.80 ^{a}	1.05	***
Average daily gain (g)	54.0 ^{b}	58.1 ^{b}	111 ^{a}	114 ^{a}	10.4	***

Means with different lowercase letters in the same row are significantly different ($p < 0.05$)
SE = standard error of the mean; * = $p < 0.05$; ** = $p < 0.01$; *** = $p < 0.001$; NS = not significant; LS = level of significance

3.4   Carcass yield and dressing percentage

The average slaughter weight of goats supplemented with 170 and 220 g MSL was 18.2 and 18.5 kg, respectively, being significantly higher than those of the control and with 120 g MSL supplemented goats (Table 4). Similarly, the hot carcass weight varied ($P < 0.01$) across the treatment diets with higher values being observed in goats supplemented with 220 g MSL followed by those of 170 g and 120 g. The results further indicated that the average dressing percentage was higher ($P < 0.05$) in goats supplemented with 220 g MSL than those reared on the control diet.

The results indicated differences ($P < 0.01$) in rib-eye area between goats supplemented with various levels of MSL. The weight of the neck also varied across treatment diets with higher values being observed in goats supplemented with 170 and 220 g MSL. The weight of the lumbar region in all supplemented goats was higher ($P < 0.05$) than those reared on the control diet. The

weight of the heart was higher ($P < 0.05$) in goats supplemented with 220 g of MSL than those reared on the control and 120 g supplemented diets. The same was true for the weight of the kidneys.

3.5   Conditionally edible parts of the carcass

As presented in Table 5, the proportional value of lungs varied ($P < 0.05$) across the goats reared on the different diets being higher ($P < 0.05$) in those supplemented with 220 g MSL than those fed on the control diet. The spleen yield was higher ($P < 0.05$) among the goats supplemented with 170 and 220 g MSL. The proportional yields of the head and trotters were higher ($P < 0.05$) in goats supplemented with 220 g MSL than those reared on the control diet. The yield of the large intestine was higher ($P < 0.05$) in goats supplemented with 220 g MSL than those reared on other treatment diets.

Table 4: *The effect of different levels of air-dried Moringa stenopetala leaf supplementation on edible carcass parts of Arsi-Bale goats*

Carcass components (kg)	Levels of Moringa leaf supplementation (g head⁻¹ d⁻¹)				SE	LS
	0	*120*	*170*	*220*		
Slaughter weight	15.8 ^{b}	16.5 ^{b}	18.2 ^{a}	18.5 ^{a}	1.30	*
Hot carcass weight	4.83 ^{c}	5.26 ^{c}	6.40 ^{b}	7.30 ^{a}	0.37	**
Dressing (%)	30.8 ^{b}	32.5 ^{b}	35.1 ^{ab}	39.7 ^{a}	4.50	*
Rib-eye area (cm2)	4.65 ^{c}	5.96 ^{b}	6.63 ^{a}	7.27 ^{a}	0.51	**
Neck	0.41 ^{b}	0.44 ^{b}	0.49 ^{ab}	0.58 ^{a}	0.10	*
Thorax	1.34 ^{c}	1.35 ^{c}	1.37 ^{b}	1.43 ^{a}	0.01	**
Rack	0.108 ^{b}	0.188 ^{b}	0.213 ^{b}	0.302 ^{a}	0.07	*
Lumbar	0.440 ^{b}	0.574 ^{a}	0.615 ^{a}	0.675 ^{a}	0.10	*
Liver	0.308	0.314	0.363	0.365	0.15	NS
Heart	0.073 ^{b}	0.078 ^{b}	0.130 ^{ab}	0.180 ^{a}	0.06	*
Kidneys	0.026 ^{b}	0.045 ^{ab}	0.053 ^{ab}	0.066 ^{a}	0.025	*

Means with different lowercase letters in the same row are significantly different ($p < 0.05$)
SE = standard error of the mean; * = $p < 0.05$; ** = $p < 0.01$; NS = not significant; LS = level of significance

Table 5: *The effect of diets supplemented with various levels of air-dried Moringa stenopetala leaf on the yield of conditionally edible parts of the carcass in Arsi Bale goats*

Organs relative to slaughter weight (%)	Levels of Moringa leaf supplementation (g head⁻¹ d⁻¹)				SE	LS
	0	*120*	*170*	*220*		
Lung	0.81 ^{b}	0.93 ^{b}	1.21 ^{b}	2.04 ^{a}	0.27	*
Testis	0.58	0.64	0.61	1.00	0.19	NS
Head	4.20 ^{b}	5.44 ^{ab}	5.77 ^{ab}	6.75 ^{a}	0.52	*
Trachea	0.74	0.73	0.81	0.81	0.02	NS
Trotters	3.42 ^{b}	4.55 ^{a}	4.40 ^{a}	4.41 ^{a}	0.26	*
Spleen	0.06 ^{b}	0.11 ^{b}	0.25 ^{a}	0.26 ^{a}	0.05	*
Rumen	3.70	3.63	3.60	3.90	0.07	NS
Large intestine	1.34 ^{b}	1.81 ^{b}	1.73 ^{b}	2.71 ^{a}	0.3	*
Small intestine	1.63	1.96	1.76	2.0	0.09	NS
Blood	3.35	3.38	3.11	3.21	0.06	NS
Gallbladder	0.13	0.16	0.20	0.50	0.08	NS
Skin	10.45	10.40	9.10	9.45	0.33	NS

Means with different lowercase letters in the same row are significantly different ($p < 0.05$)
SE = standard error of the mean; * = $p < 0.05$; NS = not significant; LS = level of significance

3.6 Chemical composition of the meat

As shown in Table 6, the contents of protein and fat were higher in the meat from goats fed on diets supplemented with the 170 and 220 g MSL than those reared on the 120 g supplemented and the control diets. The results further indicated that there were no differences among treatments for the other parameters studied.

Table 6: *Chemical composition, pH value and cooking loss of meat in Arsi-Bale goats supplemented with different levels of air-dried Moringa stenopetala leaf*

Parameter	Levels of Moringa leaf supplementation (g head^{-1} d^{-1})				SE	LS
	0	120	170	220		
Ash (%)	1.15	1.06	1.76	1.35	0.76	NS
Moisture (%)	69.0	70.4	71.8	72.9	4.06	NS
Protein (%)	19.1[b]	20.1[b]	24.0[a]	26.5[a]	2.12	*
Fat (%)	4.66[b]	5.10[b]	6.41[a]	7.60[a]	1.03	*
pH of meat	5.56	5.48	5.45	5.42	0.01	NS
Cooking loss	39.1	39.1	38.9	36.7	2.96	NS

Means with different lowercase letters in the same row are significantly different ($p < 0.05$)
* = $p < 0.05$; NS = not significant; SE = standard error of the mean; LS = level of significance

4 Discussion

4.1 Nutrient composition of Moringa stenopetala and experimental feed

The ash contents of MSL in this study are in accordance with the observations of Dechasa *et al.* (2006) and Melesse *et al.* (2011) for *M. stenopetala* leaves and those of Olugbemi *et al.* (2010a) and Melesse *et al.* (2012) for *M. oleifera* leaves. The CP contents of MSL in the current study are similar to those reported by Melesse *et al.* (2009, 2011, 2012, 2013) for the same plant species. However, higher CP contents were reported by Negesse *et al.* (2009) for *M. stenopetala* leaves which might be attributed to the stage of maturity of leaves as young leaves tend to contain higher protein contents than older ones. The EE contents are within the range of those reported by Melesse *et al.* (2009) for *M. stenopetala* leaves. The NDF and ADF contents in the present study are also comparable to those of Melesse *et al.* (2011) for leaves of the same Moringa species but higher than those of *M. oleifera* reported by Olugbemi *et al.* (2010b). The CP, DM and OM contents of the natural grass hay in the present study are consistent with those reported by Banerjee *et al.* (2013). However, the NDF and ADF contents of the natural grass are lower than those reported by Banerjee *et al.* (2013). This difference might be attributed to the stage of maturity of the grass when it was cut and the way and duration it was stored. The relatively high NDF and ADF contents of the natural grass hay shows that it has a poor nutritional potential and is not adequate to support the maintenance requirements of goats as it contains CP below the minimum level (7 %) required for microbial function (Van Soest, 1994). Consequently, such poor quality roughages must be supplemented with protein sources like leaves of *M. stenopetala* that are easily accessible and affordable for small farmer households.

4.2 Body weight and carcass characteristics

Supplementation of natural grass hay with various levels of MSL has in general resulted in enhanced final body weight and total weight gain. The findings are in good agreement with those of Gebregiorgis *et al.* (2011) who reported that there was a significant improvement in body weights of sheep reared on a basal diet of Rhodes grass hay that was supplemented with different levels of air-dried *M. stenopetala* leaves. Similar to the present findings, Aregheore (2002) also reported that goats supplemented with fresh leaves of *M. oleifera* at 20 and 50 % levels had higher live-weight gains, and higher digestibility of DM, organic matter, energy, CP and NDF than those reared on the basal diet of batiki grass alone.

Dressing percentage, which is an important trait for carcass evaluation is influenced by age, breed, sex, plane of nutrition, management system etc. In the present study, both the slaughter weight and dressing percentage values were considerably higher in goats supplemented with 170 g and 220 g MSL. The observations are in accordance with the reports of Qwele *et al.* (2013) who had observed higher slaughter weight and digestible protein values in goats reared on diets supplemented with *M. oleifera* leaves. This might be explained by the availability of more valuable nutrients, particularly crude protein, due to the supplementation of MSL to the low quality natural grass hay.

The average dressing percentage (irrespective of the treatments) in the present study indicates that it is lower than the value reported by Kebede *et al.* (2008) for adult Arsi-Bale goats. The observed differences might be due to the fact that goats in the present study were young and were still growing. Moreover, the observed differences might be associated with the management differences used by the different scholars as goats in the tropics are not always accustomed to stall feeding conditions unless they are allowed to accustom to the new management environment before the start of the actual trial.

The rib-eye area values in goats supplemented with 120, 170 and 220 g MSL are in accordance with the reports of Kebede *et al.* (2008) and Sebsibe *et al.* (2007) for Ethiopian goat breeds. However, lower rib-eye area values have been reported by Mesfin (2007) for Arsi-Bale goats. The increased values for the rib-eye area in the present study may suggest that supplementation of low quality roughages such as natural grass hay with MSL has improved the body weight of the goats as the rib-eye measurement and body weights are positively correlated traits.

The values pertaining to the weights of the neck, rack and lumbar areas were higher in goats reared on diets supplemented with 220 g MSL. These observations are in accordance with those of Kebede *et al.* (2008) for Arsi-Bale goats reared on diets containing sweet potato vines. In general, the values obtained in this study for heart, lung, head and trotters are higher in goats supplemented with MSL than those of non-supplemented, which might suggest an active body physiological activity of these organs as a result of relatively high intake of MSL with the supplemented diets.

4.3 Chemical composition of the meat and its quality

The contents of protein and fat were found to be significantly higher in the meat from goats fed on diets supplemented with the 170 and 220 g MSL than those reared on the control diets (Table 6). The study conducted by Qwele *et al.* (2013) indicated that the *M. oleifera* leaf had the antioxidative potential which was responsible for improving meat quality (chemical composition, colour and lipid stability) of goats. The *Moringa stenopetala* leaf extract have been also reported to contain an antioxidant fraction (Tebeka & Libsu, 2014), which might be responsible for enhanced protein and fat contents observed in goats supplemented with MSL. It has been suggested that these antioxidant compounds may also assist in the prevention of meat degradation by oxidation (Moyo *et al.*, 2011, 2012).

To the authors' knowledge, no literature data for Ethiopia are available to compare the chemical composition of local goat meat. Thus, comparisons were made among goats reared in other tropical countries. The chemical composition values of the meat obtained from the present study are consistent with those reported by Kadim *et al.* (2004) for the Omani goat, Babiker & Bello (1986) for the Desert goat, and Agnihotri *et al.* (2006) for Barbari goats. The CP and fat contents of the meat of goats fed on the control and 120 g MSL supplemented diets were comparable to those reported by Agnihotri *et al.* (2006).

The meat pH values in the current study are consistent with the findings of Kannan *et al.* (2001) for Spanish goats (5.96) and Kadim *et al.* (2004) for the Omani goats (5.78). However, pH values of the meat in the present study are lower than those reported by Marinova *et al.* (2001) for Bulgarian White goat. The differences between breeds, treatment diets and also the muscle fibre proportions (red or white) might explain these differences as suggested by Kannan *et al.* (2001).

Increased cooking loss percentage is a reflection of the decreased water-holding capacity (increased expressed juice) associated with meat of high ultimate pH (Bouton *et al.*, 1971). The cooking loss percentage values across the treatment diets are within the normal range for goat muscles and are in good agreements with those of Babiker *et al.* (1990) and Das & Rajkumar (2010). However, Kannan *et al.* (2001) reported lower values for these parameters which might be attributed to the age of the studied goats themselves. The ash content of the meat did not vary across the treatments and is comparable with the results reported by Agnihotri *et al.* (2006).

5 Conclusion

The supplementation with *M. stenopetala* leaf of young Arsi-Bale male goats has considerably improved body weight gains, many carcass traits as well as meat quality as compared to the non-supplemented group. Moreover, the supplementation of *M. stenopetala* leaf enhanced the protein and ether extract content of the goats indicating a better retention of these nutrients. Consequently, supplementation with Moringa leaf is a viable option for improving the productivity of livestock particularly that of small ruminants under smallholder farming conditions where conventional protein supplementation of low quality local feed resources are not accessible and affordable.

Acknowledgements

This research project was fully supported by the research fund granted by the Vice President for Research and Technology Transfer of the Hawassa University for which the authors are highly grateful. The support received from Mr. Tadesse Bekore in analysing the feed nutrient composition is highly acknowledged.

References

Agnihotri, M. K., Rajkumar, V. & Dutta, T. K. (2006). Effect of feeding complete rations with variable protein and energy levels prepared using by-products of pulses and oilseeds on carcass characteristics, meat and meat ball quality of goats. *Asian-Australasian Journal of Animal Sciences*, 19 (10), 1437–1449.

AOAC (1995). *Official methods of analysis of AOAC International.* (16th ed.). Association of Official Analytical Chemists - AOAC International, Arlington, Virginia, USA.

Aregheore, E. M. (2002). Intake and digestibility of *Moringa oleifera*-batiki grass mixtures by growing goats. *Small Ruminant Research*, 46, 23–28.

Babiker, M. S. (2012). Chemical composition of some non-conventional and local feed resources for poultry in Sudan. *International Journal of Poultry Science*, 11 (4), 283–287.

Babiker, S. A. & Bello, A. (1986). Hot cutting of goat carcasses following early post-mortem temperature ageing. *Meat Science*, 17, 111–120.

Babiker, S. A., El Khider, I. A. & Shafie, S. A. (1990). Chemical composition and quality attributes of goat meat and lamb. *Meat Science*, 28, 273–277.

Banerjee, S., K., D. & Abebe, A. (2013). Some serum biochemical and carcass traits of Arsi Bale rams reared on graded levels of *Millettia ferruginea* leaf meal. *World Applied Sciences Journal*, 28 (4), 532–539.

Bouton, P. E., Harris, P. V. & Shorthose, W. R. (1971). Effect of ultimate pH upon the water-holding capacity and tenderness of mutton. *Journal of Food Science*, 36, 435–439.

CSA (2014). *Agricultural Sample Survey, Vol. II: Report on Livestock and Livestock Characteristics.* Central Statistical Agency (CSA), Federal Democratic Republic of Ethiopia, Addis Ababa, Ethiopia. Pp. 12–25

Das, A. K. & Rajkumar, V. (2010). Comparative study on carcass characteristics and meat quality of three Indian goat breeds. *Indian Journal of Animal Sciences*, 80 (10), 1014–1018.

Debela, E. & Tolera, A. (2013). Nutritive value of botanical fractions of *Moringa oleifera* and *Moringa stenopetala* grown in the mid-Rift Valley of southern Ethiopia. *Agroforestry Systems*, 87 (5), 1147–1155.

Dechasa, J., Sonder, K., Alemayehu, L., Mekonen, Y. & Anjulo, A. (2006). Leaf yield and Nutritive value of *Moringa stenopetala* and *Moringa oleifera* accessions: Its potential role in food security in constrained dry farming agro forestry system. *In:* Proc. of the International Workshop: Moringa and Other Highly Nutritious Plant Resources: Strategies, Standards and Market for Better Impact on Nutrition in Africa, Nov. 16–18, Accra, Ghana. pp. 1–14, Moringanews network.

Gala, S. (2005). Biodiversity in goats. *Small ruminant Research*, 60, 75–81.

Gebregiorgis, F., Negesse, T. & Nurfeta, A. (2011). Feed intake and utilization in sheep fed graded levels of dried moringa (*Moringa stenopetala*) leaf as a supplement to Rhodes grass hay. *Tropical Animal Health and Production*, 44, 511–517.

Kadim, I. T., Mahgoub, O., Srikandakumar, A., Al-Ajmi, D. S., Al-Maqbaly, R. S., Al-Saqri, N. M. & Johnson, E. H. (2004). Comparative effect of low levels of dietary cobalt and parenteral injection of vitamin B12 on carcass and meat quality characteristics in Omani goats. *Meat Science*, 66, 837–844.

Kannan, G., Kouakou, B. & Gelaye, S. (2001). Color changes reflecting myoglobin and lipid oxidation in chevon cuts during refrigerated display. *Small Ruminant Researc*, 42, 67–75.

Kebede, T., Lemma, T., Tadesse, E. & Guru, M. (2008). Effect of level of substitution of sweet potato (*Ipomoea batatas* L.) vines for concentrate on body weight gain and carcass characteristics of browsing Arsi-Bale goats. *Journal of Cell and Animal Biology*, 2, 36–42.

Manaye, T., Tolera, A. & Zewdu, T. (2009). Feed intake, digestibility and body weight gain of sheep fed Napier grass mixed with different levels of *Sesbania sesban*. *Livestock Science*, 122, 24–29.

Marinova, P., Banskalieva, V., Alexandrov, S., Tzvetkova, V. & Stanchev, H. (2001). Carcass composition and meat quality of kids fed sunflower oil supplemented diet. *Small Ruminant Research*, 42, 217–225.

McDonald, P., Edwards, R. A., Greenhalgh, J. F. D., Morgan, C., Sinclair, L. A. & Wilkinson, R. G.

(2010). *Animal Nutrition.* Pearson Education Limited. Edinburgh, Great Britain. 692p

Melesse, A., Bulang, M. & Kluth, H. (2009). Evaluating the nutritive values and *in vitro* degradability characteristics of leaves, seeds and seedpods from *Moringa stenopetala. Journal of the Science of Food and Agriculture*, 89, 281–287.

Melesse, A., Getye, Y., Berihun, K. & Banerjee, S. (2013). Effect of feeding different levels of *Moringa stenopetala* leaf meal on growth performance, carcass traits and some serum biochemical parameters of koekoek chickens. *Livestock Science*, 157, 498–505.

Melesse, A., Steingass, H., Boguhn, J., Schollenberger, M. & Rodehutscord, M. (2012). Effects of elevation and season on nutrient composition of leaves and green pods of *Moringa stenopetala* and *Moringa oleifera. Agroforestry Systems*, 86, 505–518.

Melesse, A., Tiruneh, W. & Negesse, T. (2011). Effects of feeding *Moringa stenopetala* leaf meal on nutrient intake and growth performance of Rhode Island Red chicks under tropical climate. *Tropical and Subtropical Agroecosystem*, 14, 485–492.

Mesfin, T. (2007). The influence of age and feeding regimen on the carcass traits of Arsi-Bale goats. *Livestock Research for Rural Development*, 19 (4).

Moyo, B., Masika, P. J., Hugo, A. & Muchenje, V. (2011). Nutritional characterization of Moringa (*Moringa oleifera* Lam.) leaves. *African Journal of Biotechnology*, 10 (60), 12925–12933.

Moyo, B., Masika, P. J. & Muchenje, V. (2012). Effect of supplementing crossbred Xhosa lop-eared goats castrates with *Moringa oleifera* leaves on growth performance, carcass and non-carcass characteristics. *Tropical Animal Health and Production*, 44 (4), 801–809.

Negesse, T., Makkar, H. P. S. & Becker, K. (2009). Nutritive value of some non-conventional feed resources of Ethiopia determined by chemical analyses and an *in vitro* gas method. *Animal Feed Science and Technology*, 154, 204–217.

NMA (2012). National Metrological Agency (NMA), Hawassa Branch Directorate, Hawassa, Southern Nations, Nationalities and Peoples Regional State, Ethiopia.

Olugbemi, T. S., Mutayoba, S. K. & Lekule, F. P. (2010a). Effect of Moringa (*Moringa oleifera*) inclusion in Cassava based diets fed to broiler chickens. *International Journal of Poultry Science*, 9 (4), 363–367.

Olugbemi, T. S., Mutayoba, S. K. & Lekule, F. P. (2010b). *Moringa oleifera* leaf meal as a hypocholesterolemic agent in laying hen diets. *Livestock Research for Rural Development*, 22 (4).

Qwele, K., Hugo, A., Oyedemi, S. O., Moyo, B., Masika, P. J. & Muchenje, V. (2013). Chemical composition, fatty acid content and antioxidant potential of meat from goats supplemented with Moringa (*Moringa oleifera*) leaves, sunflower cake and grass hay. *Meat Science*, 93 (3), 455–462.

Sánchez, R., Spröndly, E. & Ledin, I. (2006). Effect of feeding different level of foliage of *Moringa oleifera* to Creole dairy cows on intake, digestibility, milk production and composition. *Livestock Science*, 101, 24–31.

Sarwatt, S. V., Kapange, S. S. & Kakengi, A. M. V. (2002). Substituting sunflower seed-cake with *Moringa oleifera* leaves as a supplemental goat feed in Tanzania. *Agroforestry Systems*, 56, 241–247.

SAS (2010). *SAS/STAT® 9.3 User's guide.* SAS Institute Inc., Cary, NC, USA.

Sebsibe, A., Casey, N. H., van Niekerk, W. A., Azage, T. & Coertze, R. J. (2007). Growth performance and carcass characteristics of three Ethiopian goat breeds fed grainless diets varying in concentrate to roughage ratios. *South African Journal of Animal Science*, 37 (4), 221–232.

Simela, L. & Merkel, R. (2008). The contribution of chevon from Africa to global meat production. *Meat Science*, 80 (1), 101–109.

Tebeka, T. & Libsu, S. (2014). Assessment of antioxidant potential of *Moringa stenopetala* leaf extract. *Ethiopian Journal of Science and Technology*, 7 (2), 93–104.

Thurber, M. D. & Fahey, J. W. (2009). Adoption of *Moringa oleifera* to combat under-nutrition viewed through the lens of the "Diffusion of Innovations" theory. *Ecology of Food and Nutrition*, 48 (3), 212–225.

Van Soest, P. J. (1994). *Nutritional Ecology of the Ruminant.* (2nd ed.). Cornell University Press, Ithaca.

Van Soest, P. J., Robertson, J. B. & Lewis, B. A. (1991). Methods of dietary fibre, neutral detergent fibre and non-starch polysaccharides in relation to animal nutrition. *Journal of Dairy Science*, 74, 3583–3597.

17

Evaluation of cassava leaf meal protein in fish and soybean meal-based diets for young pigs

Siaka Seriba Diarra*, Malakai Koroilagilagi , Simeli Tamani , Latu Maluhola ,
Sila Isitolo , Jiaoti Batibasila , Tevita Vaea , Vasenai Rota , Ulusagogo Lupea

School of Agriculture and Food Technology, Alafua Campus, University of the South Pacific, Apia, Samoa

Abstract

The unavailability and high cost of traditional ingredients calls for more research into alternative sources for pig feeding in the South Pacific region. The effect of replacing feed protein with cassava leaf meal (CLM) protein in weaner and growing pigs' diets was investigated in two experiments. In experiment 1, three diets in which CLM protein replaced 0, 15 and 30 % of feed protein were fed each to five replicate pens of weaner pigs. Feed intake (FI), body weight gain (BWG) and feed conversion ratio (FCR) were improved and feed cost of gain reduced ($P < 0.05$) on 30 % while dressing percentage was maximized ($P < 0.05$) on 15 % protein replacement diets. In experiment 2, three diets containing 0, 30 and 45 % CLM protein as replacement for feed protein were fed as in experiment 1 to grower pigs. FI and BWG were reduced while FCR and feed cost of gain were increased ($P < 0.05$) above 30 % protein replacement. Dressing percentage assumed the highest value ($P < 0.05$) on 30 % replacement. It was concluded that replacing 30 % of feed protein with sun-dried CLM protein will maintain growth and reduce cost of pork production. Efficient use of CLM in the diet will be an alternative way of value addition to this by-product.

Keywords: high feed cost, alternative ingredients, feed processing, pig performance

1 Introduction

The scarcity and high cost of conventional feed ingredients such as corn and soybean is a major impediment to commercial pig production worldwide. The situation is further compounded in the South Pacific region where conventional feed ingredients are not grown in the region thus imported at exorbitant prices. Ayalew (2011) reported a 56 to 100 % increase in retail price of commercial pig feeds in Papua New Guinea between 2003 and 2011. There is therefore the need to increase research into locally available feed materials of low economic value in the region.

Locally available energy and protein sources exist in the South Pacific region but their use in commercial feed production has not received much attention. Cassava (*Manihot esculenta*) is a root crop well adapted to many climatic and soil types. Cassava root is a major source of carbohydrates in human diets in many parts of the world. Cassava leaf, a by-product which may make up to 30 % of the root yield at harvest (Ravindran, 1993) is a moderate source of protein. The protein content of the leaves ranges from 16.7 to 39.9 % (Ravindran et al., 1987; Ly & Ngoan, 2007) with almost 85 % of the protein fraction being true protein (Ravindran, 1993). Cassava leaf protein is deficient in sulphur amino acids but has a good balance of other essential amino acids (Gomez & Valdivieso, 1985). The digestibility of cassava leaf amino acids is reported to be high in rats (Eggum, 1970). Cyanide is the major anti-nutritional factor in cassava products which can be reduced below toxic levels by sun-drying (Ravindran, 1993). Sweet cultivars of cas-

* Corresponding author
Email: diarra_s@usp.ac.fj or siakadi2012@gmail.com

sava which are low in cyanide content are grown in Samoa as pig feed and rarely consumed by Samoans. Currently, the leaves from these cultivars are not utilised in the country, thus making it readily available as stock feed. There are reports on the inclusion of cassava leaf meal in monogastric diets (Phuc *et al.*, 1999), but its use as replacement of feed protein in diets for young pigs is not documented. This study investigated the effects of replacing feed protein with cassava leaf protein on growth, carcass traits and gut weight of young pigs.

2 Materials and methods

2.1 Source and processing of cassava products

A sweet cultivar of cassava grown on the University of the South Pacific's School of Agriculture and Food Technology farm was harvested at 7 months after planting and used for the experiment. Cassava leaves (with petiole) and roots were chopped and sun-dried (mean temperature and relative humidity ~28° C and 58 % respectively) separately on a cemented floor for 48 and 72 hours respectively, then ground in a hammer mill to pass through a 2 mm sieve and labelled cassava leaf meal (CLM) and cassava root meal (CRM). Cassava products (CLM and CRM) and fish meal were analysed for proximate composition, CLM and fish meal for amino acid composition and CLM and CRM for hydrocyanic acid content (Table 1) and used in the formulation of the diets in experiments 1 and 2. The research protocols were approved by the animal ethics committee of the University of the South Pacific.

2.2 Experiment 1: Replacement of feed protein with cassava leaf meal protein in weaner pigs' diets (14–30 kg live weight)

2.2.1 Experimental diets

Three pig weaner diets were formulated to meet NRC (2012) recommendations (ME $\approx$ 14 MJ kg^{-1}, crude protein $\approx$ 18 %, lysine $\approx$ 1.1 % and Methionine $\approx$ 0.8 %, Table 2). All diets contained maize and cassava root meal as main energy sources. The control diet was devoid of CLM while in diets 2 and 3; feed protein was replaced with CLM protein at 15 and 30 % respectively.

2.2.2 Pigs and management

Thirty weaner crossbred pigs (Yorkshire × Landrace) were weighed individually (14 kg ± 0.18) and assigned in pairs to fifteen standard concrete floor pig pens. Each of the diets was fed to pigs in 5 pens in a completely randomized design for a period of 60 days. The diets and drinking water were provided *ad-libitum* throughout the experimental period.

2.3 Experiment 2: Replacement of feed protein with cassava leaf meal protein in growing pigs' diets (28–52 kg live weight)

2.3.1 Experimental diets

Three pig grower diets were formulated as in experiment 1 except that protein, lysine and methionine contents were reduced to 16, 1 and 0.6 % respectively (Table 3). Based on the results of experiment 1 protein replacement levels in this experiment were increased to 30 and 45 % in diets 2 and 3 respectively.

2.3.2 Pigs and management

Thirty grower crossbred pigs (Yorkshire × Landrace) were weighed individually (28 kg ± 0.21) and assigned to the same pens as in experiment 1. Each of the diets was fed to pigs in 5 pens in a completely randomized design for a period of 60 days as in experiment 1. Feed and drinking water were supplied *ad-libitum* throughout the experimental period.

2.4 Data collection

In both experiments, data were collected on growth (feed consumption, weight gain and feed conversion ratio), carcass traits and gut measurements. Feed consumption data were obtained by difference between quantities fed and left-over in each pen. Pigs were weighed at the beginning and end of the experiment and weight gain calculated by difference. Feed conversion ratio (FCR) was calculated as feed consumed to weight gained. At the end of each experiment all the pigs were stunned electrically, euthanized and dressed, and carcass weighed and expressed as percentages of the live weight. Back fat thickness of the carcass was measured on the last rib at 2 cm from the backbone using a Renco Lean-Meater (Renco Corporation, Minneapolis, MN. 55401 USA). Segments of the gut (stomach, small and large intestines) were weighed full and empty and the content calculated by difference and expressed as percentage of the live weight of the animal.

2.5 Chemical analysis

Cassava products and experimental diets were analysed for proximate composition according to AOAC (1990). Dry matter was determined after oven-dying samples at 105° C overnight. Ash was determined by

Table 1: *Proximate composition, NDF, amino acid contents (g kg^{-1} DM) and HCN content (mg kg^{-1} DM) of Cassava root meal (CRM), Cassava leaf meal (CLM) and fish meal used in the experiment*

Constituents	CRM	CLM	Fish meal	Soybean meal[†]
Dry matter	94.8	93.65	95.0	96.9
Crude protein	3.7	25.8	63.2	34.5
Ether extract	2.84	11.4	9.15	18.7
Ash	4.33	9.62	17.1	5.6
Crude fibre	4.6	13.91	1.8	16.4
NDF	10.2	29.55		13.6
Phenylalanine		2.6	1.88	1.69
Histidine		2.1	1.1	0.94
Lysine		1.04	3.02	2.2
Leucine		3.1	3.33	2.58
Valine		3.2	2.58	1.77
Methionine		0.34	1.9	0.56
Arginine		2.9	3.09	2.39
Threonine		2.7	2.14	1.48
Isoleucine		1.12	2.03	1.60
HCN[‡]	12.0	15.03		

[†] Values adapted from Devi (2016)
[‡] Fresh cassava root and leaves contained 47.8 and 61.6 mg HCN kg^{-1} DM respectively.

Table 2: *Ingredients composition (on as-fed basis) and analysed composition of the diets in experiment 1.*

Ingredients (g kg^{-1} DM)	Level of protein replacement (%)		
	0	15	30
Corn	234	227.8	222.1
CRM	350	341.7	333.1
Wheat middling	117	114	111
Fish meal	92	87	82
Soybean meal	184	174	164
CLM	0	32	64
Salt	6	6	6
Lysine HCl	3	3.3	3.5
DL-Methionine	1.5	1.7	1.8
Premix[†]	2.5	2.5	2.5
Coral sand	10	10	10
Analysed composition (% DM)			
Dry matter	91.33	90.78	90.65
Crude protein	18.08	17.97	18.03
Ether extract	11.74	11.98	11.79
Ash	10.30	9.80	9.84
Crude fibre	7.62	8.86	9.20
NDF	21.16	26.09	27.06
Calculated composition (%)			
ME (MJ kg^{-1})[‡]	14.62	14.30	14.12
Lysine	1.10	1.08	1.08
Methionine	0.79	0.77	0.78

[†] Biomix provided per kg diet: Vit A: 10,000 IU; Vit D$_3$: 2,000 IU; Vit E: 23 mg; Niacin: 27.5 mg; Vit B$_1$: 1.8 mg; B$_2$: 5mg; B$_6$: 3 mg; B$_{12}$: 0.015 mg; K$_3$: 2 mg; Pantothenic acid: 7.5 mg; Biotin: 0.06 mg; Folic acid: 0.75 mg; Choline Chloride: 300 mg; Cobalt: 0.2 mg; Copper: 3 mg; Iodine: 1 mg; Iron: 20 mg; Manganese: 40 mg; Selenium: 0.2 mg; Zinc: 30 mg; Anti-oxidant: 1.25 mg.

[‡] calculated according to Wiseman (1987) as: ME $= 3951 + 54.4 \times EE - 88.7 \times CF - 40.8 \times Ash$
Where EE: ether extract and CF: crude fibre; CRM: Cassava root meal; CLM: Cassava leaf meal

Table 3: *Ingredients composition (on as-fed basis) and analysed composition of the diets in experiment 2.*

Ingredients (g kg^{-1} DM)	Level of protein replacement (%)		
	0	*30*	*45*
Corn	254.9	240.7	233.8
CRM	318.7	361	350.6
Wheat middling	191.2	120	116.4
Fish meal	70.7	68.5	67.5
Soybean meal	141.5	137.5	134.4
CLM	0	48.8	73.5
Salt	6	6	6
Lysine HCl	3	3.3	3.5
DL-Methionine	1.5	1.7	1.8
Premix [†]	2.5	2.5	2.5
Coral sand	10	10	10
Analysed composition (% DM)			
Dry matter	93.5	92.9	93.1
Crude protein	16.04	15.98	16.01
Ether extract	8. 98	9.24	9.45
Ash	7.22	7.45	7.80
Crude fibre	8.54	8.88	8.9
NDF	25.39	27.24	28.23
Calculated composition (%)			
ME (MJ kg^{-1}) [‡]	14.17	14.07	14.05
Lysine	1.02	1.01	1.01
Methionine	0.62	0.59	0.60

[†] Biomix provided per kg diet: Vit A: 10,000 IU; Vit D$_3$: 2,000 IU; Vit E: 23 mg; Niacin: 27.5 mg; Vit B$_1$: 1.8 mg; B$_2$: 5mg; B$_6$: 3 mg; B$_{12}$: 0.015 mg; K$_3$: 2 mg; Pantothenic acid: 7.5 mg; Biotin: 0.06 mg; Folic acid: 0.75 mg; Choline Chloride: 300 mg; Cobalt: 0.2 mg; Copper: 3 mg; Iodine: 1 mg; Iron: 20 mg; Manganese: 40 mg; Selenium: 0.2 mg; Zinc: 30 mg; Anti-oxidant: 1.25 mg.

[‡] calculated according to Wiseman (1987) as: ME = $3951 + 54.4 \times EE - 88.7 \times CF - 40.8 \times Ash$
Where EE: ether extract and CF: crude fibre; CRM: Cassava root meal; CLM: Cassava leaf meal

burning the samples in a furnace at 500° C. Nitrogen was determined using the Kjeldahl apparatus and crude protein calculated as nitrogen $\times$ 6.25 (feed factor). Samples were hydrolysed in acid (6 mol L^{-1} HCl, 115° C, for 23 h) and amino acids analysed by chromatography using the AAA 400 analyser (INGOS, Prague, CZ). Fat extraction was carried out in the Soxhlet apparatus and neutral detergent fibre according to van Soest *et al.* (1994). Hydrogen cyanide (HCN) was determined by boiling and titrating samples in 0.1N AgNO$_3$ and entrapping HCN in KOH (AOAC, 1990).

2.6 Statistical analysis

Analysis of variance (ANOVA) (Steel & Torrie, 1980) was carried out on data collected using the GLM of SPSS (Statistical Package for Social Sciences, version 22). Pen was the experimental unit for feed consumption data while weight gain, carcass and gut weight were measured on individual pigs. Treatment means were compared using the Bonferoni test and significant differences between means were reported at 5 % level of probability.

3 Results

3.1 Chemical analysis

Results of chemical analysis are presented in Table 1.With the exception of methionine the protein of CLM used in the study had a good supply in essential amino acids. The concentration of HCN was higher in cassava leaf than the root. Sun-drying reduced HCN content in cassava root and leaf meals by 74.9 and 75.6 % respectively. The cost of CLM (US$ 0.14 per kg DM) was mainly from labour (harvesting and drying) and grinding.

3.2 Performance of the pigs in experiment 1

Growth, carcass and gut content data of the pigs in experiment 1 are presented in Table 4. Feed intake (FI), body weight gain (BWG) and FCR were significantly improved and feed cost of gain reduced ($P < 0.05$) on the 30 % protein replacement diet (6.4 % dietary CLM). Daily intake of essential nutrients (ME, lysine and methionine) was also increased in the group fed 30 % protein replacement diet. Crude fibre and neutral detergent fibre increased linearly as CLM level increased in the diet.

The pigs were slaughtered at an average weight of 30 and 50 kg in experiments 1 and 2 respectively. These weights are within the preferred slaughter weight range of 25 to 55 kg in Samoa. A higher dressing percentage ($P < 0.05$) was recorded on the 15 % protein replacement (3.2 % dietary CLM). Per cent dressing did not differ between the control and 30 % protein replacement diets ($P > 0.05$). There was no significant dietary effect on back fat thickness. Pigs fed the CLM-based diets recorded higher digesta content in the stomach compared to the control fed group ($P < 0.05$). Digesta in the small and large intestines of the pigs was not affected by the diet ($P > 0.05$).

3.3 Performance of the pigs in experiment 2

From the results of the performance of the pigs in experiment 2 (Table 5) feeding CLM above 30 % protein replacement or 4.88 % dietary CLM significantly reduced FI and BWG and increased FCR and feed cost of gain ($P < 0.05$). The best FCR and lowest feed cost of gain were observed on the 30 % protein replacement diet. The reduced FI on the 45 % protein replacement diet resulted in significantly lower daily intake of lysine and methionine on this diet ($P < 0.05$). There was no significant dietary effect on ME intake of the pigs ($P > 0.05$).

Dressing per cent assumed the highest value ($P < 0.05$) on 30 % replacement but did not differ (($P > 0.05$) between the control and 45 % protein replacement (7.35 % dietary CLM) diets. Back fat thickness and digesta content of the different segments of the gut were not affected by dietary treatment ($P > 0.05$).

4 Discussion

4.1 Chemical analysis

Poor methionine content in CLM has earlier been reported (Roggers & Milner, 1963; Eggum, 1970). The crude protein and crude fibre contents of cassava root

and leaf meals used in this study fall within the ranges of 1–3 % (Buitrago A., 1990; Salcedo et al., 2010) and 16.7–39.9 % (Allen, 1984) reported for cassava root and leaf meals respectively. The higher fibre content of CLM was reflected by a higher fibre in the diets containing CLM compared to the control. The high HCN in CLM compared to cassava root in this study supports earlier findings by Yeoh & Chew (1976) who reported six times more HCN in cassava leaf compared to the root. The significant reduction in HCN in the dried products (root and leaves) is in agreement with earlier reports (Gomez & Valdivieso, 1985; Ravindran, 1993; Ly & Ngoan, 2007). The latter authors reported a 51 % reduction of HCN in cassava leaves after 24 hours of sun-drying. The higher HCN reduction in CLM in this study may be attributed to the duration of sun-drying (48 hours) and chopping which might have increased the activity of endogenous linamarase on cyanogenic glycosides. HCN content of both sun-dried CRM and CLM was below the threshold level of 50 mg kg^{-1} reported for growing pigs (Bolhuis, 1954).

4.2 Performance of the pigs in experiment 1

The increased fibre content with increasing CLM was probably the reason for increased feed intake of pigs fed the CLM-containing diets to compensate for energy on these diets. Contrary to these results, Kallabis & Kaufmann (2012) reported reduced feed intake and weight gain in growing pigs fed diets containing lower fibre level (7.3 %) compared to about 9 % fibre in the diets containing CLM in this experiment (8.86–9.2 % CF). Pig age, fibre source and composition, feed processing and diet composition have all been reported to affect the utilisation of dietary fibre by pigs (Low, 1993). The higher feed intake on CLM diet resulted in higher daily intake of essential nutrients (ME, lysine and methionine) compared to the control. This pattern of amino acids intake may explain the improved growth performance of pigs fed the 30 % CLM diet. Edwards & Campbell (1993) observed a positive relationship between amino acid intake and body protein accretion in pigs. Ly et al. (2012) also reported improved performance of pigs fed ensiled cassava leaf protein diets supplemented with dl-methionine and l-lysine. Methionine is involved in growth and maintenance of body protein and also known for detoxification of HCN in-vivo (Tewe, 1992). The linear reduction in feed cost and the improved FCR with CLM inclusion were the main reasons for lower feed cost of gain observed on these diets.

The improved dressing percentage observed on the 15 % protein replacement diet could not be explained. The role of dietary methionine (Corzo et al., 2006;

Table 4: *Growth performance, carcass traits and gut content of pigs fed cassava leaf protein as replacement for feed protein (14 to 30 kg body weight)*

Parameters	Level of protein replacement (%)				
	0	15	30	s.e.m	P value
Initial weight (kg/pig)	14.2	14	14.1	0.047	0.529
Feed intake (kg/pig/day)	1.0[b]	1.06[ab]	1.15[a]	0.036	0.008
Body weight gain (kg/pig/day)	0.28[b]	0.29[ab]	0.35[a]	0.018	0.007
Feed conversion ratio (feed: gain)	3.57[a]	3.66[a]	3.29[b]	0.091	0.011
Cost of kg feed (US$)	0.62	0.52	0.47	n.a.	n.a.
Feed cost per kg gain (US$)	2.21[a]	1.90[ab]	1.5[b]	0.168	0.005
Daily nutrient intake					
ME (MJ kg^{-1})	14.62[b]	15.16[b]	16.24[a]	0.287	0.036
Lysine (g)	11[b]	11.45[ab]	12.42[a]	0.342	0.028
Methionine (g)	7.9[b]	8.2[ab]	9.0[a]	0.263	0.023
Carcass traits and gut content (% live weight)					
Dressing per cent	64.5[b]	68.3[a]	63.8[b]	1.143	0.004
Back fat thickness	11.2	10.8	11.4	0.172	0.150
Digesta in the stomach	0.8[b]	1.3[a]	1.0[a]	0.121	0.001
Digesta in the small intestine	0.5	0.5	0.5	0.008	0.063
Digesta in the large intestine	0.6	0.6	0.6	0.021	0.052

s.e.m: standard error of the mean; [a], [b]: means within the row bearing different superscripts are significantly different ($P=0.05$); n.a: not analysed

Table 5: *Growth performance, carcass traits and gut content of pigs fed cassava leaf protein as replacement for feed protein (28 to 55 kg body weight)*

Parameters	Level of protein replacement (%)				
	0	30	45	s.e.m	P value
Initial weight (kg/pig)	28.3	28.3	28.3	0.012	0.529
Feed intake (kg/pig/day)	1.6[a]	1.6[a]	1.5[b]	0.016	0.005
Body weight gain (kg/pig/day)	0.5[ab]	0.6[a]	0.4[b]	0.044	0.000
Feed conversion ratio (feed: gain)	3.4[ab]	2.8[b]	3.7[a]	0.218	0.011
Cost of kg feed (US$)	0.5	0.5	0.4	n.a.	n.a.
Feed cost of kg gain (US$)	1.7[a]	1.3[b]	1.6[a]	0.109	0.004
Daily nutrient intake					
ME (MJ kg^{-1})	23.0	23.1	21.8	0.614	0.056
Lysine (g)	16.5[a]	16.6[a]	15.7[b]	0.242	0.008
Methionine (g)	10.0[a]	9.7[ab]	9.3[b]	0.174	0.013
Carcass traits and gut content (% live weight)					
Dressing per cent	68.8[b]	79.4[a]	66.9[b]	3.176	0.018
Back fat thickness	14.5	14.4	14.5	0.023	0.101
Digesta in the stomach	1.5	1.5	1.4	0.167	0.061
Digesta in the small intestine	0.9	0.9	0.9	0.017	0.059
Digesta in the large intestine	0.9	1.0	1.1	0.052	0.064

s.e.m: standard error of the mean; [a], [b]: means within the row bearing different superscripts are significantly different ($P=0.05$); n.a: not analysed

Wang *et al.*, 2010) and lysine (Grisoni *et al.*, 1991; Nasr & Kheiri, 2011) in reducing body fat deposition and the similarity of the diets in these amino acids may be a reason for the pattern of back fat deposition observed in both experiments. Heavier digesta in the stomach of pigs fed CLM included diets may be due to slower digesta transit time due to higher fibre content. Jørgensen *et al.* (1996) observed that pigs adapt to high fibre diets by increasing gut weight.

4.3 Performance of the pigs in experiment 2

When the level of protein replacement in the diet exceeded 30% (11.38% crude fibre and 27.24% NDF) feed intake was reduced probably on account of the higher dietary fibre. Similar results have been reported by Len *et al.* (2008) who found that feed intake was reduced in Mong Cai × Yorkshire and Landrace × Yorkshire crossbred pigs when dietary fibre increased from 6.4 to 10%. Souza da Silva *et al.* (2012) also observed lower feed intake in growing pigs on high fibre diets as a result of gut fill. This reduced feed intake with the resultant lower intake of lysine and methionine may explain the lower weight gain, poorer FCR and higher feed cost of gain on the 45% compared to the 30% protein replacement diet. Low availability of CLM methionine (Eggum, 1970) could be implicated in the reduced performance above this level of protein replacement.

Like in experiment 1 the reason for the pattern of dressing percentage in this experiment was not understood. The similarities observed in digesta contents of different gut segments is probably the result of improved feed utilisation on the control and 30% replacement and low feed intake on the 45% replacement diets.

Based on the results of experiments 1 and 2, it is concluded that replacement of 30% feed protein with sun-dried cassava leaf meal protein in fish and soybean meal-based diets will maintain the growth performance of young pigs and reduce dependence on expensive feed resources. More research in the effects of age of cassava leaves, processing methods and diet composition on pig performance and pork quality is needed. In addition to feed cost reduction, the substitution will add value to cassava foliage in the study area.

Acknowledgements

Authors wish to acknowledge USP, Alafua farm for providing the cassava products and experimental pigs.

References

Allen, R. D. (1984). Feedstuffs ingredient analysis table. *Feedstuffs*, 56 (30), 25–30.

AOAC (1990). *Official Methods of Analysis 15th edition*. Association of Official Analytical Chemists (AOAC), Washington, DC.

Ayalew, W. (2011). Improved use of local feed resources for mitigating the effects of escalating food prices in PNG: a contribution for food security policy dialogue. Paper presented at the Food Security Conference on High Food Prices in Papua New Guinea, National Research Institute, Port Moresby, 6–8 September 2011.

Bolhuis, G. G. (1954). The toxicity of cassava root. *Netherlands Journal of Agricultural Science*, 2, 176–185.

Buitrago A., J. A. (1990). *La Yuca en la Alimentacion Animal*. Centro Internacional de Agricultura Tropical (CIAT), Cali, Colombia. CIAT No. 85, 446 pp.

Corzo, A., Kidd, M. T., Dozier, W. A., Shack, L. A. & Burgess, S. C. (2006). Protein expression of pectoralis major muscle in chickens in response to dietary methionine status. *British Journal of Nutrition*, 95, 703–708.

Devi, A. (2016). *Effect of diet composition on the utilization of copra meal by finishing broiler chickens*. Master's thesis, University of the South Pacific. Pp 22–23.

Edwards, A. C. & Campbell, R. G. (1993). Energy-protein interactions in pigs. *In:* Cole, D. J. A., Haresign, W. & Garnsworthy, P. C. (eds.), *Recent Developments in Pig Nutrition 2*. pp. 30–46, Nottingham University Press, Nottingham.

Eggum, B. O. (1970). The protein quality of cassava leaves. *British Journal of Nutrition*, 24, 761–769.

Gomez, G. & Valdivieso, M. (1985). Cassava foliage: chemical composition, cyanide content and effect of drying on cyanide elimination. *Journal of the Science of Food and Agriculture*, 36, 433–441.

Grisoni, M. L., Uzu, G., Larbier, M. & Geraert, P. A. (1991). Effect of dietary lysine on lipogenesis in broilers. *Reproduction, Nutrition, Development*, 31, 683–690.

Jørgensen, H., Zhao, X., Eggum, B. O. & Zhao, X. Q. (1996). The influence of dietary fibre and environmental temperature on the development of gastrointestinal tract, digestibility, degree of fermentation in the hind-gut and energy metabolism in pigs. *British Journal of Nutrition*, 75, 365–378. doi: 10.1079/BJN19960140.

Kallabis, K. E. & Kaufmann, O. (2012). Effect of a high-fibre diet on the feeding behaviour of fattening pigs. *Archiv für Tierzucht*, 55 (3), 272–284.

Len, N. T., Lindberg, J. E. & Ogle, B. (2008). Effect of dietary fibre level on the performance and carcass traits of Mong Cai, F1 crossbred (Mong Cai × Yorkshire) and Landrace × Yorkshire pigs. *Asian-Australasian Journal of Animal Science*, 21, 245–251. doi:10.5713/ajas.2008.60598.

Low, A. G. (1993). Role of dietary fibre in pig feeds. *In:* Cole, D. J. A., Haresign, W. & Garnworthy, P. C. (eds.), *Recent Developments in Pig Nutrition 2.* pp. 137–162, Nottingham University Press, Nottingham.

Ly, N. T. H. & Ngoan, L. D. (2007). Evaluation of the economic efficiency of using cassava leaves (variety KM 94) in diets for pigs in Central Vietnam. *Journal of Science and Technology of Agriculture*, 12, 275–284.

Ly, N. T. H., Ngoan, L. D., Verstegen, M. W. A. & Hendriks, W. H. (2012). Pig performance increases with the addition of dl-methionine and l-lysine to ensiled cassava leaf protein diets. *Tropical Animal Health and Production*, 44 (1), 165–172. doi: 10.1007/s11250-011-9904-3.

Nasr, J. & Kheiri, F. (2011). Effect of different lysine levels on Arian broiler performances. *Italian Journal of Animal Science*, 10, 170–174.

Phuc, B. H. N., Ogle, B. & Lindberg, J. E. (1999). Effect of replacing soybean protein with cassava leaf protein in cassava root meal based diets for growing pigs on digestibility and N retention. *Animal Feed Science and Technology*, 83, 223–235.

Ravindran, V. (1993). Cassava leaves as animal feed: Potential and limitations. *Journal of the Science of Food and Agriculture*, 61 (2), 141–150.

Ravindran, V., Kornegay, E. T. & Rajaguru, A. S. B. (1987). Influence of processing methods and storage time on the cyanide potential of cassava leaf meal. *Animal Feed Science and Technology*, 17, 227–234.

Roggers, D. J. & Milner, M. (1963). Amino acid profile of manioc leaf protein in relation to nutritive value. *Economic Botany*, 17 (3), 211–216.

Salcedo, A., Valle, A. D., Sanchez, B., Ocasio, V., Ortiz, A., Marquez, P. & Siritunga, D. (2010). Comparative evaluation of physiological post-harvest root deterioration of 25 cassava (*Manihot esculenta*) accessions: visual vs. hydroxycoumarins fluorescent accumulation analysis. *African Journal of Agricultural Research*, 5, 3138–3144.

Souza da Silva, C., van den Borne, J. J. G. C., Gerrits, W. J. J., Kemp, B. & Bolhuis, J. E. (2012). Effects of dietary fibres with different physicochemical properties on feeding motivation in adult female pigs. *Physiology and Behaviour*, 107, 218–230. doi: 10.1016/j.physbeh.2012.07.001.

van Soest, P. J., Robertson, J. B. & Lewis, B. A. (1994). Methods for dietary fibre, neutral detergent and non-starch polysaccharides in relation to animal nutrition. *Journal of Dairy Science*, 74, 3583–3597.

Steel, R. G. D. & Torrie, J. H. (1980). *Principles and Procedures of Statistics. A biometrical Approach.* (2nd ed.). McGraw Hills Book Co., New York, USA.

Tewe, O. O. (1992). Detoxification of cassava products and effects of residual toxins on consuming animals. *In:* Machin, D. & Nyvild, S. (eds.), *Roots, Tubers, Plantains and Bananas in Animal Feeding.* pp. 81–98, FAO, Rome, Italy. 301 p.

Wang, Z. Y., Shi, S. R., Zhou, Q. Y., Yang, H. M., Zou, J. M., Zhang, K. N. & Han, H. M. (2010). Response of growing goslings to dietary methionine from 28 to 70 days of age. *British Poultry Science*, 51, 118–121.

Wiseman, J. (1987). Feeding of Non-Ruminant Livestock. *In:* Wiseman, J. (ed.), *Meeting nutritional requirement from available resources.* Butterworth and Co. Ltd. 370 p.

Yeoh, H. H. & Chew, M. Y. (1976). Protein content and amino-acid composition of cassava leaf. *Phytochemistry*, 15, 1597–1599.

18

Effect of pre-plant treatments of yam (*Dioscorea rotundata*) setts on the production of healthy seed yam, seed yam storage and consecutive ware tuber production

Abiodun O. Claudius-Cole [a,c,*], Lawrence Kenyon [b], Daniel L. Coyne [c]

[a]*Department of Crop Protection and Environmental Biology, University of Ibadan, Ibadan, Nigeria*
[b]*AVRDC – The World Vegetable Center, Shanhua, Tainan, Taiwan*
[c]*International Institute of Tropical Agriculture (IITA), Oyo Road, PMB 5320, Ibadan (Oyo State), Nigeria*

Abstract

Numerous pests and diseases of yams are perpetuated from season to season through the use of infected seed material. Developing a system for generating healthy seed material would disrupt this disease cycle and reduce losses in field and storage. The use of various pre-plant treatments was evaluated in field experiments carried out at three sites in Nigeria. Yam tubers of four preferred local cultivars were cut into 100 g setts and treated with pesticide (fungicide + insecticide mixture), neem extract (1 : 5 w/v), hot water (20 min at 53 °C) or wood ash (farmers practice) and compared with untreated setts. Pesticide treated setts sprouted better than all other treatments and generally led to lower pest and disease damage of yam tubers. Pesticide treatment increased tuber yields over most treatments, depending on cultivar, but effectively doubled the production as compared to the control. Pesticide and hot water treated setts produced the healthiest seed yams, which had lower storage losses than tubers from other treatments. These pre-treated seed yams produced higher yields corresponding to 700 % potential gain compared to the farmers usual practice. Treatments had no obvious influence on virus incidence, although virus-symptomatic plants yielded significantly less than non-symptomatic plants. This study demonstrated that pre-plant treatment of setts with pesticide is a simple and effective method that guarantees more, heavier and healthier seed yam tubers.

Keywords: hot water treatment, mancozeb, neem, pesticide dip, seed health, yam tubers

1 Introduction

The development of sustainable seed systems, which can consistently supply seed of high quality that farmers can rely upon and trust, is a necessary foundation for improving yam (*Dioscorea* spp.) productivity. In particular, seed material is commonly infected with viruses and plant-parasitic nematodes, which affect seed viability. This in turn reduces sprouting and plant vigour,

leading to missing plants and reduced yields of plants that have sprouted (Degras, 1993).

Pests and pathogens play a major role in yam losses, which are incurred both in the field and during storage. Mealybugs, scale insects and nematodes (*Scutellonema bradys*, *Pratylenchus* spp. and *Meloidogyne* spp.) will also exaggerate fungal (e.g. *Botryodiplodia* spp., *Aspergillus* spp., *Fusarium* spp.) and bacterial (e.g. *Erwinia* spp.) pathogen tuber infection in the field (Amusa *et al.*, 2003; Ogaraku & Usman, 2008). Damage that occurs during storage, leads to reduced quality and quantity of food and planting material (Emehute *et al.*, 1998;

---
* Corresponding author
Email: b.claudiuscole@gmail.com; bi_cole@yahoo.com

Bridge *et al.*, 2005). Most losses originate from pre-harvest invasion or infection and/or damage during harvest and transportation (Morse *et al.*, 2000). In Nigeria, storage losses in yams are estimated at ca. 30 % due to nematodes and at ca. 50 % due to fungi, (Wood *et al.*, 1980; Amusa *et al.*, 2003).

Poor quality planting material leads to the perpetuation of disease cycles, returning inoculum from the store back to the field. This adversely affects crop establishment, yield and storability of harvested tubers, ensuring a continued negative impact on quality, especially of highly susceptible cultivars. Yams are usually cultivated mainly for the market as ware yam, while seed yams are produced intentionally as planting material. Ware yams are fairly large tubers traditionally weighing between 2–10 kg while seed yams, used for the production of the ware yams, typically weigh 200 g to 1 kg. Traditional methods by farmers for supplying seed includes reserving smaller tubers or 'milking' their ware yam ahead of plant senescence for seed, by removing most of the ware yam and leaving a small portion behind to produce small sized tubers for seed (Asumugha *et al.*, 2007; Nchinda *et al.*, 2009). However, such smaller sized tubers may be a consequence of inherent disease infection, which has suppressed production and tuber size, while the regrowth of milked tubers is prone to enhanced levels of seed infection borne from a mature mother plant. Seed yam production using the minisett technology was introduced in Ghana and Nigeria, however, the small (25 g) minisett size required special care and the resulting tubers were often too small to plant as seed yams. These proved to be key obstacles to farmers towards adopting the technique (Onyenweaku, 1991; Langyintuo, 1996). The aim of the current study was to identify a suitable and acceptable system that could provide a basis for sustainable production of healthy, affordable, whole seed yam. Simple, pre-plant treatments of cut yam setts of were assessed for their effect on yield and health of seed yams from the field and during storage. The study further assessed the extended impact of the resultant seed through to a second season of ware yam productivity.

2 Materials and methods

2.1 Site and experimental details

Field trials were carried out in Nigeria at three sites located in three ecological zones: the Guinea savanna (Idah, Kogi State), the forest zone (Aramoko, Ekiti State) and the forest transition zone (Ibadan, Oyo State). Seed yams of two locally popular cultivars of *D. rotundata* were sourced from markets at each location: cv. *Imola* and cv. *Akpaji* were planted at Idah, while cv. *Sogbe* and cv. *Ajimokun* were planted at Aramoko and Ibadan. Yam setts were planted in plots of 10×10 m (400 plants per plot) in Idah and 5×10 m (200 plants per plot) in Aramoko and Ibadan, with five replications per treatment at each location. The experiments were laid out in a 5×5 Latin square design in Idah and a randomized complete block design in Ekiti and Ibadan. Plants were spaced at 25 cm within rows and 1 m between rows. The variation in plot size between sites was due to the deterioration of the planting material at Ekiti and Ibadan as a result of poor quality seed.

2.2 Pre-treatment for seed yam production

The five pre-planting seed sett treatments were: pesticide (fungicide + insecticide) dip; hot water (HW) treatment; coating in wood ash (farmer practice); coating in neem leaf slurry; and untreated control. Yam setts were cut from whole tubers (~ 500–1000 g) into ~ 100 g pieces prior to treatment, except for tubers treated with HW, which were cut following treatment. The pesticide treatment was prepared at the rate of 100 g of fungicide (Mancozeb [Maneb®] a.i. concentration 6 mg g^{-1}) and 70 ml of insecticide (Diazinon [Basudin® 600EC] a.i. concentration 600 g L^{-1}) in 10 L of water. The pesticide mix was prepared in a 30 L plastic container into which yam setts were dipped for 5 min then set aside to drain. Neem leaves were collected fresh from trees in Ibadan, air dried in a glasshouse and ground to powder. At planting, a slurry was prepared by mixing 1.0 kg of powdered neem leaves in 5 L of water, which was sufficiently thick to provide a thin coat on the yam setts when dipped. Coated yam setts were spread out to dry. For the hot water (HW) treatment whole tubers were fully submerged in water, heated to 53 °C, for 20 min, then set aside to dry and cool before cutting into setts. Treatment of cut setts with wood ash was undertaken by rolling cut setts in wood ash in a large nylon bag until all setts were covered, this represented the farmers' usual practice. Another farmers' practice of cutting and allowing setts to dry overnight without treatment represented the control. All treatments were undertaken one day ahead of planting.

2.3 Crop growth and damage parameters measured

Percentage germination (sprouting) was assessed at 4 and 8 weeks after planting (WAP) from all plants per plot. Foliar disease assessment was conducted at 8 and 12 WAP per plot and percentage incidence was calculated. Plants exhibiting symptoms of virus were labelled with a tag to be separated at harvest for data collection. At harvest, approx. 7 months after planting, tuber yield and number per plant and plot were recorded.

At harvest, twenty randomly selected tubers per plot were assessed for nematode damage (galls, cracking and flaking), insect damage (termite tunnels, beetle holes, presence of scale insect and mealy bugs) and rots using a rating scale of 1 to 3, where 1 = absence, 2 = mild to moderate and 3 = severe damage. The 20 tubers were then placed in nylon net sacks and stored for four months on raised shelves in a well-ventilated barn after recording fresh weight. Tubers were again scored for damage and weighed at four months after storage and percentage fresh weight loss calculated. These tubers were also assessed for nematode population density. Tubers were peeled using a kitchen peeler, chopped finely and a 5 g sub-sample per tuber removed for nematode extraction over 48 h using a modified Baermann method (Coyne et al., 2007). Nematode suspensions were reduced to 10 ml and nematodes counted from 3×1 ml aliquots of the suspension using a Leica Wild M3C stereomicroscope.

2.4 Ware production

Seed yams produced from the previous season were used as planting material, following storage, for the production of ware yams in the following year at two sites, Idah and Aramoko. Tubers, which remained viable after storage, were selected from each of the respective treatments for ware production. Whole, uncut tubers without additional treatment were planted in plots of 50 tubers (~ 200 g). The experiment was laid out in a randomized complete block design with five replications (plots) per treatment. One cultivar per site was selected: Imola at Idah and Ajimokun at Aramoko. Tubers were planted on mounds spaced 1×1 m and harvested seven months after. Data was collected in the same manner as for seed yam above both at harvest and after storage for 4 months.

2.5 Statistical treatment of data

Differences among treatment means were compared with ANOVA using SAS, version 9 (SAS, 2001) and means separated using the Fisher's protected least significant difference test (LSD) at 5 % probability level or standard errors (SE). Nematode population density data was normalized using $\log_{10}(x + 1)$ transformation, while percentage data was transformed using arcsine of x prior to analysis.

Table 1: *Percentage incidence of virus-affected plants of four yam cultivars 12 weeks after planting at three field sites in Nigeria, following pre-plant treatments of ~ 100 g setts.*

Treatment[†]	Idah		Ibadan and Aramoko[‡]	
	Akpaji	Imola	Ajimokun	Sogbe
Pesticide	79.0 [a]	4.0 [c]	5.0 [bc]	0.0 [b]
HW	76.0 [ab]	2.0 [cd]	7.0 [a]	0.0 [b]
Neem	65.0 [c]	12.0 [a]	4.0 [c]	0.5 [a]
Wood ash	73.0 [b]	8.0 [b]	6.5 [ab]	0.0 [b]
Control	67.0 [bc]	6.0 [bc]	2.0 [d]	0.0 [b]

Figures with the same letter within a column for each cultivar are not significantly different at ($P \le 0.05$) using LSD.
[†] Pesticide = fungicide (mancozeb) + insecticide (diazinon);
HW = hot water; control = untreated yam setts.
[‡] Data combined for Ibadan and Aramoko sites.

3 Results

3.1 Seed production

Pre-planting waste and discard of planting material was high at both Ibadan and Ekiti as a consequence of rots and nematode damage, resulting in fewer setts planted per plot than at Idah. Data from Ibadan and Aramoko seed yam production in the seed yam trial were combined as the source of seed yam was the same and a similar trend was observed between sites with respect to treatments (F value = 0.14; P value = 0.71).

Virus incidence was relatively higher for Akpaji at Idah than for other cultivars, with a similar trend at 8 and 12 WAP (12 WAP data only presented; Table 1). Virus incidence in cv. Akpaji -treated pesticide mix was similar to the incidence observed on plants treated with hot water. Significantly lower virus incidence was observed in the untreated plants, neem and wood ash-treated plants. In other cultivars the virus incidence was variable with treatments. Percentage sprouting was higher ($P \le 0.05$) at 4 and 8 WAP for setts pre-treated with the pesticide mixture, than for all other treatments, no cultivar effect was found (8 WAP data only presented; Tables 2, 3). However, percentage sprouting for control setts, HW, neem and wood ash treated setts varied depending on site and cultivar. Neem and HW treatment appeared to reduce sprouting of some cultivars. Seed yam tuber yield per plot was consistently higher from setts treated with the pesticide than other treatments ($P \le 0.05$) (Table 3). Seed tuber weight per plant was higher for plants from the pesticide treatment of cv. Sogbe, and neem and pesticide treatment for Ajimokun, although differences varied considerably by cultivar and site. Tubers from HW treated setts were lower in weight and number per plot for cv. Akpaji, as compared to

Table 2: *Effect of pre-plant treatments of yam setts (~ 100 g) on sprouting and tuber yield at harvest of two yam cultivars at a field site in Idah, Nigeria.*

Treatment [†]	Percent sprouting at 8 WAP [‡]	No. of tubers per plot	Tuber weight per plot (kg)	Tuber weight per plant (g)
	cv. Akpaji			
Pesticide	73.6 [a]	299.8 [a]	29.2 [a]	101.3 [a]
HW	38.9 [d]	146.0 [c]	8.8 [c]	57.5 [b]
Neem	57.0 [b]	263.0 [ab]	25.0 [ab]	118.4 [a]
Wood ash	45.2 [cd]	189.4 [bc]	18.5 [b]	99.9 [ab]
Control	52.7 [bc]	213.0 [bc]	20.5 [b]	97.9 [ab]
	cv. Imola			
Pesticide	75.2 [a]	269.6 [a]	27.7 [ab]	135.7 [ab]
HW	58.5 [b]	219.8 [ab]	24.8 [abc]	110.7 [b]
Neem	28.3 [d]	116.4 [c]	18.4 [c]	167.4 [a]
Wood ash	52.8 [bc]	205.6 [b]	28.5 [a]	94.5 [b]
Control	41.1 [c]	184.6 [b]	20.2 [bc]	123.1 [ab]

Figures with the same letter within a column for each cultivar are not significantly different at ($P \leq 0.05$) using LSD.
[†] Pesticide = fungicide (mancozeb) + insecticide (diazinon); HW = hot water; control = untreated yam setts; [‡] WAP = weeks after planting; means separation undertaken on arcsin($\sqrt{x}$) transformed data with non-transformed data presented.

Table 3: *Effect of pre-plant treatments of cut yam setts (~ 100 g) on sprouting and tuber yield of two yam cultivars at two field sites (Ibadan and Aramoko combined) in Nigeria.*

Treatment [†]	Percent sprouting at 8 WAP [‡]	No. of tubers per plot	Tuber weight per plot (kg)	Tuber weight per plant [§] (g)
	cv. Ajimokun			
Pesticide	33.3 [a]	63.4 [a]	31.9 [a]	684.6 [ab]
HW	16.9 [b]	29.7 [b]	12.7 [bc]	423.7 [b]
Neem	15.7 [bc]	37.0 [bc]	16.9 [b]	1047.7 [a]
Wood ash	12.8 [bc]	19.5 [c]	10.8 [c]	514.5 [b]
Control	10.4 [c]	19.2 [c]	9.3 [c]	504.5 [b]
	cv. Sogbe			
Pesticide	12.8 [a]	16.1 [a]	9.4 [a]	737.2 [a]
HW	6.5 [b]	14.8 [a]	5.5 [abc]	474.6 [ab]
Neem	4.5 [b]	11.4 [a]	4.6 [bc]	390.7 [ab]
Wood ash	5.9 [b]	15.5 [a]	6.3 [ab]	723.2 [a]
Control	0.3 [c]	0.8 [b]	1.5 [c]	135.7 [b]

Figures with the same letter within a column for each cultivar are not significantly different at ($P \leq 0.05$) using LSD.
[†] Pesticide = fungicide (mancozeb) + insecticide (diazinon); HW = hot water; control = untreated yam setts; [‡] WAP = weeks after planting; means separation undertaken on arcsin($\sqrt{x}$) transformed data with non-transformed data presented; [§] Tuber weight per plant = total weight of tubers per plot divided by the total number of tubers per plot.

other treatments. The number of tubers per plot varied by site and cultivar, but was relatively higher in the Idah site, than in Ibadan and Aramoko. At harvest, the number of tubers between non-symptomatic and virus-infected plants differed ($P \leq 0.05$) within treatments, with fewer symptomatic plants in neem and HW treatments (Table 4).

Virus-infected plants, however, consistently produced smaller ($P \leq 0.05$) tubers for all treatments across the experiments (5.0 kg per plot) than non-symptomatic plants (10.3 kg per plot) (Table 4). Percentage rot on seed tubers was high for both *Ajimokun* and *Sogbe* cvs., particularly for the control, which had >65 % increase in rot after 4 months of storage, as compared with 27 %

Table 4: *Yam tuber yield from virus symptomatic and non-symptomatic plants from three field trials in Nigeria, following pre-plant treatments of ~ 100 g cut yam setts* [†].

Treatment [‡]	No. of tubers from non-symptomatic plants per plot	No. of tubers from symptomatic plants per plot	SE	Tuber weight from non-symptomatic plants per plot (kg)	Tuber weight from symptomatic plants per plot (kg)	SE	Tuber weight per non-symptomatic plant (g)	Tuber weight per symptomatic plant (g)	SE
Pesticide	58.8	61.6	1.4	15.6	7.7	4.0	265.3	125.0	70.9
HW	43.8	34.4	4.7	9.6	3.5	3.1	219.2	101.7	59.3
Neem	42.3	35.3	3.5	10.2	3.4	3.4	241.1	96.3	73.1
Wood ash	31.5	35.3	1.9	8.5	4.5	2.0	269.8	127.5	71.9
Control	43.7	52.9	4.6	7.8	5.9	1.0	178.5	111.5	33.8
Mean	44.0	43.9		10.3	5		234.8	112.4	
SE	4.4	5.7		1.4	0.8		17.0	6.3	

[†] Data are means from three locations each with two cultivars; [‡] Pesticide = fungicide (mancozeb) + insecticide (diazinon); HW = hot water; control = untreated control; SE = Standard error ($P \leq 0.05$).

rot increase for the cv. *Ajimokun* on pesticide-treated setts (Fig. 1). Tubers from pesticide and HW treated plants stored better than other treatments in Aramoko and Ibadan. Storage rot in Idah was relatively low in comparison to the other storage sites.

Nematodes found in tubers after four months storage were mostly *S. bradys* (data not shown). The initial number of nematodes (Pi) at planting was 122 taken from sampled tubers. Tubers from HW and neem treatments were least ($P \leq 0.05$) infected, with 14 and 91 nematodes per 5 g of tuber cortex respectively, while tubers from wood ash-treated setts showed highest ($P \leq 0.05$) densities (186 per 5 g). *Meloidogyne* spp. were observed only in tubers from the wood ash treatment, but at barely detectable levels of 4 nematodes per 5 g of cortex. Dry rot damage was generally low (data not shown). Only the cv. *Akpaji*, showed that up to 5 % of the untreated control tubers were dry rot damaged ($P \leq 0.05$), while for cv. *Imola* just 1 % of wood ash treated tubers were affected. Tubers from HW and pesticide-treated setts had, in general, the lowest dry rot damage after storage ($P \leq 0.05$). Although present at harvest, yam beetle damage (3 % in Idah only) and termite infestation (41.5 % across locations and cultivars) did not progress during storage (data not shown). There was no evidence of the yam scale insect (*Aspidiella hartii*) infestation from either the field or storage. Mealybug infestation increased during storage, especially in Idah on cv. *Imola*; pesticide and HW pre-treated plants were least affected ($P \leq 0.05$) (Fig. 2). Generally the HW and pesticide treatment tubers hosted fewer insect pests.

Following four months storage, percentage weight loss of seed yam tubers was higher ($P \leq 0.05$) for cultivars *Ajimokun* and *Sogbe* than for *Imola* and *Akpaji* (Fig. 3). Differences between treatments occurred across cultivars with greatest differences observed for the cvs. *Ajimokun* and *Sogbe*. Less ($P \leq 0.05$) tuber weight loss occurred in the pesticide and HW treatments (9.5 % and 10.2 %, respectively), than for the control (e.g. 86.6 % and 62.8 % for cvs. *Sogbe* and *Ajimokun*, respectively). Total combined loss per treatment, as a result of poor sprouting in the previous season and storage weight loss was also lower for the pesticide and HW treatments, which reduced availability of planting material for the most affected treatments for the following ware yam production.

3.2 Ware production

More tubers were lost during storage from plots with previously untreated tubers and wood ash pre-treated tubers compared to tubers pre-treated with the pesticide mix, HW and neem slurry. Although plants pre-treated with neem had more losses compared to the either pesticide or HW treated plants (Table 5). Ware yam produced from seed that originated from the control treatments in the previous seed yam trial, yielded less than seed arising from pesticide treated setts ($P \leq 0.05$) for cv. *Ajimokun*, while *Imola* cv. tubers from the HW treatment also yielded better than the control ($P \leq 0.05$) (Table 5). This occurred even though seed material used for production of ware yam was selected only from the viable material that remained following storage, which was of relatively even quality and size (~ 200 g) for all treatments. Taking into account the difference in production of seed material between control and pesticide treatments in the previous season, combined with tuber losses during storage, and further to the difference in production of the ware yam from the remaining seed material (in the seed yam production season), the pesticide pre-plant treatment of setts (best treatment) led to an overall 214 % greater yield for *Imola* and 700 % for *Ajimokun* compared to the control (Table 5). This equates to an increase in ware yield from 7.4 to

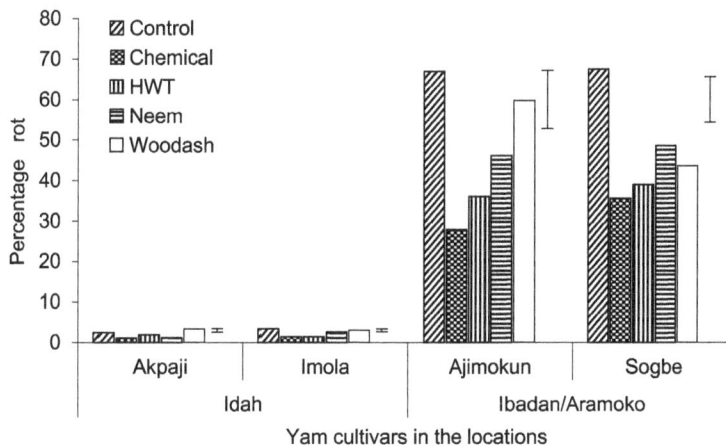

Fig. 1: *Percentage increase in tuber rot incidence of yam tubers following pre-plant treatments, between harvest and storage for four months in Nigeria. Error bars = Standard errors; means separation undertaken on* $\arcsin(\sqrt{x})$ *transformed data with back-transformed data presented. N = 100 per cultivar.*

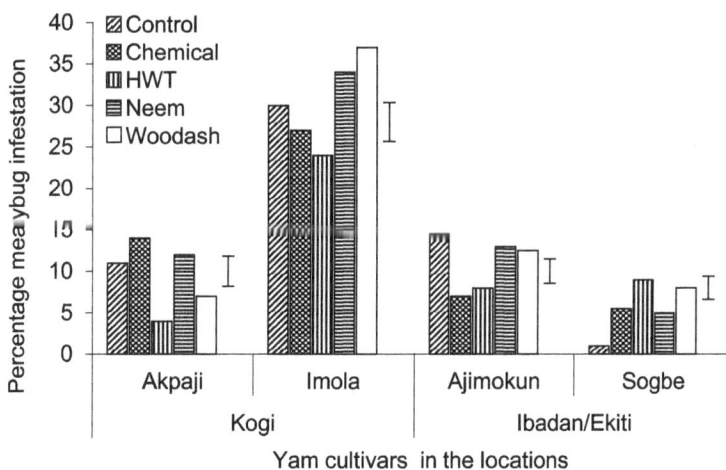

Fig. 2: *Percentage increase in mealybug infestation incidence of yam tubers following pre-plant treatments, between harvest and storage for four months in Nigeria. Error bars = Standard errors; means separation undertaken on* $\arcsin(\sqrt{x})$ *transformed data with back-transformed data presented. N = 100 per cultivar.*

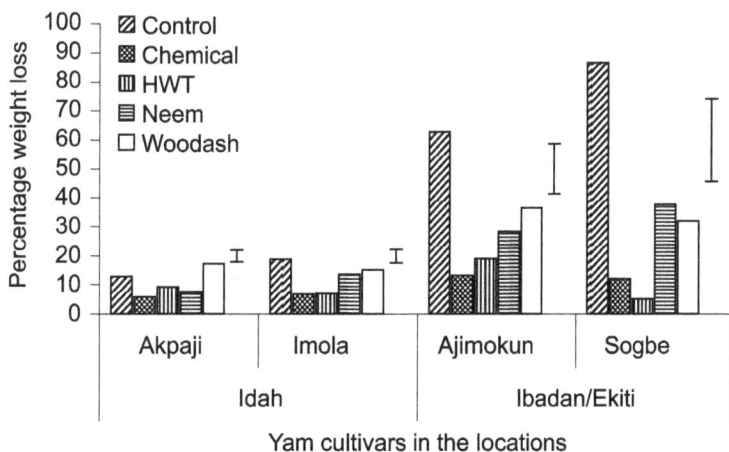

Fig. 3: *Weight loss of harvested yam tubers following pre-plant treatments, during storage for four months in Nigeria. Error bars = Standard errors; means separation undertaken on* $\arcsin(\sqrt{x})$ *transformed data with back-transformed data presented. N = 100 per cultivar.*

Table 5: *Potential ware yam yield for two cultivars produced from seed yams arising from different pre-plant treatment of 100 g seed setts.*

Treatment	No. of tubers available for planting [†]	Yield of ware yam per plot (kg)	Yield increase of ware yams (%) [‡]	Actual yield (t ha^{-1}) [§]	Potential yield per plot (kg) [¶]	Percentage increase in potential yield [‖]	Potential yield increase (t ha^{-1}) [**]
cv. Imola							
Control	82.3 (55.4)	26.2	0	5.2	43.1	0.0	0.0
Chemical	182.7 (32.2)	37.1	41.6	7.4	135.6	214.4	15.9
HW	133.9 (36.9)	33.8	29	6.8	90.5	109.9	7.5
Neem	63.6 (45.4)	24.6	−6.1	4.9	31.3	−27.5	−1.3
Wood ash	96.8 (52.9)	31.4	19.8	6.3	60.8	41.0	2.6
Mean	111.9	30.6	16.9	6.1	72.3	67.5	4.9
SE	21.2	2.3	8.9	0.5	18.8	43.5	3.1
cv. Ajimokun							
Control	7.4 (61.5)	10.5	0	2.1	1.5	0.0	0.0
Chemical	41.4 (31.7)	15	17.3	3.0	12.4	700.9	21.0
HW	22.2 (36.1)	12.3	7.2	2.5	5.5	252.3	6.3
Neem	15.3 (48.4)	13.5	11.6	2.7	4.1	167.1	4.5
Wood ash	8.7 (55.4)	13.5	11.5	2.7	2.4	52.0	1.4
Mean	19.0	13.0	9.5	2.6	5.2	234.5	6.7
SE	6.2	0.8	2.9	0.6	1.9	125.0	3.8

[†] Number of tubers remaining for planting after the storage period (number at harvest - number lost in storage) with percentage number lost in parenthesis; [‡] Ware yield difference of treatments over untreated control; [§] Yield of ware yams from pre-treated seed yams calculated per ha; [¶] Yield of ware yams provided all available tubers per treatment were planted; [‖] Percentage increase of treatments over control provided all available tubers were planted; [**] Increase in potential yield (provided all tubers were planted) of other treatments over control.

23.3 t ha^{-1} for *Imola* and 3.0 to 24 t ha^{-1} for *Ajimokun*. The potential increase in ware yam yield from HW treated seed was also significantly higher ($P \leq 0.05$) than from other treatments (neem and wood ash) for *Imola*.

4 Discussion

The poor state of yam planting material in Nigeria was emphasized during the current study, with large proportions of the planting material purchased for the study in Ekiti State becoming unusable and discarded before planting. Production of seed yam from the remaining suitable material then resulted in poor quality and quantity of seed from untreated setts, with low germination rates, high incidence of tuber diseases and ultimately high levels of loss during storage, culminating in marked low yam production potential over two seasons. The traditional practice of using wood ash before planting reduced tuber losses during storage to a small extent but in comparison to the pesticide treatment, yields remained low while losses during storage were high. Consequently, the current study provided a clear insight of the general state and quality of seed yam material available for farmers in Nigeria, reaffirming earlier reports (Asiabaka *et al.*, 2001). The dual combination of pesti-

cides used on pre-plant setts led to greater production of seed yam and, along with HW, yielded healthier seed tubers than traditional practices. HW treatment is therefore effective at producing healthy material, as previously reported (Meerman & Speijer, 2001), but showed serious shortcomings on its initial effect on sett viability and sprouting. The success rate of the HW treatment is cultivar dependent, possibly due to the number of eye buds and cortex thickness (Coyne *et al.*, 2010). It is also of limited attraction due to its burdensome application, reducing its adoption by individual farmers (Asiabaka, 1994; Agbaje & Oyegbami, 2005). Although neem is well recognized for its pest and disease control properties (Onalo *et al.*, 2001; Okigbo, 2004; Okigbo *et al.*, 2009), its effect was not consistent in the current study. The neem treatment may have proved better using seed extracts or commercial products, for more consistent concentration of the active ingredient azadirachtin (Schmutterer, 1990). The systemic nature of azadirachtin has been reported to lead to phytotoxicity in crops, particularly by stressed plants or those with limited root mass (Oetting *et al.*, 1990; Schmutterer, 1990). This may have contributed to the poor sprouting of neem-treated setts of some cultivars, particularly as yam setts produce roots only a few weeks after planting (Orkwor & Ekanayake, 1998).

When assessing the relative performance of seed sett treatment over the two cycles of seed-through-ware yam production, the pesticide treatment consistently provided heavier seed tubers, which were healthier and stored better than other treatments. Following storage, this seed produced heavier and healthier ware yam, compared with other treatments. It is important to note, however, that ware yam was produced only from the seed tubers surviving the storage period, which were inevitably healthier and less infected than those from other treatments and subject to lower losses. Taking these losses into account therefore vastly exaggerates the overall potential yield difference between the pesticide treatment and traditional treatment, emphasizing the negative "knock-on" effect of using poor seed yam. For the two cultivars, *Imola* and Ajimokum a sufficient quantity of tubers remained after storage in order to potentially increase ware yam production by a staggering 212 and 700 %, respectively. This difference in potential yield could be achieved through the use of a simple pesticide dip on seed yam setts in the previous season, as compared to the no-treatment control, a regular farmer practice. Some of the treatments assessed in the current study are recognized and have previously been studied (Bridge *et al.*, 2005); pesticide and HW for example are not "new" or innovative (Meerman & Speijer, 2001). They have both proved effective treatments, with remarkable rates of return for the pesticide treatment. The current study provides information that further supports previous studies, but builds on this further through following the assessment over two crop cycles and the intervening storage period. Our study underscores the necessity for the development of a sustainable healthy seed yam system, but furthermore highlights the colossal differences that such simple techniques can provide. The benefits of a healthy seed yam system are not under question, but rather how to systematically develop, implement and sustain it.

The use of pesticides can be contentious where availability of good, unadulterated products is often limited or inconsistent (Neuenschwander, 2004; Ngowi *et al.*, 2007). However, the products used in this study were selected based on suitable sized packages for small scale farmers, ease of availability, affordability and were in their original (sealed) packages. The method of application by dipping could be applied with relative simplicity by farmers. The sett dipping application helps to prevent adverse effect on non-target organisms and general environmental contamination. A further consideration to promoting the use of pesticides for seed yam production, is that their use is targeted as a pre-treatment of seed yam, which can then be used without any further

treatment for ware yam production. The application of pesticide at least two seasons prior to consumption thus limits residue contamination to the consumer. Additionally, diazinon is a contact, non-systemic compound, apparent from the increased incidence of mealybugs during storage. The combined fungicide and insecticide sett treatment improved yields substantially (62 %), more so than the wood ash treatment, resulting in healthier seed yam tubers, with improved storability and thus improved quality and health for ware production in the following season.

The cost of seed yam equates to approximately 50 % of the total cost of yam production (Nweke *et al.*, 1991; Aighewi *et al.*, 2003). An effective and simple treatment of setts, that guarantees more, heavier and healthier seed yam tubers, that preserve longer and store better than the farmers' usual practices, implies substantial benefits to yam productivity. The higher quality of tubers additionally attracts premium prices. Planting healthy, superior quality seed tubers for ware yam production ensures fewer empty mounds, reducing labour wastage and leads to better ware yam yields. Healthy and sustainable seed systems remain the platform upon which to improve crop productivity. The technique is referred to as the adapted yam minisett technique (AYMT) and has been demonstrated in farmer managed trials (Morse & McNamara, 2017). The simple, cost effective method demonstrated here is ideal for small and large-scale yam farmers alike in West Africa, providing that good quality products are used and that safe and proper pesticide practices are employed and observed.

Acknowledgements

This work was funded by DFID through NRI in collaboration with the University of Reading UK and Diocesan Development Services (DDS), Kogi State, Nigeria. The contributions made by Nora McNamara, Steve Morse, Moses Acholo and Lava Kumar to this study are much appreciated.

References

Agbaje, G. O. & Oyegbami, A. (2005). Survey on the adoption of yam minisett technology in South-Western Nigeria. *Journal of Food Agriculture and Environment*, 3 (2), 134–137.

Aighewi, B. A., Akoroda, M. O. & Asiedu, R. (2003). Seed yam (*Dioscorea rotundata* Poir) production, storage and quality in selected yam zones of Nigeria. *African Journal of Root Tuber*, 5, 20–23.

Amusa, N. A., Adegbite, A. A., Muhammed, S. & Baiyewu, R. A. (2003). Yam diseases and its management in Nigeria. *African Journal of Biotechnology*, 2, 497–502.

Asiabaka, C. C. (1994). Factors influencing the adoption of yam minisett technology among farmers in Nigeria. *In:* Ofori, F. & Hahn, S. K. (eds.), *Symposium on tropical root crops in a developing economy held in Accra Ghana, 1 November 1994*. pp. 376–380, Acta Horticulturae (ISHS) 380.

Asiabaka, C. C., Morse, S. & Kenyon, L. (2001). *The development, dissemination and adoption of technologies directed at improving the availability of clean yam planting material in Nigeria and Ghana*. DFID Crop Protection Programme (Za0478), Study Mission Report (54 pp.) (11–22 June 2001), Natural Resources Institute (NRI) Chatham, Kent, UK.

Asumugha, G. N., Ugwu, B. O., Aniedu, O. C., Orkwor, G. C. & Amegbeto, K. (2007). Minisett technique of seed yam production in two major yam producing states of Nigeria: a function of input availability and production objective. *In:* Mahungu, N. M. & Manyong, V. M. (eds.), *Advances in root and tuber crops technologies for sustainable food security, improved nutrition, wealth creation and environmental conservation in Africa. Proceedings of the 9th Triennial Symposium of the International Society for Tropical Root Crops – African Branch (ISTRC-AB), Mombasa, Kenya, 1–5 November, 2004*. pp. 209–214, IITA, Ibadan, Nigeria.

Bridge, J., Coyne, D. & Kwoseh, C. K. (2005). Nematode parasites of tropical root and tuber crops. *In:* Luc, M., Sikora, R. & Bridge, J. (eds.), *Plant parasitic nematodes in subtropical and tropical agriculture. Revised 2nd Edition*. pp. 221–258, CAB International, UK.

Coyne, D., Nicol, J. & Claudius-Cole, B. (2007). *Practical plant nematology: field and laboratoty guide*. IITA, Ibadan, Nigeria. 82 pp.

Coyne, D. L., Claudius-Cole, A. O., Kenyon, L. & Baimey, H. (2010). Differential effect of hot water treatment on whole tubers versus cut setts of yam (*Dioscorea* spp.). *Pest Management Science*, 66, 385–389.

Degras, L. (1993). *The yam: a tropical root crop*. The Macmillan Press Ltd., London. 408 pp.

Emehute, J. K. U., Ikotun, T., Nwauzor, E. C. &

Nwokocha, H. N. (1998). Crop protection. *In:* Orkwor, G. C., Asiedu, R. & Ekanayake, I. J. (eds.), *Food yams: Advances in Research*. pp. 187–214, IITA and NRCRI, Nigeria.

Langyintuo, A. S. (1996). *Economic evaluation of the use of in vitro plantlets for healthy seed yam production. Report on: Collaborative study on technologies for germplasm conservation and distribution of pathogen-free Dioscorea yams to national root crop research programs*. Savanna Agric. Res. Inst., Ghana/IITA/ODA.

Meerman, J. C. & Speijer, P. R. (2001). Perspectives for large scale distribution of nematode disinfested yam planting material in Southern Nigeria. *In:* Proceedings of the 7th Triennial Symposium of the International Society for Tropical Root Crops-Africa Branch, 11–17 October, 1998. pp. 620–622, IITA Cotonou, Benin.

Morse, S., Acholo, M., McNamara, N. & Oliver, R. (2000). Control of storage insects as a means of limiting yam tuber fungal rots. *Journal of Stored Product Research*, 36, 37–45.

Morse, S. & McNamara, N. (2017). Impact of the adapted yam minisett technique on ware yam (*Dioscorea rotundata*) production under farmer-managed conditions in Nigeria. *Experimental Agriculture*, 53 (1), 131–143.

Nchinda, V. P., Njualen, D. K., Ngassam, S. B., Che, M. A. & Nkwate, S. P. (2009). Adoption of minisett technology in the agroecological zones of high guinea savannah and western highlands of Cameroon. *In:* Nkamleu, B., Annang, D. & Bacco, N. M. (eds.), *Securing livelihoods through yams. Proceedings of a technical workshop on progress in yam research for development in West and Central Africa. Held in Accra, Ghana, 11–13 September 2007*. pp. 230–240, IFAD, IITA publication.

Neuenschwander, P. (2004). Biological pest control can benefit the pocket, health and the environment – Commentary. *Nature – Harnessing nature in Africa*, 432, 801–802.

Ngowi, A. V. F., Mbise, T. J., Ijani, A. S. M., London, L. & Ajayi, O. C. (2007). Smallholder vegetable farmers in Northern Tanzania: Pesticides use practices, perceptions, cost and health effects. *Crop Protection*, 26, 1617–1624.

Nweke, F. I., Ugwu, B. O., Asadu, C. L. A. & Ay, P. (1991). Production cost in yam based cropping system of South Western Nigeria. Resource and Crop Management Division Research Monograph No. 6. IITA Ibadan, Nigeria. 29 pp.

Oetting, R. D., Sandersom, K. C. & Smith, D. A. (1990). Treatment of cuttings before shipment with neem. *In:* Locke, J. C. & Lawson, R. H. (eds.), *Proceedings of a workshop on neem's potential in pest management programmes.* pp. 113–117, USDA-ARS.

Ogaraku, A. O. & Usman, H. O. (2008). Storage rot of some yams (*Dioscorea* spp.) in Keffi and environs, Nasarawa State, Nigeria. *Protection Agriculture and Technology*, 4, 22–27.

Okigbo, R. N. (2004). A review of biological control methods for postharvest yams in storage in southern Nigeria. *KMITL Science Journal*, 4 (1), 207–215.

Okigbo, R. N., Putheti, R. & Achusi, C. T. (2009). Post-harvest deterioration of cassava and its control using extracts of *Azadirachta indica* and *Aframomum melegueta*. *E-Journal of Chemistry*, 6, 1274–1280.

Onalo, J., Asiedu, R. & Adesiyan, S. (2001). Control of the yam nematode (*Scutellonema bradys*) with neem fruit powder. *In:* Proceedings of the 7th Trien-nial Symposium of the International Society for Tropical Root Crops-Africa Branch, 11–17 October, 1998. Cotonou. pp. 634–637, IITA, Ibadan, Nigeria.

Onyenweaku, C. E. (1991). Factors associated with the adoption of seed yam 'minisett' technique by farmers in Imo State. *Nigeria Journal of Agricultural Science and Technology*, 2, 23–28.

Orkwor, G. C. . & Ekanayake, I. J. (1998). Growth and Development. *In:* Orkwor, G. C., Asiedu, R. & Ekanayake, I. J. (eds.), *Food Yams. Advances in Research.* pp. 39–62, IITA and NRCRI, Nigeria.

SAS Institute (2001). *Statistical Analysis Software (SAS) user's guide.* SAS Institute Inc, Cary, NC.

Schmutterer, H. (1990). Future tasks of neem research in relation to agricultural needs worldwide. *In:* Locke, J. C. & Lawson, R. H. (eds.), *Proceedings of a workshop on neem's potential in pest management programmes.* pp. 15–22, USDA-ARS.

Wood, T. G., Smith, R. W., Johnson, R. A. & Komolafe, P. O. (1980). Termite damage and crops loss studies in Nigeria – Pre-harvest losses to yams due to termites and other soil pests. *Tropical Pest Management*, 26, 355–370.

19

Activities and influence of veterinary drug marketers on antimicrobial usage in livestock production in Oyo and Kaduna States, Nigeria

Olufemi Ernest Ojo [a,*], Olajoju Jokotola Awoyomi [b],
Eniola Fabusoro [c], Morenike Atinuke Dipeolu [b]

[a] *Department of Veterinary Microbiology and Parasitology, College of Veterinary Medicine,*
Federal University of Agriculture Abeokuta, Abeokuta, Nigeria
[b] *Department of Veterinary Public Health and Reproduction, College of Veterinary Medicine,*
Federal University of Agriculture Abeokuta, Abeokuta, Nigeria
[c] *Department of Agricultural Extension and Rural Development, College of Agricultural Management and Rural Development,*
Federal University of Agriculture Abeokuta, Abeokuta, Nigeria

Abstract

Antimicrobial usage in animals contributes to the emergence of antimicrobial resistant bacterial strains. Investigations were carried out on how the characteristics, knowledge, attitude and practices of antimicrobial marketers influenced antimicrobials usage in animal production in Oyo and Kaduna States, Nigeria. Focus group discussions, in-depth interviews and structured questionnaires were used to gather information about the characteristics and activities of antimicrobial marketers. Overall, 70 (56.9 %) of 123 marketers had post-secondary education while 76 (61.8 %) were trained on the use of antimicrobials. Eighteen (14.6 %) of the marketers were licensed veterinarians. Only 51 (41.5 %) marketers displayed adequate knowledge about antimicrobials and antimicrobial usage. Sixty-seven (54.6 %) marketers requested a prescription before selling antimicrobials while 113 (91.9 %) marketer recommended antimicrobials for use in animals. Two-third of the marketers (66.7 %) prescribed antimicrobials without physically examining sick animals but based their prescriptions on verbal reports of clinical signs by farmers and on their personal experience. Marketers with higher educational qualification displayed more adequate knowledge of antimicrobials and antimicrobial usage than those with basic education background only. More years of experience in antimicrobial marketing did not translate to better knowledge on antimicrobial usage. Only 45 (36.6 %) respondents were aware of the existence of regulatory agencies monitoring the use of antimicrobials in animals. Farmers ignored the services of veterinarians in the diagnosis and control of animal diseases but resorted to drug marketers for help. Effective communication of existing legislations on antimicrobial usage, improved access to veterinary services and strict enforcement of regulatory policies are recommended for checking non-judicious use of antimicrobial agents in animal production. Sales of antimicrobial agents for animal use without veterinarian's prescription must be prohibited.

Keywords: antimicrobial resistance, attitude, injudicious antimicrobial usage, knowledge, socio-cultural practices

1 Introduction

The challenge posed by infectious disease is a major limiting factor to the development of livestock industry in Africa (Rweyemamu *et al.*, 2006). Recurrent

* Corresponding author
Email: ojooe@funaab.edu.ng

morbidity and high mortality that characterise bacterial infections make animal production a high-risk business (Pritchett *et al.*, 2005). To mitigate against economic losses associated with disease outbreaks, farmers rely heavily on antimicrobial agents for the prevention and treatment of infections (Ojo *et al.*, 2016). Antimicrobials are also used as growth promoters to enhance performance and boost feed conversion efficiency (Dhama et al., 2014). Sometimes, farmers use antimicrobial agents to cover for lapses such as poor management, inadequate resources and substandard facilities that could predispose animals to infections (Ojo et al., 2016).

With the increasing demands for antimicrobials in Nigeria, marketing of these drugs has become an attractive and lucrative business (Adesokan et al., 2015). There is high turnover rate, good profit and low risk. Poor regulation, high demand and relatively easy start-up have led to great increase in the activities of drug marketers with influx of different brands of antimicrobial agents into the market for animal consumption. With more participation and increasing competition, marketers have devised strategies to outwit competitors and maximise profits. Some of these marketing strategies are unwholesome as they are geared towards profit making with little or no consideration for the preservation of antimicrobial efficacy. Thus, antimicrobials are readily accessible to the general public without restrictions leading to injudicious use of antimicrobial with grave consequences on the efficacy of antimicrobials for the treatment of human and animal diseases.

Okeke et al. (1999) observed that in Africa, many interwoven socioeconomic and behavioural factors contribute to the emergence and spread of antimicrobial resistance. The use of antimicrobial agents especially in animals has been implicated as a major contributory factor to the increasing emergence and dissemination of antimicrobial resistance among bacteria (Chantziaras et al., 2014). Marketing may have direct and indirect influence on buyers and the pattern of antimicrobial usage in animals. There is a need for better regulation of the sales and distribution of antimicrobial agents in any attempt to promote judicious use of antimicrobials.

We hypothesised that proliferation of veterinary drug marketing and ready availability of antimicrobial agents promote the excessive use of antimicrobials in livestock production in Nigeria. Furthermore, the activities of antimicrobial marketers may have profound impact on antimicrobial usage pattern including the level of antimicrobial usage, choice of antimicrobial agents, mode of administration and consequently the outcome of antimicrobial therapy. Hence, knowledge, attitude and practices of antimicrobial marketers are relevant to antimicrobial usage in animal production and should be considered among the factors that contribute to the emergence of antimicrobial resistance.

In view of this, the present study investigated some factors associated with the activities of antimicrobial marketers in the use of antimicrobial agents in animal production. Emphasis was placed on the characteristics, knowledge, attitudes and practices of marketers in relation to antimicrobial usage in food animal production in Oyo and Kaduna States of Nigeria.

2 Materials and methods

The present study examined the knowledge, practices and attitude of antimicrobial marketers in Kaduna and Oyo States of Nigeria and assessed the influence of the activities of antimicrobial marketers on antimicrobial usage in animal production. With huge human populations and booming commercial activities, Oyo and Kaduna States are the socioeconomic hubs of western and northern Nigeria, respectively. The favourable climatic conditions of the States support profitable agriculture and encourage high levels of agricultural activities including animal husbandry. There are many large-, medium- and small-scale livestock farms in these states thus creating a huge demand for veterinary drugs and readily available markets for veterinary drugs merchants.

Data were collected through focus group discussions (FGD), in-depth interviews and structured questionnaires. The structured questionnaire comprised of 38 questions divided into two sections. In the first section, data were obtained on the demographic attributes of respondents while the second section was used to assess the knowledge, attitude and practices of the respondents.

Information was obtained from 123 marketers of antimicrobial drugs (AMD) in 12 communities within 12 Local Government Areas (LGAs) in both States (Table 1). Sixty-two marketers were sampled from seven communities within seven LGAs in Kaduna State while 61 marketers were sampled from five communities within five LGAs in Oyo State (Table 1). The survey targeted marketers of antimicrobial agents including drug importers, major distributors of antimicrobial agents, wholesalers and drug retailers (including local drug vendors and veterinary drug outlets) (Table 2).

3 Data analysis

Data were analysed using statistical package software (Statistical Package for the Social Sciences, SPSS version 16). The level of knowledge among participants

Table 1: *Locations of sampled antimicrobial drug marketers in Oyo and Kaduna States.*

State	Local Government Area	Communities Selected	Number of marketers selected
Oyo	Egbeda	Ibadan	21
	Atiba	Oyo township	12
	Saki West	Saki	12
	Atisbo	Ago-Are	6
	Iseyin	Iseyin	10
Subtotal	5	5	61
Kaduna	Zango Kataf	Zonkwa	9
	Jema	Kafanchan	10
	Soba	Maigana	6
	Zaria	Zaria	13
	Birnin Gwari	Birnin Gwari	8
	Kaduna North	Kaduna	8
	Lere	Saminaka	8
Subtotal	7	7	62
Overall Total	12	12	123

Table 2: *Business status of antimicrobial marketers.*

Ownership status of marketers	Oyo State ($n = 61$; %)	Kaduna State ($n = 62$; %)	Total ($n = 123$; %)
Manufacturer	3.3	0	1.6
Importer	0	1.6	0.8
Major distributor	14.8	1.6	8.1
Wholesaler	24.6	16.1	20.3
Retailer	57.4	80.6	69.1

was examined by ten questions. The questions were graded with allotment of scores. One point each was allotted for correct answer to the ten questions. Respondents with total score of seven points and above were considered to have good knowledge while those with total score of six points and below were categorised to have poor knowledge. Other attributes in terms of attitude and practices were expressed with descriptive statistics and frequency distributions, which were used to generate summarized results. Pearson Chi Square and Fisher's exact test were used to test for association between responses from Kaduna and Oyo States and levels of knowledge. Inferences were interpreted as being statistically significant at $p \leq 0.05$. Relationship between personal attributes of respondents (in terms of level of education and years of experience in marketing of antimicrobial drugs) and level of knowledge about antimicrobial drug usage was assessed using Spearman Correlation Coefficient.

4 Results

4.1 Characteristics of marketers of antimicrobial drugs

Small to medium scale retailers dominated antimicrobial marketing in the study areas. Overall, 69 % of all marketers were retailers (Table 2). Out of the 123 marketers in both states, 106 (86 %) were male (Table 3).

More than 90 % of all marketers had at least primary/basic education (Table 3). The primary occupations of the marketers were much related to agriculture and relevant to marketing of antibiotic drugs. A significantly higher ($p < 0.001$) percentage of marketers in Oyo State (77 %) than those in Kaduna State (37 %) attained post-secondary education. Sixty-six (54 %) marketers had less than 10 years of experience in the business. This indicated an expansion of the veterinary drug marketing over the last 10 years (Table 3). Sixty-seven (54 %) drug marketers were involved in trade associations.

Table 3: *Personal and business characteristics of marketers of antibiotics drugs.*

	Oyo State (n = 61; %)	Kaduna State (n = 62; %)	Total (n = 123; %)
Sex			
Male	80.3	91.9	86.2
Female	19.7	8.1	13.8
Age categories (years)			
Less than 30	21.3	19.4	20.3
30–60	73.7	74.2	74.0
Above 60	4.9	6.5	5.7
Educational status			
No formal education	3.3	9.7	6.5
Primary education	3.3	17.7	10.6
Secondary education	16.4	35.5	26.0
Post-secondary education	77.0*	37.1*	56.9
Years of experience			
Less than 10	47.5	59.7	53.7
Above 10	52.5	40.3	46.3
Mean values	12.7	13.1	12.9
Primary occupation			
Veterinary doctors	26.2*	3.2*	14.6
Livestock producers	37.7*	6.5*	21.9
Agrochemical input dealers	3.3*	41.9*	22.8
Civil servant	9.9*	22.6*	16.2
Crop Farming	22.9	25.8	24.4
Membership of association			
None	45.9	66.1	54.0
Yes	54.1	33.9	43.9

* : values are significantly different ($p < 0.001$)

4.2 Activities of marketers of antimicrobial drugs

The number of antimicrobial brands sold by individual marketer per day ranged from 1 to 25. On a daily basis, marketers in Oyo State sold an average of 11 commercial antimicrobial brands while in Kaduna State, the average number of antimicrobial brands sold by marketers was 10. An average of 18 people in Oyo State and 23 people in Kaduna State requested for antimicrobials from each marketer daily. Marketers in both states sold imported and local brands of antimicrobial agents. Most of the antimicrobial agents marketed for use in animals were imported to the country. Local brands of antimicrobial agents were manufactured overseas for local companies that packaged, branded and marketed the drugs in Nigeria. The antimicrobial stocks came predominantly from Asia (China, India) and the Middle East (Jordan). Other sources were Europe (The Netherlands and the United Kingdom) and North America (Canada).

Overall, 55 % of marketers asked for a prescription before selling. This number was significantly higher ($p < 0.005$) in Oyo State ($n = 41$; 67 %) than in Kaduna State ($n = 26$; 42 %) Many marketers (81 %) asked questions relating to the purpose of antimicrobial purchase before selling to farmers. Marketers also recommended antimicrobial agents to farmers for use in animals. However, findings showed that majority of the marketers had no formal training on antimicrobial usage and were not licensed to prescribe antimicrobials for animal use. In most cases, antimicrobials agents were prescribed without physical examination of sick animals and collection of clinical samples for laboratory investigations such as bacterial isolation and antimicrobial susceptibility testing (Table 4). A significantly higher proportion ($p < 0.001$) of marketers in Oyo State (57 %) than in Kaduna State (29 %) followed up with farmers after sales of antimicrobial agents. Most (94.5 %) of the marketers that followed up cases with farmers in Oyo State were veterinarians.

Table 4: *Activities of marketers in handling of antimicrobials.*

S/No	Activities	Oyo State (n = 61; %)	Kaduna State (n = 62; %)	Total (n = 123; %)
1.	Ask for prescription before selling antibiotics	67.2*	41.9*	54.5
2.	Ask for reasons for purchasing antibiotics from buyers	75.4	85.5	80.5
3.	Volunteer information on the use of antibiotics	93.4	90.3	91.9
4.	Buyers asking for guidance on the use of antibiotics	90.2	88.7	89.4
5.	Buyers ask for drug prescription for their animals	93.4	95.2	94.3
6.	Prescribe drugs for buyers	90.2	93.2	91.9
7.	See/examine animals before drug prescription	39.3	27.4	33.3
8.	Submit samples for bacteria isolation and antimicrobial sensitivity before prescribing drugs	9.8	8.1	8.9
9.	Based prescription on past experience	86.9	91.9	89.4
10.	Receive complaints from farmers about drug failures	54.1	69.4	61.8
11.	Had training on the use of antibiotics	65.6	58.1	61.8

*: values are significantly different ($p < 0.05$)

4.3 Perceptions and knowledge of marketers on antimicrobial usage

Only 37 % of the total marketers in both States knew about the existence of agencies regulating drug marketing activities and sales of antimicrobial agents (Table 5). Regulatory agencies known to marketers included National Agency for Food Drug Administration and Control (NAFDAC), Veterinary Council of Nigeria (VCN), Departments of Veterinary Services in State Ministries of Agriculture as well as Local Government Veterinary Departments.

Most of the drug marketers interviewed agreed that the use of antimicrobial agents required special training. Types of training included formal training through school system leading to award of diplomas and degrees, training through seminars and workshops, on-the-job training/apprenticeship, informal training through membership of associations, trade groups and other sources. The proportion of marketers that got the correct description of an antimicrobial agent is significantly higher ($p < 0.001$) in Oyo State (36 %) than in Kaduna State (13 %). About 93 % of the marketers from both States believed that the sales of antimicrobial agents should be by prescription.

This study showed a positive correlation between level of education and adequate knowledge of antimicrobial usage and not year of experience, sex and age. This indicated that knowledge was a function of training and level of education but not years of experience, sex, and age. More marketers in Oyo (48 %) than in Kaduna (36 %) had adequate knowledge of antimicro-

bial usage. However, the difference in the level of knowledge between the two states was not statistically significant.

4.4 Experience of drug failure among marketers of antimicrobial drugs

In Kaduna State, 69 % of antimicrobial marketers and 54 % in Oyo State reported that farmers do complain about failure of antimicrobial chemotherapy (Table 6). When farmers complained about drugs, many of the antimicrobial marketers (57 % in Kaduna State and 53 % in Oyo State) would simply recommend another drug. Most veterinary drug marketers (91 %) knew that the use of antimicrobial agents contribute to the incidence of antimicrobial resistance in bacterial pathogens (Table 6).

5 Discussion

Overdependence on antimicrobial usage in food animals and non-compliance with withdrawal period before processing edible animal products for human consumption are responsible for the high level of antimicrobial resistance and presence of antimicrobial residues in animal-source foods (Olatoye & Ehinmowo, 2010; Mensah *et al.*, 2014). Every form of antimicrobial usage has the tendency to precipitate antimicrobial resistance in exposed bacteria (Davies & Davies, 2010). However, the increasing pace of development and widespread dissemination of antimicrobial resistance is aggravated by excessive and indiscriminate use of antimicrobial agents in food animal production (Founou *et*

Table 5: *Perception of marketers on the use of antimicrobials.*

Perception	Oyo State (n = 61; %)	Kaduna State (n = 62; %)	Total (n = 123; %)
Aware of regulatory bodies controlling antimicrobial drug administration	44.3	29.0	36.6
Use of antibiotics requires special training	88.5	95.2	91.9
Sales of antibiotics require prescriptions	86.9	98.4	92.7
Stricter measures are necessary to restrict of antibiotics use on animals	80.3	91.9	86.2
Use of antibiotics in animals could impact human health	90.2	93.5	91.9
Correct description of antibiotics	36.1*	9.7*	22.8

* : values are significantly different ($p < 0.001$)

Table 6: *Experience on drug failure among marketers.*

Perception	Oyo State (n = 61; %)	Kaduna State (n = 62; %)	Total (n = 123; %)
Do farmers complain about drug failures			
Yes	54.1	69.4	61.8
No	45.9	30.6	38.2
Frequency of complaints			
Occasionally	63.9	64.5	64.2
Often	1.6	1.6	1.6
Very often	0	3.2	1.6
Action taken in case of drug failure			
Referred to veterinarian	29.5	41.9	35.8
Referred to laboratory	16.4	9.7	13.0
Changed the drug	52.5	56.5	54.5
Visited the farm	23.0	19.4	21.1
Took no action	1.6	3.2	2.4
Antibiotic use contributes to incidence of antibiotic resistance	88.5	93.5	91.1
Follow up after sales	57.4	29.0	43.1

al., 2016; Ojo *et al.*, 2016). High level of exposure to infectious agents, poor accessibility to veterinary services from qualified professionals and inadequate implementation of regulatory policies on antimicrobial usage contribute to overdependence on antimicrobials for the prevention and treatment of diseases in food animal production (Grace, 2015). There are not enough reliable and up-to-date information on antimicrobial usage in many developing countries (Grace, 2015). Many developed countries have put in place organised antimicrobial resistance monitoring and surveillance programmes as parts of concerted effort to identify the sources of threat, understand the extent of the problems and develop preventive strategies against antimicrobial resistance (Founou *et al.*, 2016). Regrettably, in developing countries, there are no monitoring and surveillance pro-

grammes (Grace, 2015). Insufficient data from developing countries on antimicrobial usage and antimicrobial resistance is a major limitation in the assessment of the actual situation regarding antimicrobial resistance. Thus, there is no strong basis for the development of realistic and sustainable preventive measures (ibid).

Sales of antimicrobial agents for use in animal is on the increase in Nigeria (Adesokan *et al.*, 2015). Results from the present study showed there is the possibility of greater accessibility and more usage of antimicrobials in Oyo State relative to Kaduna State due to presence of several major distributors. The increasing use of antimicrobials in animals is due largely to a system that allows unrestricted access to antimicrobial agents (Grace, 2015; Ojo *et al.*, 2016). More men than women were involved in marketing of antimicrobial agents probably because men may have more capital to invest in business than women. Sociocultural practices in the study areas favour men being in business than women (Danjuma *et al.*, 2013). More young people (below 50 years old) than old people (above 50 years old) participated in marketing of antimicrobial drugs. This implies that most of the marketers are within the active age category. The youthfulness of the marketers should influence positively their innovativeness and opportunities for improving their capacity of acquiring information towards effective adoption and compliance with new regulations (Sebba *et al.*, 2009; Adegbidi *et al.*, 2012).

More educated people were involved in handling of livestock drugs in Oyo State than in Kaduna State. Improved educational status promotes good practices in handling of drugs and in provision of useful information to end-users. Low level of education may promote quackery and substandard service delivery among antimicrobial retailers thus jeopardizing the health of the nation's livestock and consequently public health. Marketers join trade associations to establish networks among themselves. This is important as a form of internal self-regulatory mechanisms for coordinating the activities of the marketers (FAO, 2005). These associations also serve as channels of disseminating information and for capacity building. Government interventions in monitoring and regulating antimicrobial sales and usage in animals can be routed through these associations. The higher involvement of marketers in trade associations in Kaduna State than in Oyo State could be a sign of greater interdependency and mutual cooperation among the traders in Kaduna State. Marketers in Kaduna State probably rely more on one another for support and information because of their lower level of education. Lower level of involvement in trader association by marketers in Oyo State suggests that marketers are more independent-minded probably because of higher level of educational attainment.

The years of experience in drug marketing did not translate to increased knowledge; many of the experienced marketers displayed inadequate knowledge about antimicrobials and antimicrobial usage. As a business with implications for public health, the experience and knowledge of the marketers (in drugs handling and provision of correct information and advice to users) are critical factors in the correct handling of antimicrobials. The availability of different brands of antimicrobial agents reflects the many sources of antimicrobial agents arriving at the markets for use in animal production (Ojo *et al.* 2016). This study showed that most of the antimicrobial agents marketed for use in animals were imported to the country.

The number of people requesting for antimicrobial drugs from marketers was relatively high. This signifies the high demand for antimicrobial agents and high consumption rate (Alo & Ojo, 2007; Ojo *et al.*, 2016). Antimicrobial agents are widely used by farmers without prescription and supervision by veterinarians (Grace 2015). In most livestock operations, bacterial infections are common due to unsanitary conditions of animal husbandry, inadequate biosecurity measures and poor vaccination programmes (European Food Safety Authority, 2007; Leung *et al.*, 2011). Many marketers showed interest in the welfare of the animals and made efforts in providing good service to the purchasers of antimicrobial drugs by asking questions that could provide useful clues to the clinical conditions of the animals on farms. However, these marketers were mostly non-professional with poor knowledge of diagnosis and treatment of animal diseases. Although, marketers provided guidance to farmers based on experience, they were not officially authorised to prescribe and dispense antimicrobial agents. The practice of prescribing antimicrobial agents by unauthorised persons should be discouraged (Leung *et al.*, 2011; Office International des Épizooties, 2016). Asking questions about the clinical conditions of animals cannot substitute for physical examination of the animals and laboratory analysis of clinical samples for more accurate diagnosis. Laboratory investigations are very important aids for accurate disease diagnosis and crucial to making informed decision on the choice of appropriate antimicrobial agent in the treatment of bacterial infections (Leekha *et al.*, 2011; Leung *et al.*, 2011; De Briyne, *et al.* 2013, Office International des Épizooties, 2016). Use of antimicrobial agents without microbiological analysis of clinical samples and antimicrobial sensitivity testing could lead to misuse thereby promoting the emergence of antimicrobial resistant strains of bacteria (Berild *et al.*, 2006).

Previous studies have shown that antimicrobial agents that have been prohibited for use in both humans and animals were widely marketed for use in Nigerian food animal production system (Adebowale *et al.*, 2016; Ojo *et al.*, 2016). Moreover, many antimicrobial brands marketed for use in animal production have similar active agents as those used in humans (Lander *et al.*, 2012; Ojo *et al.*, 2016). The abuse of these drugs in animals can accelerate the emergence and widespread dissemination of antimicrobial resistance (Woolhouse *et al.*, 2015). It can lead to the presence of antibiotic residues in animal products which when consumed by humans can cause toxicity, allergy and malignancies. The transfer of antimicrobial resistance from animals to humans is of global public health importance with potentially devastating socioeconomic consequences (Cosgrove & Carmeli, 2003).

Drug marketers suggested antimicrobial agents to farmers, advised on the usage of these drugs and inquired about the outcome of antimicrobial therapy in treated animals. Veterinarians normally should carry out these activities. Marketers provided these important services to farmers without additional costs to the price of antimicrobial agents. Thus, by patronising antimicrobial marketers, farmers did not need to pay for veterinary services that could increase the total cost of production and impinge on profitability. In cases where farmers do not have ready access to veterinary services from qualified professional, the marketers are available as (reliable) alternatives to bring reprieve to farmers and their animals. However, the services provided by the marketers could be uninformed, insufficient and sometimes could complicate the case at hand leading to more losses. Marketers reported that farmers do complain about failure of antimicrobial therapy. Non-response to antimicrobial therapy could be due to many factors including wrong diagnosis and involvement of antimicrobial resistant bacterial strains in disease conditions (Finch & Chalmers, 2014). Therefore, farmers should be encouraged to seek help from qualified professionals. Moreover, Government intervention is required to make veterinary services more readily accessible and affordable to farmers.

Most of the marketers agreed that handling and use of antimicrobials required special training. Furthermore, majority concurred that antimicrobial agents should be sold based on presentation of a prescription from qualified and authorised professionals. The unrestricted sales and uncontrolled access to antimicrobials is possible because of poor regulation and inadequate monitoring by regulatory agencies. Most marketers were not aware of the presence of regulatory agencies and were unfamiliar with regulatory processes. Although there are good policies to regulate sales and usage of antimicrobials in humans and animals, these policies are poorly implemented (Grace, 2015). Lack of communication on legislations, low level of awareness on health hazards and socioeconomic implications of antimicrobial resistance as well as poor access to veterinary services have been identified as major constrains to policy implementation (Adebowale et al., 2016; Grace, 2015). In the present study, marketers showed their dislike for introduction of stricter measures in marketing and use of antimicrobial agents. However, they believe stricter measures are necessary to safeguard the continued efficacy of antimicrobial agents for the protection of animal welfare and public health. Aversions to regulatory measure could be due to perceived loss of patronage and possible loss of business with stricter control of antimicrobial usage in animals.

6 Conclusion and recommendations

The study showed that antimicrobial agents are widely distributed and poorly regulated in the study areas. The activities of antimicrobial marketers influenced antimicrobial usage in animal production. The existing regulatory policies on distribution and use of antimicrobial agents are ineffective therefore implying inadequacies of regulatory agencies in the administration and control of antimicrobial drugs.

There should be restricted access to antimicrobial drug through effective communication and enforcement of existing policies and regulations (Office International des Épizooties, 2016). Antibiotics should only be sold on the basis of a veterinary prescription. All antimicrobial marketers should be made to register with appropriate government regulatory agency and obtain an operating licence. The licence should be renewed periodically after demonstration of satisfactory compliance with regulatory standards. All marketers should operate within the limits of their roles in dispensing antimicrobial agents without infringing on the duties of veterinarians. Only veterinarians should prescribe and supervise antimicrobial usage in animals. Marketers should keep detailed records of every transaction involving antimicrobial sales. These records must be made available to regulatory agencies for monitoring purposes. Implementation of antimicrobial legislations through adequate monitoring and penalisation of offenders will help to promote compliance. Government can employ 'stick and carrot' approach whereby there are rewards and compensations for good practices and punishments for unacceptable practices. Moreover, local government authorities should employ more veterinarians to make veterinary services readily accessible to farmers at subsidised cost (Leung et al., 2011). The ensuing increase in the involvement of qualified veterinarians in disease diagnosis, treatment and consultancy services will curb excessive and non-judicious use of antimicrobials in food animals. Adequate supervision of antimicrobial usage by veterinarians at farm level would ensure adherence to dosage regimen. Furthermore, regulatory agencies can utilise the operations of the trade associations to disseminate information, create awareness and publicize regulatory policies by providing training workshops aimed at reorienting the marketers towards responsible antimicrobial stewardship and best practices in antimicrobial dispensing. Better control of drug manufacturing, importation and distribution in Nigeria will limit the influx of antimicrobial agents into the market. This will help to check the increasing trend of antimicrobial usage in animals.

Authors' contributions:

OOE, FE and DMA. were responsible for project conceptualisation, design and implementation. OOE and FE were involved in sample collection and data collation. OOE, AOJ and FE were responsible for data analysis and interpretation of results. OOF. prepared the manuscript. AOJ, FE and DMA were involved in manuscript review. DMA was the project leader.

Competing interests:

The authors declare that they have no conflict of interest.

Acknowledgement:

This study was part of a project commissioned by the Ministerial Codex Committee of the Federal Ministry of Agriculture and Rural Development, Nigeria.

References

Adebowale, O. O., Adeyemo, O. K., Awoyomi, O., Dada, R. & Adebowale, O. (2016). Antibiotic use and practices in commercial poultry laying hens in Ogun State Nigeria. *Revue d'Élevage et de Médecine Vétérinaire des Pays Tropicaux*, 69 (1), 41–45.

Adegbidi, A., Mensah, R., Vidogbena, F. & Agossou, D. (2012). Determinants of ICT use by rice farmers in Benin: from the perception of ICT characteristics to the adoption of the technology. *Journal of Research in International Business and Management*, 2 (11), 273–284.

Adesokan, H. K., Akanbi, I. O., Akanbi, I. M. & Obaweda, R. A. (2015). Pattern of antimicrobial usage in livestock animals in southwestern Nigeria: The need for alternative plans. *Onderstepoort Journal of Veterinary Research*, 82 (1), #816. doi: 10.4102/ojvr.v82i1.816.

Alo, O. S. & Ojo, O. (2007). Use of antibiotics in food animals: A case study of a major veterinary outlet in Ekiti State, Nigeria. *Nigerian Veterinary Journal*, 28 (1), 80–82.

Berild, D., Mohseni, A., Diep, L. M., Jensenius, M. & Ringertz, S. H. (2006). Adjustment of antibiotic treatment according to the results of blood cultures leads to decreased antibiotic use and costs. *Journal of Antimicrobial Chemotherapy*, 57, 326–330.

Chantziaras, I., Boyen, F., Callens, B. & Dewulf, J. (2014). Correlation between veterinary antimicrobial use and antimicrobial resistance in food-producing animals: a report on seven countries. *Journal of Antimicrobial Chemotherapy*, 69, 827–834.

Cosgrove, S. E. & Carmeli, Y. (2003). The impact of antimicrobial resistance on health and economic outcomes. *Clinical Infectious Disease*, 36, 1433–1437.

Danjuma, S. K., Muhammad, Y. A. & Alkali, L. F. (2013). Factors militating against women economic empowerment end poverty reduction in African countries. *IOSR Journal of Business and Management*, 13 (6), 47–51.

Davies, J. & Davies, D. (2010). Origins and Evolution of Antibiotic Resistance. *Microbiology and Molecular Biology Review*, 74 (3), 417–433.

De Briyne, N., Atkinson, J., Pokludová, L., Borriello, S. P. & Price, S. (2013). Factors influencing antibiotic prescribing habits and use of sensitivity testing amongst veterinarians in Europe. *Veterinary Record*, 173 (19), #475. doi:10.1136/vr.101454.

Dhama, K., Tiwari, R., Khan, R. U., Chakraborty, S., Gopi, M., Karthik, K., Saminathan, M., Desingu, P. A. & Sunkara, L. T. (2014). Growth promoters and novel feed additives improving poultry production and health, bioactive principles and beneficial applications: the trends and advances–A Review. *International Journal of Pharmacology*, 10, 129–159.

European Food Safety Authority (EFSA) (2007). Scientific opinion of the panel on animal health and welfare on a request from the commission on animal health and welfare aspects of different housing and husbandry systems for adult breeding boars, pregnant, farrowing sows and unweaned piglets. *The European Food Safety Authority Journal*, 572, 1–13.

FAO (2005). *Associations of market traders: Their roles and potential for further development*. Food and Agriculture Organization of the United Nations (FAO), Publishing Management Service, Information Division, Viale delle Terme di Caracalla, 00100 Rome, Italy.

Finch, S. & Chalmers, J. D. (2014). Brief clinical review: non-responding pneumonia. *European Medical Journal, Respiratory*, 2, 104–111.

Founou, L. L., Founou, R. C. & Essack, S. Y. (2016). Antibiotic Resistance in the Food Chain: A Developing Country-Perspective. *Frontiers in Microbiology*, 7, #1881. doi:10.3389/fmicb.2016.01881.

Grace, D. (2015). Review of evidence on antimicrobial resistance and animal agriculture in developing countries. Evidence on Demand, UK. , iii + 39 pp.

Lander, T. F., Cohen, B., Wittum, T. E. & Larson, E. L. (2012). A Review of Antibiotic Use in Food Animals: Perspective, Policy, and Potential. *Public Health Reports*, 127, 4–22.

Leekha, S., Terrell, C. L. & Edson, R. S. (2011). General principles of antimicrobial therapy. *Mayo Clinic Proceedings*, 86 (2), 156–167.

Leung, E., Weil, D. E., Raviglione, M. & Nakatani, H. on behalf of the World Health Organization World Health Day Antimicrobial Resistance Technical Working Group (2011). The WHO policy package to combat antimicrobial resistance. *Bulletin of the World Health Organization*, 89, 390–392. doi: 10.2471/BLT.11.088435.

Mensah, S. E., Koudandé, O. D., Sanders, P., Laurentie, M., Mensah, G. A. & Abiola, F. A. (2014). Antimicrobial residues in foods of animal origin in Africa: public health risks. *Revue Scientifique et Technique de l'Office International des Épizooties*, 33 (3), 987–996.

Office International des Épizooties (2016). Responsible and prudent use of antimicrobial agents in veterinary medicine. *In:* OIE – Terrestrial Animal Health Code, 25th edition. ISBN 978-92-95108-01-1. World Organisation for Animal Health 12, Rue de Prony, 75017 Paris, France.

Ojo, O. E., Fabusoro, E., Majasan, A. A. & Dipeolu, M. A. (2016). Antimicrobials in animal production: usage and practices among livestock farmers in Oyo and Kaduna States of Nigeria. *Tropical Animal Health and Production*, 48, 189–197.

Okeke, I. N., Lamikanra, A. & Edelman, R. (1999). Socioeconomic and behavioral factors leading to acquired bacterial resistance to antibiotics in developing countries. *Emerging Infectious Diseases*, 5 (1), 18–27.

Olatoye, I. O. & Ehinmowo, A. A. (2010). Oxytetracycline residues in edible tissues of cattle slaughtered in Akure, Nigeria. *Nigerian Veterinary Journal*, 31 (2), 93–102.

Pritchett, J., Thilmany, D. & Johnson, K. (2005). Animal disease economic impacts: A survey of literature and typology of research approaches. *International Food and Agribusiness Management Review*, 8 (1), 23–45.

Rweyemamu, M., Otim-Nape, W. & Serwadda, D. (2006). *Foresight. Infectious Diseases: preparing for the future. Africa*. Office of Science and Innovation, London.

Sebba, J., Griffiths, V., Luckock, B., Hunt, F., Robinson, C., Flowers, S., Farlie, J., Mulmi, R. & Drew, N. (2009). *Youth-led innovation – Enhancing the skills and capacity of the next generation of innovators*. NESTA, 1 Plough Place, London EC4A 1DE, UK.

Woolhouse, M., Ward, M., van Bunnik, B. & Farrar, J. (2015). Antimicrobial resistance in humans, livestock and the wider environment. *Philosophical Transactions of the Royal Society B*, 370, #20140083. doi: 10.1098/rstb.2014.0083.

20

The effect of feeding restriction with cassava flour on carcass composition of broilers

Youssouf Toukourou [a,*], Dassouki Sidi Issifou [a], Ibrahim Traore Alkoiret [a],
Armand Paraïso [a], Guy Appolinaire Mensah [b]

[a]*Department of Animal Production, Faculty of Agronomy, University of Parakou, Parakou, Benin Republic*
[b]*National Agricultural Research Institute of Benin (NARIB), Agricultural Research Centre Agonkanmey, 01 Cotonou, Benin Republic*

Abstract

In order to promote poultry farming in resource-limited rural areas, the effects of feeding restriction with cassava flour on the carcass composition of broilers was studied. After three weeks on a restrictive diet (step 1), the broilers were re-fed during four weeks according to their physiological needs (step 2). In total, 75 four-weeks old chicks were randomly divided into three lots of 25 subjects. Lot I (control) is fed without cassava flour. The lots II and III are fed with diets containing respectively 10 and 30 % of cassava flour, with energetic and protein density of 85 and 70 % of the control. Eight broilers of each lot have been randomly selected and slaughtered at the end of each step. At the end of the restrictive step, the carcass yields and the weights of the digestive tracts are 67.1, 66.3, and 64.7 % and 178.5, 170.0, and 113.3 g respectively for the lots I, II, and III with a significant difference ($p \leq 0.05$) between lot I and III and then between lots II and III. After 4 weeks of re-feeding, the lots I, II, and III had respectively 69.9, 73.2, and 67.7 % of carcass yield as well as digestive tract weights of 178.3, 180.8, and 156.0 g. The carcass yield had been entirely made up ($p \geq 0.05$) to the broilers previously submitted on a restrictive diet. However, the weight of the empty cold carcass was not fully compensated ($p \leq 0.05$).

Keywords: broiler, cassava flour, carcass productivity, compensatory growth

1 Introduction

The breeding of short cycle animal species, such as poultry, has remarkably increased over the last decades because of its socio-economic importance and mainly its obvious dietary and nutritional value in underdeveloped countries such as the Benin Republic. A growing number of households, particularly in rural areas, are devoted to small size animal breeding, mainly in poultry farming. Poultry farming contributes a great extent to the provision of highly nutritious diet values and helps satisfy numerous social obligations (consumption, savings, schooling, health care, clothing, etc.). In order to improve the very low zoo-technical performance of local broilers, particularly the carcass yields, many traditional farmers and households do not hesitate to introduce improved and genetically efficient broilers strains. Two difficulties which farmers face in such a process of efficiency improvement are sanitary and dietary requirements of these improved progenitors. These broilers are not able to express their production and reproduction potentials unless their husbandry sanitary and dietary conditions are sufficiently improved. Broilers feed in order to be efficient requires both quantitative and qualitative norms. Indeed, broilers bred to grow quickly need a diet that gives them all the necessary nutrients in sufficient quantity. The raw materials used in the feeding of broiler chickens must satisfy the requirement of digestive and metabolic efficiency, because the digestive

* Corresponding author
Email: ytoukourou@gmail.com

tract of poultry is shorter than that of other domestic animals. According to Rougière (2010), total transit is relatively rapid in chickens compared to monogastric mammals. On average, the average residence time in the digestive tract varies between 5 and 9 hours. So, to compensate the relatively short digestive tract and rapid digesta transit time, high-performing birds need easily digestible nutrient-dense diets (Phocas *et al.*, 2014). Poultry farming thus becomes a very uncertain activity for these rural households because of its high cost. According to several authors (Detimmerman *et al.*, 1992; Picard *et al.*,1993; Agreste Conjoncture, 2015) the feeding of monogastric animal species, especially poultry occupies more than $^2/_3$ of the total cost of production. This diet is all the more onerous because it uses grains of cereals that usually are concurrently a part of the human diet. In Benin, one of the most utilised essential raw materials in broilers diets is maize, which can be used as whole kernels or maize meal in feed. For the production of a poultry diet, farmers generally mix 50 to 70 % of maize as an energy source into the diet (Salami & Odunsi, 2003; Teguia *et al.*, 2004; Ukachukwu, 2008). However, maize is a staple food for the population in Benin. Therefore, the periodical unavailability and inaccessibility of this diet ingredient limit its permanent use as raw material in the formulation of broiler diets. As a result, periodic feed restriction as a feeding strategy is necessary because of the increasing scarcity of food resources for humans and livestock. Indeed, climate change is becoming increasingly unfavourable for agricultural activities, especially for the production of food crops. According to Baudoin (2010), the Benin agriculture is mainly rain-fed, losses or yield reductions are a common risk exacerbated by climate change. The challenge for local traditional breeders is to be able to adapt to this new situation. The strategy of periodic feed restriction in broiler chicken production offers the alternative of a more efficient management of feed resources at the village level. Moreover, it helps to avoid metabolic disorders, excessive fat deposition and therefore makes the production profitable (Sahraei, 2012). The objective of the study is to be able to partially replace maize with cassava flour in the feed of genetically improved broiler chickens. The aim of this feeding strategy is to reduce the usual high production costs of traditional local poultry farmers who do not have enough resources. Because cassava as a source of food energy is more available and more accessible, it is expected that the use of cassava flour in poultry diets will lead these poultry producers to be less dependent on maize. Many studies have been carried out on the feeding of broilers based on cassava flour with sometimes contradictory results (Tada *et al.*, 2004; Zanu & Dei, 2010; Ngiki *et al.*, 2014; Motielal *et al.*, 2016). Since cassava flour is more fibrous than maize (Ngiki *et al.*, 2014), its incorporation into the diet of monogastric animal species, especially poultry, certainly affects zoo-technical performance, specifically carcass yield and carcass composition (Zanu *et al.*, 2017; Okosun & Eguaoje, 2017). The level and duration of the incorporation of cassava flour into the broiler feeding thus determine the extent of the consequences on expected performance. One of the consequences is undoubtedly the increase in feed consumption which can be economically compensated by a decrease in the nutritional value of the ration. It has been shown that poultry primarily seek to cover their dietary energy needs (Hossain *et al.*, 2012). This explains the fact that these animals tend to consume more feed with low energy density (Leeson *et al.*, 1996). Short-term feed and nutritional deprivation are likely to induce a bulimic reaction in many animal species that may result in compensatory growth (Sainz *et al.*, 1995; Toukourou & Peters, 1999; Hoch *et al.*, 2003; Więcek *et al.*, 2008). This phenomenon, whose bio-physiological mechanism is still poorly understood, can be used as an effective strategy for the efficient management of food resources. Several studies showed the possibility to replace 10 to 80 % maize with cassava flour in the broilers diets without any significant difference in their production efficiencies and other zoo-technical parameters (Eshiett *et al.*, 1980; Gomez & Valdivieso 1983; Garcia & Dale 1999). Cassava flour is more available and accessible than maize in almost all areas in Benin. It may be strategic to partially replace maize with cassava flour in broiler diets therefore limiting protein and energy content of the diet. The research question is the following: Can the broilers previously submitted to a restrictive diet made up of cassava flour reconstitute their carcass once their food condition is normalized?

2 Materials and methods

2.1 Framework

The test is conducted at the Faculty of Agronomy Application and Research Farm of the University of Parakou. The township of Parakou is located in the centre of the Benin Republic between 19° 21′ North latitude and 36° 2′ East longitude. With an average altitude of 350 m, it covers an area of 441 km², 30 km² of which is urbanized. The climate is Sudano-Guinean type with an annual average rainfall that varies between

Table 1: *Composition of feed rations for the broilers.*

Ingredients	Starting regime	Finishing-growth regime			
	Regime I	Regime II (control)	Regime III	Regime IV	Refeeding diets
Maize (%)	65	70	30	20	70
Cottonseed cake (%)	7	0	4	6	0
Soybean meal (%)	8	10	5	3	10
Fishmeal (%)	6	4	4	2	4
Cassava flour (%)	0	0	10	30	0
Malt brewery residues (%)	0	0	6.5	6.5	0
Maize bran(%)	0	0	25	20	0
Wheat bran (%)	8	12	12	9	12
Oystershell (%)	1.2	1.2	1.2	1.2	1.2
Table Salt (%)	0.25	0.25	0.25	0.25	0.25
Ironsulphate(%)	0.05	0.05	0.05	0.05	0.05
CMV (%)	4.5	2.5	2	2	2.5
Total	100	100	100	100	100
ME (Kcal/kg DM)	3000.00	3100.00	2650.00	2170.00	3100.00
CP (%)	20.00	18.76	16.40	13.10	18.76

CMV: Concentrated Mineral Vitamin; ME: Metabolisable Energy; CP: Crude Protein.

1000 and 1500 mm and two seasons that alternate as follows: a rainy season from mid-April to mid-October and a dry season from mid-October to mid-April. In this part of Benin, the temperature oscillates between 28 and 35 °C.

2.2 Diet Compositions

The following four types of diet are prepared (Table 1). Diet 0 (starting diet) is given to all chicks up to the age of 4 weeks. Diet I (growth-finishing diet) is served to the chickens in lot I (control) during growth-finishing period which lasts seven weeks. Diet II is served to chickens in lot II during dietary restriction period and diet III is served to chickens in lot III during the same period. All diets are composed and made from the following raw materials: maize grain, wheat bran, soybean meal, cottonseed meal, fish meal and cassava flour, which are commonly available and accessible at the local markets. In order to obtain the energy and protein levels required in the rations of lots II and III (regime III and IV in Table 1), certain raw materials, in particular, fish meal, soybean meal, cottonseed cake, maize bran and wheat bran, have been readjusted (Table 1). The goal is to restrict protein and energy to a specific level using cassava flour essentially. During the refeeding phase, all lots were subsequently fed at the same level as the control.

2.3 Animal materials and experimental design

On the whole, 75 chicks of *hubbard* strain are used. On reception, the chicks receive various usual sanitary treatments (vaccination against *Newcastle*, antibiotic treatment, vitamins, and vaccination against *Gumboro* disease anti-coccidian treatment). The chicks are raised in a room preheated at 35 °C with charcoal during the first week. They are installed on litter made of wood chips at a density of 20 chicks per m^2. The starting phase corresponds to the first four weeks. From the fifth week, the chicks are randomly grouped in 3 lots of 25 subjects at the rate of 20 chicks per m^2. The feed is served twice daily at fixed hours (8.30 am and 5.30 pm) in wooden linear feeders. Drinking water is always available in five litre semi-automatic plastic containers. All the chicks are individually identified by a numbered plastic ring secured to their feet. The experimental design is a *Fisher's* random block design whose treatments are described in Table 2.

2.4 Broiler slaughtering and carcass dissecting

Eight broilers are randomly chosen from each lot after each dietary phase and are fasted for twelve-hours prior to slaughtering but were given water during this period. The animals were not stunned prior to slaughter. The slaughtering is executed by jugular vein severing at the neck level. Then, they are soaked in 60 °C hot water

Table 2: *Experimental design of the study.*

	Test phases							
Criteria for study	Phase 1	Phase 2			Phase 3			
	Starting	Control	Restriction		Control	Refeeding		
		Lot I	Lot II	Lot III	Lot I	Lot II	Lot III
Test duration (weeks)	4	3	3	3	4	4	4
Number of chicks at the beginning of each phase	75	25	25	25	17	17	17
Cassava flour in the diet (%)	0	0	10	30	0	0	0
Physiological needs of energy and protein in the diet (%)	100	100	85	70	100	100	100
Number of slaughtered chickens at the end of each phase	0	8	8	8	8	8	8
Number of chicks at the end of each phase	75	17	17	17	9	9	9

and plucked manually. Their body weight is noted before slaughtering. The carcass cutting is done according to the methods described by Ricard *et al.* (1967). The feet are severed at the *tibiotarsis-metatarsia* articulation and the head is dissociated from the neck at the skull-atlas junction. The abdominal fat and all the viscera were then removed in order to obtain the eviscerated carcass. By abdominal fat, we mean the fatty tissue which covers the abdominal wall and surrounds the gizzard. The broiler residues and offal (heart, liver and gizzard) are weighed individually just after slaughter by way of 100 ± 1 g. The dressed carcass was weighed hot, just after slaughter, and cold after 24 hours *post mortem* in the refrigerator at $4\,°C$. The carcass yield was determined by making the ratio between the weight of the carcass eviscerated without head and feet and live weight before slaughter.

2.5 Data analysis

Statistical analysis of the collected data is performed with SAS (Statistical Analysis System) 9.2 software. The dependent variables such carcass yields and carcass compositions were previously tested for their normal distribution. The statistical model used for the analysis of variance (ANOVA) was as follows:

$$Y_{ijk} = \mu + a_i + b_j + e_{ijk},$$

where:

Y_{ijk}: is an observed value of the dependent variable of interest Y;

μ: the average of the dependent variable of interest Y;

a_i: the fixed effect of dietary level ($i = 1, 2, 3$) ;

b_j: the fixed effect of the live weight of the chicks at the end of the initial period (continuous variable);

e_{ijk}: variance residue.

3 Results

The average live body weight of the broilers at slaughtering, after three weeks of feed restriction (seven weeks of age), was 1264.5, 1174.9, and 802.1 g respectively for the control lot (I), lot II, and III. The average live body weights were 1.57 and 1.46 times significantly ($p \leq 0.05$) heavier for broilers in lot I and II than those in lot III, respectively. The substitution of maize with cassava flour at 30 % in the diet affected the dressed cold carcass weight and the carcass yield of broilers after three weeks of restrictive feeding. The broilers in lot I (control) shown respectively 38.85 and 2.42 % more ($p \leq 0.05$) dressed cold carcass weight and carcass yield, than those of lot III and 8.25 % and 0.85 % ($p \geq 0.05$) than those in the lot II (Table 3). At the end of seven weeks of testing, the full digestive tract weighed 178.5, 170.0, and 113.3 g respectively for broilers in lot I (control), II, and III, with lot I and lot II being significantly different ($p \leq 0.05$) from lot III. Between the two phases of feeding, the gastrointestinal tract experienced a weight increase in subjects previously subjected to a nutritional restriction. It was nearly 38 % among the subjects of lot III against 6.33 % in those of lot II. Subjects in the control group had almost no change (-0.11 %) in their digestive tract. The heart, gizzard, liver, head, feet and abdominal fat have been particularly affected by the dietary restriction with the partial substitution of maize with cassava flour in the diet. Chickens of lot I (without cassava flour in the diet) show an average weight of 8.4, 45.5, 30.8, 36.4, 54.9, and 32.7 g respectively for the heart, gizzard, liver, head, feet and abdominal fat (Table 3). Some of these values are close to those obtained by Ricard (1988) on chickens produced by crossing between *Cornish* and *White Rock* types reared under confined conditions at a density of 10 subjects per m^2 and slaughtered at seven weeks of

Table 3: *Carcass composition of broilers after the period of feed restriction and refeeding.*

Carcass characteristics	Lot I (control)			Lot II			Lot III		
	Step 1	Step 2	Difference (%)	Restriction (Step 1)	Refeeding (Step 2)	Difference (%)	Restriction (Step 1)	Refeeding (Step 2)	Difference (%)
Slaughter weight (g)	1264.5^a ±49.3	1894.4^a ±99.6	49.82	1174.9^a ±48.9	1811.1^a ±99.9	54.15	802.1^b ±49.1	1341.4^a ±99.6	67.23
Dressed cold carcass (g)	848.8^a ±37.6	1323.8^a ±76.3	55.96	778.8^a ±37.3	1325.2^a ±76.5	70.17	519.1^b ±37.4	908.4^b ±76.3	75.01
Carcass yield (%)	67.1^a ±9.8	69.9^a ±8.3	4.10	66.3^a ±9.8	73.2^b ±9.0	10.40	64.7^b ±10.1	67.7^a ±9.5	4.64
Full gastrointestinal Tract (g)	178.5^a ±8.8	178.3^a ±6.5	-0.11	170.0^a ±8.9	180.8^a ±6.5	6.33	113.3^b ±8.8	156.4^b ±6.5	37.99
Heart (g)	8.4^a ±0.6	11.0^a ±0.9	31.54	7.3^a ±0.6	9.97^{ab} ±0.90	35.83	5.7^b ±0.6	8.0^b ±0.9	41.77
Gizzard (g)	45.5^a ±1.7	62.2^a ±3.2	36.50	46.4^a ±1.7	62.3^a ±3.2	34.27	37.1^b ±1.7	54.2^a ±3.2	46.14
Liver (g)	30.8^a ±1.3	36.7^a ±1.9	19.25	30.4^a ±1.3	34.9^a ±1.95	14.69	17.95^b ±1.3	25.7^a ±1.9	43.12
Head (g)	36.4^a ±1.1	52.4^a ±3.7	44.18	33.1^a ±1.1	46.7^a ±3.7	41.28	28.4^b ±1.1	43.6^a ±3.7	53.32
Feet (g)	54.9^a ±2.1	80.0 ±5.8	45.75	46.7^b ±2.1	71.7^a ±5.8	53.60	37.4^c ±2.1	64.3^b ±5.8	71.75
Abdominal fat (g)	32.7^a ±3.1	43.5^a ±7.6	33.01	20.4^b ±3.8	46.7^a ±7.6	129.08	16.7^b ±3.8	31.1^b ±7.6	86.37

The values on the same line during the same period (step 1or step 2) with the same letters are not significantly different at the 5 % threshold.

age. Indeed, the author has registered 7.6 g for the heart, 29.8 g for the liver and 33.3 g for the abdominal fat. The weight of these organs significantly declined ($p \leq 0.05$) for lot III compared to lot I. It should be noted that organs such as feet and abdominal fat were already negatively affected ($p \leq 0.05$) by the 15 % of the energy and protein's restriction, corresponding to the treatment in lot II.

At the end of the refeeding phase, which lasted 4 weeks, the broilers showed on average, the dressed carcass weights of 1323.8, 1325.2 and 908.4 g, respectively, for lot I (control), II and III with a significant ($p \leq 0.05$) gap between broilers in lot III and the others. The carcass yield at the end of the refeeding phase was 69.9 %, 73.2 % and 67.7 % respectively in lot I, II and III. This significant superiority of the carcass yield observed in the subjects of lot II compared to those of lot I (control) is certainly due to better food efficiency, due to the phenomenon of compensatory growth. The extent of this efficacy was less remarkable in the subjects of lot III previously subjected to a much more severe food restriction. However, no significant differences in carcass yield were noted between chickens in lot I (control) and those in lot III. Outside the gizzard and head, all other organs listed, showed a significantly lower weight by chickens in lot III compared to other lots (Table 3).

4 Discussion

The use of cassava flour in the diet of broilers submitted to dietary restriction has clearly affected the development of the carcass and its composition. This shows that the energy and protein density of the diet is one of the determinants of the level of protein synthesis in animal growth. Like most species, body composition of poultry depends on the energy and protein density of the diet (Lebret *et al.*, 2015). The body weight average at slaughtering, the dressed carcass weight average and the broilers' carcass yield progressed along with an increase in the energy and protein density of the diet, which negatively corresponds with the level of cassava flour in the diet during the period of feeding restriction. With 30 % cassava flour in the diet, the broilers in the lot III show a significantly lower body weight at slaughtering than those in lot I and II. The diet containing 10 % cassava flour for broilers in lot II did not significantly affect the body weight at slaughtering, the dressed carcass weight and carcass yield after three weeks of feed restriction. Only broilers in lot III submitted to a dietary restriction containing 30 % of cassava flour in substitution of maize have a significant decrease in the development of their carcass. The subjects in lot II, regardless of their 15 %

reduced level of energy and protein restriction compared to the control and a substitution of 10 % of maize with cassava flour in the diet, have shown any change in the carcass yield. According to McMillan & Dudley (1941) and Vogt (1966) cited by Morgan *et al.* (2016), cassava in poultry diets reduced the performances. Protein and specific amino acids, such as lysine and methionine which are necessary in broilers diets, are very low in cassava flour. However in the study conducted by Ngandjou *et al.* (2011), the 50 % substitution of maize with cassava flour led to a significant rise in broiler carcass yield after seven weeks of testing. This result is certainly due to the high dietary protein content of the diet. Indeed the author incorporated 18.6 % of crude protein in the diet, compared with 13.1 % in the present study. Above the 50 % introduction rate, the author has remarked no significant difference among the broilers in different lots of testing. These results clearly show that cassava flour can validly replace maize in broiler feed, if measures to increase the protein content of the ration are taken. The use of cassava residues in substitution of maize in the broilers feed has a significant effect on all carcass parameters except the heart in the study conducted by Kana *et al.* (2014). Carcass yields were definitely higher with the broilers fed a diet lacking cassava residues compared to the one in which maize has been replaced by 50, 75 and 100 % cassava residues in the diet. This is explained by the fact that cassava residues that are more fibrous than maize are much less valorised by poultry. Other raw materials rich in fibre, such as cereal bran may, from a certain proportion, negatively affect the performance of poultry. The introduction of 20 % of rice bran in broilers diet resulted in a significant reduction of hot carcass weight in the study conducted by Deniz *et al.* (2007). The weak development of the digestive tract of broilers in this study, particularly in lot III, is the result of high dietary restriction (70 % of control diet). Pokniak *et al.* (1984) have made the same remarks on the broilers submitted to a quantitative restriction of 45 % between the 14th and 28th day. All the registered organs, in particular the heart, the stomach, the liver, the head, the feet and the abdominal fat have been more or less severely affected by the various treatments to which the broilers were submitted. The abdominal fat has considerably declined. Unlike the other organs, a substitution of 10 % maize with cassava flour resulted in a reduction of over 37 % of abdominal fat. This reduction has passed over 50 % for a 30 % substitution of maize with cassava flour. Rations containing cassava flour are found to be less nutritious in energy and protein. This results in a relatively low abdominal fat deposition proportional to the degree of maize substitution with cassava flour. This shows clearly that in a situation of under-feeding, the adipose reserves are initially mobilized to meet the metabolic requirement.

The refeeding phase has been marked by a significant rise in all carcass components as well as carcass yield in broilers in all lots and more particularly in broilers previously submitted to a dietary restriction containing cassava flour. Carcass yield was 69.9, 73.2, and 67.7 % respectively for lots I, II, andIII. This carcass yield, in particular in the control lot is almost identical to that of 69.56 % obtained by Malher *et al.* (2015) on chickens of Ross strain slaughtered at a live weight of 1.9 kg. The weight increase of cold carcass of broilers in lot I (control) was 14.21 % less than broilers in lot III and 19.05 % less than of broilers in lot II. Carcass yields deduced from the ratio between live weight at slaughter and cold dressed carcass weight recorded an increase in all lots of broilers and in particular by broilers in lots II andIII, previously subject to a restrictive diet with cassava flour. The more energy and protein restriction in the diet has been severe, the more broilers have known an improvement of their carcass weight after the refeeding phase. Broilers of lot II previously subjected for three weeks to a diet containing 10 % cassava flour with an energetic and protein density of 85 % showed, after the refeeding phase which lasted four weeks, a carcass yield significantly higher than those of lot I (control).Such a reaction can be explained by better use of nutrients within the limit of the substitution rate of maize by cassava flour in the diet. Indeed, a periodic feed restriction is usually accompanied by a better nutritional efficiency that can persist beyond the restriction period. According to Van Eenaeme *et al.* (1998), Rossi *et al.* (2001) cited by Hoch *et al.* (2003), re-feeding results in a sharp increase in protein synthesis, degradation, and protein gain throughout the body and in myofibrillar muscle proteins. This increase in protein may be, according to the same authors, 50 % higher in re-fed animals compared to the control fed *ad libitum*. Broilers in lot III, despite a spectacular rise in dressed carcass weight, have not succeeded in fully compensating the delay registered during the dietary restriction. Consequently, the four weeks scheduled for the refeeding period was insufficient to ensure complete compensation. However, the organs such as the stomach, and head did not present any difference compared to those in lot I and II. The spectacular increase of the weight of the digestive system, particularly in broilers, previously subjected to severe food restriction (lot III) clearly illustrates the extent of the metabolic activity that led to the intensification of the synthesis of organic tissues. At this rhythm, it is certainly not excluded that these subjects know full compensation if the period of refeeding was prolonged. According to Leeson & Zubair (1997), it is mainly digestive organs such as gizzard, pancreas, crop and liver that improve food consumption and thus promote compensatory growth. Susbilla *et al.* (1994) found no difference in liver weights in broiler chicks subjected to 75 and 50 % dietary restriction between 5 and 11 days of age. The whole full digestive tract with an average approximately weight of 178g has not undergone any variation in broilers of control lot between the two dietary phases. This value is closer to the observation made by Ricard & Rouvier (1967). They obtained a total weight of 167.9 g from broilers of *Bress* strain aged 11 weeks.

5 Conclusion

A short period of qualitative food restriction with cassava flour in poultry was found to be an alternative for local breeders with insufficient maize corn. Being able to partially replace, over a period of three weeks, ingredients, such as maize, with cassava flour, a more available and accessible ingredient, without unduly affecting the zoo-technical performance of poultry can be considered as an alternative to mitigate the financial burden inherent in broilers breeding. This can contribute to the fight against poverty. The substitution of 10 % maize with cassava flour associated with an about 15 % energy and protein restriction in the diet over a period of three weeks may be considered in broilers aged four weeks, without any significant change in their final carcass productivity. A level of substitution of 30 % associated to a restriction of 25 % turns out to be too high and does not allow for complete compensation even after four weeks of refeeding.

Acknowledgements

Our thanks are addressed to the Competitive Fund of the University of Parakou who financially supported this work.

References

Agreste Conjoncture (2015). Agreste Synthèses – Moyens de production – Mars 2015 – n° 2015/261. Ministère de l'Agriculture, de l'Agroalimentaire et de la Forêt, Paris, France.

Baudoin, M.-A. (2010). Climate change adaptation in Southern Benin: A study comparing international policy and local needs. *Geo-Eco-Trop.*, 34, 155–169.

Deniz, G., Orhan, F., Gencoglu, H., Eren, M., Gezen, S. S. & Turkmen, I. I. (2007). Effects of different levels of rice bran with and without enzyme on performance and size of the digestive organs of broiler chickens. *Revue de Médecine Vétérinaire*, 158 (7), 336–343.

Detimmerman, F., Buldgen, A., Dimi, R. & Compere, R. (1992). Utilisation de la graine d'arachide dans l'alimentation des poulets de chair au Sénégal. *Tropicultura*, 10 (3), 93–97.

Eshiett, N. O., Ademosun, A. A. & Omole, T. A. (1980). Effect of feeding cassava root meal on reproduction and growth of rabbits. *Journal of Nutrition*, 110 (4), 697–702.

Garcia, M. & Dale, N. (1999). Cassava roots meal for poultry. *Journal of Applied Poultry Science*, 8, 132–137.

Gomez, G. & Valdivieso, M. (1983). Cassava meal for baby pig feeding. *Nutrition Reports International*, 28 (3), 547–558.

Hoch, T., Begon, C., Cassar-Malek, I., Picard, B. & Savary-Auzeloux, I. (2003). Mécanismes et conséquences de la croissance compensatrice chez les ruminants. *INRA Productions Animales*, 16 (1), 49–59.

Hossain, M. A., Islam, A. F. & Iji, P. A. (2012). Energy Utilization and Performance of Broiler Chickens Raised on Diets with Vegetable Proteins or Conventional Feeds. *Asian Journal of Poultry Science*, 6, 117–128.

Kana, J. R., Tadjong, R. N., Kuietche, H. M., Tefack, Y., Zambou, H. & Teguia, A. (2014). Valorisation des résidus de manioc en substitution du maïs dans la ration alimentaire du poulet de chair. *Livestock Research for Rural Development*, 26, Article 48. Available at: http://www.lrrd.org/lrrd26/3/kana26048.htm (accessed on: 23 July 2016).

Lebret, B., Prache, S., Berri, C., Lefèvre, F., Bauchart, D., Picard, B., Corraze, G., Médale, F., Faure, J. & Alami-Durante, H. (2015). Qualités des viandes : influences des caractéristiques des animaux et de leurs conditions d'élevage. *INRA Productions Animales*, 28 (2), 151–168.

Leeson, S., Caston, L. & Summers, J. D. (1996). Broiler Response to Energy or Energy and Protein Dilution in the Finisher Diet. *Poultry Science*, 75, 522–528.

Leeson, S. & Zubair, K. (1997). Nutrition of the broiler chicken around the period of compensatory growth. *Poultry Science*, 76, 992–999.

Malher, X., Coudurier, B. & Redlingshöfer, B. (2015). Les pertes alimentaires dans la filière poulet de chair. *Innovations Agronomiques*, 48, 161–175.

McMillan, A. M. & Dudley, F. G. (1941). Potato meal, tapioca meal and town waste in chicken rations. *Harper Adams Utility Poultry Journal*, 26, 191–194.

Morgan, N. K. & Choct, M. (2016). Cassava: Nutrient composition and nutritive value in poultry diets. *Animal Nutrition*, 2 (4), 253–261.

Motielal, M., Homenauth, O. & DeGroot, P. (2016). Utilization of Cassava in Poultry Feed in Guyana. *Greener Journal of Agricultural Sciences*, 6 (3), 121–126. doi:10.15580/GJAS.2016.3.022616047.

Ngandjou, H. M., Teguia, A., Kana, J. R., Mube, H. K. & Diarra, M. (2011). Effet du niveau d'incorporation de la farine de manioc dans la ration sur les performances de croissance des poulets de chair. *Livestock Research for Rural Development*, 23, Article 76. Available at: http://www.lrrd.org/lrrd23/4/mafo23076.htm (accessed on: 23 July 2016).

Ngiki, Y. U., Igwebuike, J. U. & Moruppa, S. M. (2014). Utilization of Cassava Products for Poultry Feeding: A Review. *The International Journal of Science & Technology*, 2 (6), 48–59.

Okosun, S. E. & Eguaoje, S. A. (2017). Growth performance, carcass response and cost benefit analysis of cockerel fed graded levels of cassava (Manihot esculenta) grit supplemented with Moringa (Moringa oleifera) leaf meal. *Animal Research International*, 14 (1), 2619–2628.

Phocas, F., Agabriel, J., Dupont-Nivet, M., Geurden, I., Médale, F., Mignon-Grasteau, S., Gilbert, H. & Dourmad, J.-Y. (2014). Le phénotypage de l'efficacité alimentaire et de ses composantes, une nécessité pour accroître l'efficience des productions animales. *INRA Productions Animales*, 27 (3), 235–248.

Picard, M., Sauveur, B., Fenardji, F., Angulo, I. & Mongin, P. (1993). Ajustements technico-économiques possibles de l'alimentation des volailles dans les pays chauds. *INRA Productions Animales*, 6 (2), 87–103.

Pokniak, J. A., Avaria, M. S. & Cornejo, S. B. (1984). Performances zootechniques et modifications de composition des carcasses de poulets de chair subissant une restriction alimentaire protéique et énergétique initiale et une réalimentation ultérieure. *Nutrition Reports International*, 30 (6), 1377–1383.

Ricard, F. H. (1988). Influence de la densité d'élevage sur la croissance et les caractéristiques de carcasse des poulets élevés au sol. *Annales de Zootechnie*, 37 (2), 87–98.

Ricard, F. H. & Rouvier, R. (1967). Étude de la composition anatomique du poulet. I- Variabilité de la répartition des différentes parties corporelles chez des coquelets Bresse-pile. *Annales de Zootechnie*, 16 (1), 23–39.

Rossi, J. E., Loerch, S. C., Keller, H. L. & Willett, L. B. (2001). Effects of dietary crude protein concentration during periods of feed restriction on performance, carcass characteristics, and skeletal muscle protein turnover in feedlot steers. *Journal of Animal Science*, 79, 3148–3157.

Rougière, N. (2010). *Étude comparée des paramètres digestifs des poulets issus des lignées génétiques D+ et D− sélectionnées pour une efficacité digestive divergente*. Ph.D. thesis, Université François Rabelais, Tours, France.

Sahraei, M. (2012). Feed Restriction in Broiler Chickens Production: A Review. *Global Veterinaria*, 8 (5), 449–458.

Sainz, R. D., de la Torre, F. & Oltjen, J. W. (1995). Compensatory Growth and Carcass Quality in Growth-Restricted and Re-fed Beef Steers. *Journal of Animal Science*, 73, 2971–2979.

Salami, R. I. & Odunsi, A. A. (2003). Evaluation of Processed Cassava Peel Meals as Substitutes for Maize in the Diets of Layers. *International Journal of Poultry Sciences*, 2 (2), 112–116.

Susbilla, P. J., Frankel, T. L., Parkinson, G. & Gow, C. B. (1994). Weight of internal organs and carcass yield of early food restricted broilers. *British Poultry Science*, 35, 677–685.

Tada, O., Mutungamiri, A., Rukuni, T. & Maphosa, T. (2004). Evaluation of performance of broiler chicken fed on cassava flour as a direct substitute of maize. *African Crop Science Journal*, 12 (3), 267–273.

Teguia, A., Endeley, H. N. L. & Beynen, A. C. (2004). Broiler performance upon dietary substitution of cocoa husks for maize. *International Journal of Poultry Science*, 3 (12), 779–782.

Toukourou, Y. & Peters, K.-J. (1999). Auswirkungen restriktiver Ernährung auf die Wachstumsleistung von Ziegenlämmern. *Archives Animal Breeding (Archiv Tierzucht, Dummerstorf)*, 42 (3), 281–293. doi:10.5194/aab-42-281-1999.

Ukachukwu, S. N. (2008). Effect of composite cassava meal with or without palm oil and/or methionine supplementation on broiler performance. *Livestock Research for Rural Development*, 20, Article 53. Available at: http://www.lrrd.org/lrrd20/4/ukac20053.htm (accessed on: 23 July 2016).

Van Eenaeme, C., Evrard, M., Hornick, J. L., Baldwin, P., Diez, M. & Istasse, L. (1998). Nitrogen balance and myofibrillar protein turnover in double muscled Belgian Blue bulls in relation to compensatory growth after different periods of restricted feeding. *Canadian Journal of Animal Science*, 78, 549–559.

Vogt, H. (1966). The use of tapioca meal in poultry rations. *World Poultry Science Journal*, 22, 113–126.

Więcek, J., Skomiał, J., Rekiel, A., Florowski, T., Dasiewicz, K. & Woźniak, A. (2008). Compensatory growth of pigs: Quantitative and qualitative parameters of pork carcass of fatteners administered a mixture with linseed oil in the re-alimentation period. *Polish Journal of Food and Nutrition Science*, 58 (4), 451–455.

Zanu, H. K., Azameti, M. K. & Asare, D. (2017). Effects of dietary inclusion of cassava root flour in broiler diets on growth performance, carcass characteristic and haematological parameters. *International Journal of Livestock Production*, 8 (3), 28–32. doi: 10.5897/IJLP2015.0222.

Zanu, H. K. & Dei, H. K. (2010). Evaluation of Processed Cassava Flour and Blood (PCB) in feed for Broiler Chickens. *African Journal of Food Science and Technology*, 1 (5), 98–101.

21

Production and milk marketing strategies of small-scale dairy farmers in the South of Rio Grande do Sul, Brazil

Aline dos Santos Neutzling [a,*,**], Luc Hippolyte Dossa [a,b,*], Eva Schlecht [a,***]

[a] *Animal Husbandry in the Tropics and Subtropics, University of Kassel and Georg-August-Universität Göttingen, Steinstrasse 19, 37213 Witzenhausen, Germany*

[b] *Faculté des Sciences Agronomiques, Université d'Abomey-Calavi, 03 BP 2819 Cotonou Jéricho, République du Bénin*

Abstract

Milk production is a socio-economically relevant activity for many small-scale family farms in southern Brazil. The objective of this study was to analyse their production and marketing strategies. A questionnaire was administered to 199 farm households in Rio Grande do Sul State to collect information on farm assets and activities, and particularly on the contribution of milk sale to farm income. Through categorical principal component analysis and two-step clustering, farmers were classified into three types: farmers selling only milk (M); farmers selling cash crops and milk (CM); farmers selling cash crops and surplus milk (Cm). Cattle herd (heads) and size of pasture land were larger on M farms (114 ± 71.9; 51 ± 49.4 ha) than on CM (31 ± 13.4; 9 ± 8.9 ha) and Cm (12 ± 7.5; 5 ± 8.1 ha) farms. Livestock husbandry contributed 71, 59 and 16 % to family income on M, CM and Cm farms, respectively. Daily milk production of the individual cow depended on the area cultivated with fodder maize (ha per cow; $p \leq 0.001$), on sale of milk to cooperatives or to private companies ($p \leq 0.01$), on summer pasture area (ha per cow; $p = 0.001$) and on daily amount of concentrates offered (kg per cow; $p \leq 0.01$). These results indicate that the area available for fodder cultivation is a key factor for milk production on small-scale dairy farms in southern Brazil, while concentrate feeding plays a less important role even for highly market-oriented farms. This must be accounted for when exploring options for strengthening the regional small-scale milk production, in which dairy cooperatives do play an important role.

Keywords: dairy cattle, dairy companies, dairy cooperatives, family farms, farm income, pasture area, resilience

1 Introduction

Currently, almost 50 % of the global cow milk production is concentrated in the USA, India, China, Brazil, Germany, Russia, France and New Zealand (Gerosa & Skoet, 2012). In 2012, Brazil ranked fourth and produced 32.3 million tons of milk from around 22.8 million cows in lactation, with an average annual production of 1417 litres per cow and year (FAOSTAT, 2015). Milk is the sixth most important agricultural commodity in Brazil, and the dairy value chain plays an important role for food supply, job creation and income generation. The sector employs about 4 million people (Fundação Banco do Brasil, 2011) and for each 1 Brazilian real (R$) increase in the value of the milk supply chain an increase of 5 R$ can be expected in the Brazilian Gross Domestic Product (Vilela *et al.*, 2001). Nevertheless, Brazil's dairy sector is of surprisingly low profitability for producers (Fundação Banco do Brasil, 2011). This is in part due to the heterogeneity of production systems, since only 2.3 % of the farms that are specialised in milk production operate modern produc-

---

* The first and second author have contributed equally to this article.

** Current affiliation:
Program de Pós Graduacao, Faculdade de Enfermagem, Universidade Federal de Pelotas, Rio Grande do Sul, Brazil

*** Corresponding author
Email: tropanimals@uni-kassel.de

tion systems, whereas 90 % of the milk producers are small-scale farmers with low production volumes, low productivity per cow and limited use of modern technology (*ibid.*).

Until the late 1980s, most milk processing plants were run by farmer cooperatives, producing cheese, powdered milk and UHT milk. The dairy sector in Brazil massively changed in the 1990s when governmental intervention ended, the national currency stabilised, the MERCOSUL/MERCOSUR (Southern Common Market) was created (1991) and the country joined the World Trade Organization (1995). The competition of the private sector with national and multinational companies resulted in an increase in milk production, partly also as a reaction to increasing milk demand. Data for the country's twelve major dairy companies indicate that the number of suppliers reduced by 28 % whereas the milk supply per farm increased by 37 % between 1996 and 1998 (Costales *et al.*, 2008). However, after market liberalisation, many of the local cooperatives could not compete with multinational dairy companies and were thus sold to the latter (Farina, 2002). By the second half of the 1990s, the market share of cooperatives had dropped to 60 %. Despite these developments, by 2008 the cooperatives still collected around 40 % of all milk processed in the country (Costales *et al.*, 2008) and involved more than 1.3 million of mostly small-scale farmers (OCB, 2013).

To increase milk quality, a new national policy was implemented in the early 1990s, which obliged producers to invest in on-farm milk cooling and storage in order to meet international standards. This policy, the general market liberalisation and the increasing competition on the milk market were important causes for the decrease in the number of small-scale dairy farms from 1.8 million in 1996 to 1.4 million in 2003 (Matthey *et al.*, 2004; Costales *et al.*, 2008). Even despite these developments, 80 % of the dairy farmers manage 50 % of the national dairy herd until today, all keeping a maximum of ten cows. Only about 10 % of the dairy farmers keep more than 30 cows – they manage 30 to 35 % of the national dairy herd and produce 30 to 40 % of the national milk yield (Costales *et al.*, 2008).

Milk production within the country is geographically concentrated in six of the 27 federal states: Minas Gerais, Rio Grande do Sul, Goias, Sao Paulo, Parana and Santa Catarina. Accounting for about 12 % of the national milk production, Rio Grande do Sul ranks second among the milk producing states (Zilli & Candaten, 2016). Of the milk produced in Rio Grande do

Sul, 57 % comes from small-scale farms, where land property is less than 100 ha, family labour prevails over hired labour and work is supervised by the farmer (Wagner *et al.*, 2004; MDA, 2009; MDA, 2013). Between 1996 and 2000, 26.8 % of these small-scale farmers abandoned milk production (Wagner *et al.*, 2004). However, this dropout was considered relatively moderate, partly explained by the fact that most small-scale farmers in Rio Grande do Sul are organised in cooperatives which secure market access and milk commercialisation (DESER, 2009). Cooperatives guarantee that farmers can sell any amount of milk without a minimum quantity required, because they are able to deal with private companies and larger markets more effectively than a single farmer.

Since dairy production still contributes significantly to household income of small-scale farmers in Rio Grande do Sul, this study aimed at analysing their current situation, thereby focusing on how milk production strategies of more specialised or more integrated farms and milk marketing via cooperatives or private dairy companies, are related to the size of the dairy herd, the area of land devoted to pastures, fodder or cash crops, and to inputs into the dairy unit. This analysis should provide insights into the future development of milk production as part of the income generating portfolio of the state's small-scale farms.

2   Materials and methods

2.1   Study location

The state of Rio Grande do Sul is located in the South of Brazil, comprising an area of 281,748 km^2. According to the 2010 census, the state's population amounted to 10.7 million with an average population density of 38 people per km^2 and its Human Development Index evolved from 0.49 in 1991 to 0.73 in 2010 (IBGE, 2010). The region's Gross Domestic Product (GDP) was 277.658 billion R\$ (135.5 million US\$) and the annual per capita income averaged 25,779 R\$ (12,584 US\$) in 2012 (IBGE, 2014). Agriculture plays an important economic role and in 2009 accounted for 10.1 % of the state's GDP (FEE, 2015). The main cities in the South of Rio Grande do Sul are Cangucu (31°23′ S, 52°40′ W), Pelotas (31°46′ S, 52°21′ W) and Sao Lourenco do Sul (31°22′ S, 51°59′ W). The low-lying region (7–500 m a.s.l.) is characterised by a humid subtropical climate with warm summers (December–February). A regular dry season is not observed and annual precipitation ranges from 1250–1600 mm with rainfall concentrated in the winter months (June–August, 350–500 mm per month; Defesa Civil do Rio Grande do Sul, 2011).

Dairy farming is the main livestock activity in the region, and little mechanised small farms operate alongside modernised and mechanised larger scale farms, all cultivating to different degrees tobacco, rice, soybeans, black beans and wheat, vegetables and fruits (Alonso & Bandeira, 1994; Sacco dos Anjos, 1995).

2.2 Baseline survey

A baseline survey addressing a total of 199 family farms was conducted from February to April 2010 in the rural surroundings of Pelotas ($n = 76$), Sao Lourenco do Sul ($n = 63$) and Cangucu ($n = 60$). The interviewees were chosen through snowball sampling whereby an initially interviewed farmer supplied names of three colleagues at the end of the interview; amongst these one person was then randomly selected for the next interview. Qualitative and quantitative information regarding family and farm size, household composition, education, crop cultivation, livestock activities, labour endowment, milk marketing strategies, off-farm employment, and membership in cooperatives were collected during the interviews with a pre-tested structured questionnaire.

2.3 Data analysis

The dataset originating from the survey contained ordinal, nominal and scale variables. Data was first subjected to categorical principal component analysis (CATPCA), in which categorical variables were simultaneously quantified while the dimensionality of the data was reduced. Thus, the original set of 204 variables was reduced to a set of 92 uncorrelated components that represented most of the information of the original variables.

After the CATPCA, a two-step cluster analysis was applied, which selected cattle herd size (n), amount of milk ($L \, day^{-1}$) produced per lactating cow at interview time, total pasture area (ha) and contribution of livestock income to overall family income (%) as determinant variables to classify dairy farms in the study region. The emerging three farm types were then compared for relevant variables using t-test for normally and Wilcoxon test for non-normal distributed variables. Results depict means and standard deviations (±), significance was declared at $p \leq 0.05$. All statistical analyses were computed in SPSS 19.0 (IBM Corporation, 2010).

Backwards stepwise multiple linear regression analysis was applied to test the impact of different independent variables on milk production, and the binary logistic

regression model was used to predict the variables that affect farmers' decision to sell their milk to cooperatives or to private companies. The models were tested for linearity by inspecting the scatter graph, for normality of residues by the Kolmogorov-Smirnov test on unstandardized residuals, for residual auto-correlation by the Durbin-Watson test, and for homoscedasticity by the Pesaran-Pesaran test. Multi-collinearity of independent variables was also verified (Garcia, 2005). Outliers were excluded and missing values were replaced by the mean of that variable across the specific farm type. Variables in the model showed a non-normal distribution, presenting in all cases asymmetric distribution with positive skewedness. To correct asymmetric distribution, variables were ln transformed (*ibid.*). The fit of the final model was assessed by the model Chi-square (Model χ^2) and the goodness-of-fit test of Hosmer and Lemeshow (Archer *et al.*, 2006). Well-fitting models showed significance ($p \leq 0.05$) on the Model χ^2 and non-significance ($p > 0.05$) on the goodness-of-fit test.

3 Results

3.1 Farm types and characteristics of the dairy unit

From the cluster analysis three different types of farmers were identified, namely farmers selling milk only (M, $n = 7$), farmers selling cash crops and milk (CM, $n = 74$) and farmers selling cash crops and surplus milk (Cm, $n = 118$). Milk producers were mainly located near Pelotas (57%), followed by Sao Lourenco do Sul (29%) and Cangucu (14%). CM farms were especially present in Sao Lourenco do Sul (43%), followed by Pelotas (39%) and Cangucu (18%), while Cm farms dominated in Cangucu (43%), followed by Pelotas (31%) and Sao Lourenco do Sul (26%).

Cattle herd size on M, CM and Cm farms averaged 114, 31 and 12 animals. Milk producers (M) owned more female cows aged > 24 months compared to CM ($p = 0.01$) and Cm farmers ($p = 0.007$). Significant differences between groups M, CM and Cm were also observed for the number of young females (≤ 24 months, $p = 0.029$), young males (≤ 12 months, $p = 0.002$) and adult males (> 12 months, $p = 0.04$). Average milk production of lactating cows at the time of interview varied from 5 to 11 litres per day, with insignificant differences between farm types (Table 1).

Cattle herd sizes on M, CM and Cm farms had increased by 2, 3 and 8% during the 12 months preceding the interview, mainly because female calves were kept in the herd. Animals sold during the preceding 12 months

Table 1: *Size and composition of the cattle herd and milk production on three types of small-scale dairy farms in southern Rio Grande do Sul State.*

Variable	Farm type					
	M (n = 7)		CM (n = 74)		Cm (n = 118)	
	Mean	SD	Mean	SD	Mean	SD
Herd size (n)	114	71.9	31	13.5	12	7.5
Females > 24 months (n)	70	40.5	24	11.6	8	5.1
Females ≤ 24 months (n)	25	41.3	5	4.3	2	2.8
Males > 12 months (n)	14	11.1	1	2.3	1	1.8
Males ≤ 12 months (n)	5	11.2	1	2.3	1	1.8
Milk production (L cow^{-1} day^{-1})	11	10.3	7	4.1	5	4.3
Total milk per farm (L day^{-1})	705	641.3	172	125.4	34	32.8

Farm types: M = only selling milk; CM = selling cash crops and milk; Cm = selling cash crops and surplus milk.

Table 2: *Pasture area (ha per head of cattle) across three types of small-scale dairy farms in southern Rio Grande do Sul State.*

Farm type	Total pasture land		Winter pasture area		Summer pasture area	
	Mean	SD	Mean	SD	Mean	SD
M (n = 7)	0.45	0.212	0.30	0.124	0.10	0.004
CM (n = 74)	0.20	0.087	0.20	0.105	0.10	0.005
Cm (n = 118)	0.42	0.225	0.35	0.259	0.08	0.003

Farm types: M = only selling milk; CM = selling cash crops and milk; Cm = selling cash crops and surplus milk.

accounted for 15, 13 and 8 % of the cattle on M, CM and Cm farms.

Of the farmers (household heads, 83 % male and 17 % female) belonging to type M, CM and Cm, respectively, 29, 40 and 44 % had completed primary school, and 14, 4 and 3 % had completed secondary school. Twenty two and five percent of M and Cm farmers held a bachelors' degree, whereby education level was generally higher in women than in men. The family size was equal (4 ±1.7) on CM and Cm farms and slightly lower on M farms (3 ±1.4).

Milk producers managed larger areas (ha) of land (70 ±88.6) than CM (27 ±22.5) and Cm farmers (19 ±14.5; p = 0.006). Hectares of pasture land (Table 2) differed (p = 0.008) between farm types and averaged 51 ±49.4 (M), 9 ±8.9 (CM) and 5 ±8.1 (Cm), whereby a significant correlation was observed between cattle herd size and pasture area (r = 0.88 for M; r = 0.51 for CM; r = 0.58 for Cm). Pasture land was divided in winter and summer areas; the former was sown with ryegrass (*Lolium multifolium* Lam.) and black oat (*Avena strigosa* Schreb.), and the latter with pearl millet (*Pennisetum glaucum* (L.) R.Br.). With areas (ha) of

34 ±39.8 (M), 6 ±6.4 (CM) and 4 ±4.6 (Cm), the winter pastures were larger (p = 0.006) than the summer pastures of 11 ±19.1 (M), 3 ±3.1 (CM), and 1 ±3.9 (Cm).

Fodder was cultivated by 57, 80 and 62 % of M, CM and Cm farmers. The area (ha) used to crop fodder – mainly maize (*Zea mays* L.) for silage making or feeding fresh – was larger (p = 0.004) on M (21 ±22.1) than on CM (6 ±5.3) and Cm (4 ±4.8) farms. Cattle feed was mainly derived from the fodder maize area and the cultivated winter and summer pastures, but on 114 farms also at least partly from native pastures. Across farm types, the animals' daily grazing time averaged 8 hours (range 3 to 12 hours). In addition to grazing, cattle were supplemented with concentrate feed and cereal grains on 86, 96 and 75 % of M, CM and Cm farms. The total amount of concentrate (kg/month) bought by M farmers (4590 ±4515.7) was higher (p = 0.001) than the amounts purchased by CM (1650 ±3402.8) and Cm farmers (270 ±330.3). The daily amount of concentrate and grains (kg/head) offered across young and adult cattle was 1.6 (±2.18) on M farms, 1.6 (±3.11) on CM and 1.5 (±1.98) on Cm farms (p > 0.05).

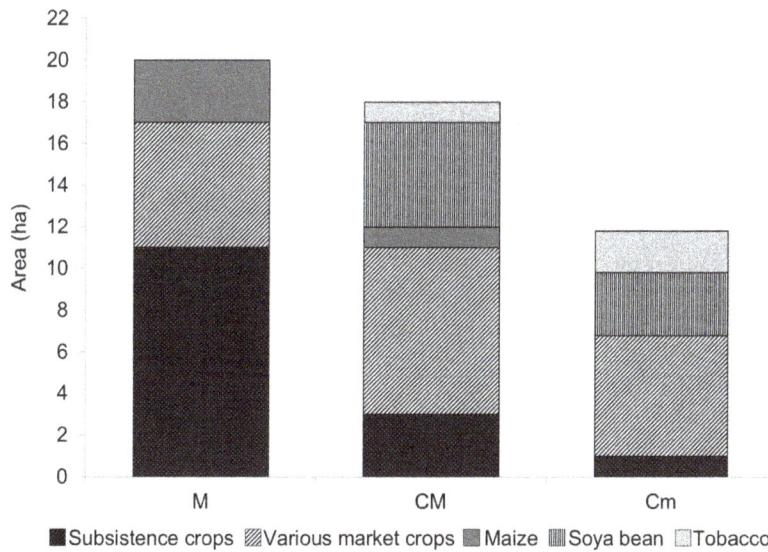

Fig. 1: *Average area per farm dedicated to different cash and subsistence crops for human use across three types of small-scale dairy farms in southern Rio Grande do Sul State (Farm types: M = only selling milk; CM = selling cash crops and milk; Cm = selling cash crops and surplus milk).*

3.2 Crop cultivation and labour allocation

In terms of area cropped, soybeans and tobacco were the most important cash crops on CM and Cm farms (Fig. 1), but a number of other crops that were partly self-consumed and partly marketed were also grown, such as lettuce, cabbage, black beans, tomatoes and strawberries. Selling of tobacco was an important source of cash for 20 % of CM and 43 % of Cm farmers, whereas selling of soybean was important for 15 and 5 % of CM and Cm farmers. Cash crops were mostly sold to traders (CM: 21 %; Cm: 30 %) and private companies (CM: 28 %; Cm: 53 %). Irrespective of farm type, maize was mainly grown for feeding animals (63.6 % of farms).

When asked about the most time consuming activity, all farmers mentioned the dairy unit, whereby milking ranked first for 86, 75 and 67 % of M, CM and Cm farmers. Milking was carried out by both partners on CM farms, only by the woman on Cm farms and only by the man on M farms. Feeding the animals was mainly a task of the male farmer on M farms (42 %), of the female farmer on Cm farms (35 %) and of both partners on CM farms (32 %). Hired labour was employed on 43, 35 and 25 % of M, CM and Cm farms, and was mainly performing general work on M and CM farms (100 and 46 %, respectively), or working in the tobacco plantations (Cm, 70 %). Only 29 % of M farms also charged hired labour with tasks of feeding, milking and pasturing the animals.

3.3 Milk marketing

Ninety-eight farmers (out of 199) were selling their milk to a cooperative; thereof 4 % belonged to cluster M, 36 % to CM and 60 % to Cm. The most important factors determining the choice of a milk cooperative were the area cultivated with subsistence crops, the pasture area per animal and investments in livestock-supporting structures during the last 10 years. Such investments have been undertaken by 14, 21 and 18 % of farmers in clusters M, CM and Cm. Among the influential variables, pasture area per animal was more important (odds ratio = 2.75) than the other predictors (Table 3).

The decision to sell milk to a private company was positively influenced by the area devoted to subsistence crops (odds ratio = 2.20) and years of education of household members, which was calculated as the average of schooling years of all household members aged > 16 years. Of the 65 farmers selling their milk to private companies, 5 % belonged to cluster M, 65 % to CM and 30 % to Cm.

3.4 Factors influencing milk production

To determine the influence of different variables on the daily milk output per farm, multiple linear regression models were run separately for farms selling their milk to cooperatives and private companies, respectively, without accounting for farm type (Table 4).

Table 3: *Parameters of the binary logistic regression analysis for variables predicting the choice of the milk marketing channel (cooperative or private company) across 199 small-scale dairy farmers in southern Rio Grande do Sul State.*

Variable	B	SE B	Wald's χ^2	df	p	Odds ratio
Dairy cooperative						
Constant	1.217	0.642	3.591	1	0.048	n.a.
Tobacco area (ha)	0.142	0.081	3.040	1	0.081	1.152
Fodder maize area (ha)	0.060	0.037	2.612	1	0.106	1.062
Subsistence crop area (ha)	−0.557	0.133	17.473	1	0.000	0.573
Pasture area per animal[†] (ha)	1.010	0.479	4.452	1	0.035	2.746
Average duration of education of household members (years)	−0.120	0.074	2.642	1	0.104	0.887
Investment in livestock activities in the last 10 years (yes/no)	−1.134	0.451	6.315	1	0.012	0.322
Overall model evaluation (Model χ^2)			69.219	6	0.001	
Goodness-of-fit[‡]			7.249	8	0.510	
Dairy company						
Constant	−3.550	0.738	23.160	1	0.001	n.a.
Subsistence crop area (ha)	0.787	0.153	26.380	1	0.000	2.197
Average duration of education of household members (years)	0.148	0.068	4.756	1	0.029	1.160
Investment in livestock activities in the last 10 years (yes/no)	1.180	0.614	3.687	1	0.055	3.253
Overall model evaluation (Model χ^2)			87.332	3	0.001	
Goodness-of-fit[‡]			20.509	7	0.457	

n.a. = not applicable; [†] animal = cattle, irrespective of age and physiological status; [‡] Goodness-of-fit test.

On all farms, total daily milk output of course depended on herd size ($p < 0.001$). In addition, on farms selling their milk to cooperatives, total daily milk output depended on area cropped with fodder maize ($p \leq 0.01$), and on farms selling their milk to private companies, the amount of concentrates fed per animal and day ($p < 0.001$) and the cultivated summer pasture area ($p \leq 0.05$) played a decisive role. In contrast, group Cm classification negatively affected the choice of private companies for milk sale ($p \leq 0.01$; Table 5).

As far as daily milk production per lactating cow is concerned, on farms selling the milk to cooperatives this variable was only and positively influenced by the fodder maize area ($p \leq 0.01$; Table 6). On farms selling their milk to private companies, variables positively influencing the production of an individual cow were the total amount of concentrate feeds offered per animal and day ($p \leq 0.01$) and the cultivated summer pasture area per animal ($p \leq 0.01$). Again, being part of group Cm had a negative impact on the milk production of an individual cow ($p \leq 0.05$).

3.5 Contribution of dairy farming to family income

Off-farm income contributed 20, 24, and 41 % to household income of groups M, CM and Cm, while crop sales contributed 9 % to total income on M farms, 17 % on CM farms and 43 % on Cm farms. The contribution of livestock husbandry to family income was highest ($p \leq 0.05$) on M farms (0.71 ±0.338), followed by CM (0.59 ±0.251) and Cm farms (0.16 ±0.141). Livestock-based income originated exclusively from milk production for 85 % of the farms (M: 100 %; CM: 78 %, Cm: 86 %), followed by milk production combined with egg sales in 13 % of farms (CM: 18 %, Cm: 12 %), and by milk production, egg and chicken sales in 2 % of farms (CM: 4 %, Cm: 2 %), whereby revenues from milk sales were responsible for at least 95 % of the overall livestock-based income in all cases. The price per litre of milk sold varied between R$ 0.51 and 0.62 (1 R$ = 0.56 US$ at the time of study; World Bank, 2011) depending on the total volume of produced milk. The amount of milk marketed (L day^{-1}) was higher ($p \leq 0.05$) on M farms (702 ±639.4) compared to CM (165 ±124.4) and Cm farms (24 ±31.7). Milk producers and CM farmers sold their milk mainly to cooperatives (M: 57 %; CM: 45 %) and private companies (M: 43 %; CM: 54 %). Cm farmers also sold their milk mainly to cooperatives (50 %) or on a personal basis to neighbours (33 %), but only rarely to private companies (17 %).

Table 4: *Characteristics of small-scale dairy farms selling their milk to cooperatives and to private companies, respectively, in southern Rio Grande do Sul State.*

Variable	Unit	Cooperative		Private company	
Presence of hired labour	(% affirmative)	27		34	
Belonging to group Cm	(% affirmative)	60		30	
		Mean	SD	Mean	SD
Cultivated winter pasture	(ha animal $^{-1}$)†	0.3	3.12	0.2	1.25
Cultivated summer pasture	(ha animal $^{-1}$)	0.1	1.15	0.2	2.41
Total pasture area	(ha animal $^{-1}$)	0.4	1.56	0.4	4.25
Fodder maize area	(ha animal $^{-1}$)	0.2	1.27	0.2	2.59
Herd size	(animals)	22	23.2	32	30.2
Concentrate feed offered	(kg animal $^{-1}$ day^{-1})	1.2	3.56	1.4	1.33

† animal = cattle, irrespective of age and physiological status.

The contribution of dairy production to the overall income of farmers selling milk to cooperatives (Table 4) was negatively affected by the use of hired labour ($p < 0.01$) and membership in group Cm ($p < 0.001$), while the area (per cattle head) of cultivated winter pasture had a positive impact ($p < 0.01$). Considering the same dependent variable for farmers selling milk to private companies, their membership in group Cm ($p < 0.001$) and area of cultivated summer pasture per cattle head ($p \leq 0.05$) were influential in a negative and a positive way, respectively (Table 7).

4 Discussion

4.1 Characteristics of small-scale dairy farms

Keeping dairy cattle is still a viable strategy to enhance income in rural areas of developing countries (Sraïri, 2005; Somda *et al.*, 2005; Radeny *et al.*, 2012). Even in Brazil this activity is essential to numerous small-scale farmers, providing 58 % of the country's milk and contributing to household income and self-consumption (Guilhoto *et al.*, 2006). The establishment of a farm typology has often been used as a tool to verify how different socio-economic and production circumstances affect farmers' management decisions and income (Daskaloupolou & Petrou, 2002; Tavernier & Tolomeu, 2004; Emtage & Harrison, 2006; Toleubayev *et al.*, 2010; Huynh *et al.*, 2014; Cortez-Arriola *et al.*, 2015).

Using labour, management and milk production as classification criteria, Wagner *et al.* (2004) also distinguished three types of small-scale dairy farmers in Rio Grande do Sul State: modern conventional producers, transition producers and traditional producers.

For the modern producers, who keep more than 10 dairy animals and have a cumulative milk production of > 68 L day^{-1}, milk production is the main source of farm income and requires the bulk of the labour force; the production is market-oriented and animal management is very good. Transition producers combine milk and cash crop production; they keep 5–10 cows and produce a total of 50–68 L day^{-1} of milk which is sold (*ibid.*). According to these authors, he or she may develop into a modern producer, or discontinue dairy activities, depending on the economic performance of the dairy unit and alternative income opportunities. Traditional producers manage < 5 cows and produce < 50 L day^{-1} of milk, which is not significantly contributing to farm income. Hence, labour input is restricted to the minimum time needed to maintain the dairy system. New investment in milk production is rare; machinery and equipment, if present, are in a poor state, animal management is not specialised and feeding is not sophisticated in most cases (*ibid.*). Carried out about 10 years later, our study shows new elements with respect to the previous classification. As far as milk production priority, importance of income from milk sales, herd size, cow productivity, animal management and feeding strategies are concerned, farmers of type M resemble the "modern conventional producer". However, farmers of type CM (cash crop and milk producers) and Cm (cash crop producers with surplus milk marketing) manage larger herds than the "transition" and "traditional" farmers. This finding shows that in the first decade of the 21st century, the transition farmers, in contrast to what was predicted by Wagner *et al.* (2004), had not yet dropped out of milk production nor, on the other hand, transformed into specialised dairy farmers. Whereas these authors found that 36 % of the small-scale farmers were

Table 5: *Parameters of the multiple linear regression on milk output per farm ($L\,day^{-1}$; dependent variable) of farms selling their milk to cooperatives or private companies, respectively.*

Variable	β_0	$SE\,\beta_0$	β_i	t	$p \leq$
Selling to cooperatives (Model 1)					
(Constant)	0.686	0.163		4.213	0.001
ln(HS)	1.055	0.127	0.898	8.323	0.001
ln(MA)	0.296	0.097	0.331	3.067	0.004
$r = 0.82$; $r^2 = 0.684$, Adj. $r^2 = 0.664$					
Selling to private companies (Model 2)					
(Constant)	1.339	0.29		4.619	0.001
ln(HS)	0.764	0.156	0.518	4.902	0.001
ln(CG)	0.476	0.124	0.325	3.847	0.001
Cm	−0.364	0.105	−0.365	−3.475	0.002
ln(SP)	0.329	0.121	0.232	2.722	0.012
$r = 0.91$; $r^2 = 0.837$; Adj. $r^2 = 0.811$					

Model 1: $\ln(y) = \beta_0 + \beta_1 \ln(HS) + \beta_2 \ln(MA) + \mu$
Model 2: $\ln(y) = \beta_0 + \beta_1 \ln(HS) + \beta_2 \ln(CG) + \beta_3 Cm + \beta_4 \ln(SP) + \mu$
ln = natural logarithm; y = milk output of the farm ($L\,day^{-1}$); HS = herd size (animals[†]);
MA = fodder maize area (ha animal^{-1}); CG = amount of concentrate and grains offered
(kg animal^{-1} day^{-1}); Cm = membership in group Cm (yes/no, i.e. 1/0); SP = cultivated summer
pasture area (ha animal^{-1}).
[†] animal = cattle, irrespective of age and physiological status.

Table 6: *Parameters of the multiple linear regression on milk offtake per cow ($L\,day^{-1}$; dependent variable) of farms selling their milk to cooperatives or private companies, respectively.*

Variable	β_0	$SE\,\beta_0$	β_i	t	$p \leq$
Selling to cooperatives (Model 3)					
(Constant)	0.939	0.071		13.256	0.001
ln(MA)	0.313	0.088	0.528	3.575	0.001
$r = 0.58$; $r^2 = 0.524$; Adj. $r^2 = 0.484$					
Selling to private companies (Model 4)					
(Constant)	1.169	0.119		9.844	0.001
ln(CG)	0.477	0.127	0.526	3.766	0.001
ln(SP)	0.424	0.125	0.480	3.402	0.002
Cm	−0.212	0.086	−0.344	−2.458	0.021
$r = 0.71$; $r^2 = 0.553$; Adj. $r^2 = 0.457$					

Model 3: $\ln(y) = \beta_0 + \beta_1 \ln(MA) + \mu$
Model 4: $\ln(y) = \beta_0 + \beta_1 \ln(CG) + \beta_2 \ln(SP) + \beta_3 Cm + \mu$
ln = natural logarithm; y = milk output of the farm ($L\,day^{-1}$); MA = fodder maize area
(ha animal^{-1}); CG = concentrate and grains offered (kg animal^{-1} day^{-1}); SP = cultivated
summer pasture area (ha animal^{-1}); Cm = membership in group Cm (yes/no, i.e. 1/0).
animal = cattle, irrespective of age and physiological status.

exclusively relying on dairy production, in our study M type farmers accounted for only 4 % of the sample. Apparently the combination of cash crop and milk production is still the most secure livelihood strategy for many small-scale farmers in the region, even though this is highly labour demanding. Yet, the shift from subsist- ence to market-oriented dairy production requires more skilled labour, and men are more likely than women to engage in professionalised dairy production (Vascon- celos Dantas *et al.*, 2016). This might explain the ob- served domination of men in large M farms compared to the two other farm types. This finding is consistent

Table 7: *Parameters of the multiple linear regression on the contribution of the dairy unit to overall income (%; dependent variable) of farms selling their milk to cooperatives or private companies, respectively.*

Variable	β_0	$SE\beta_0$	β_i	t	$p \leq$
Selling to cooperatives (Model 5)					
(Constant)	2.059	0.098		21.015	0.001
Cm	−0.532	0.073	−0.768	−7.316	0.001
ln(WP)	0.332	0.118	0.292	2.812	0.008
HL	−0.166	0.077	−0.219	−2.142	0.040
$r = 0.81$; $r^2 = 0.665$; Adj. $r^2 = 0.634$					
Selling to private companies (Model 6)					
(Constant)	1.948	0.076		25.687	0.001
Cm	−0.312	0.055	−0.737	−5.639	0.001
ln(SP)	0.195	0.079	0.322	2.465	0.020
$r = 0.74$; $r^2 = 0.556$; Adj. $r^2 = 0.523$					

Model 5: $\ln(y) = \beta_0 + \beta_1 CM + \beta_2 \ln(WP) + \beta_3 HL + \mu$
Model 6: $\ln(y) = \beta_0 + \beta_1 Cm + \beta_2 \ln(SP) + \mu$
ln = natural logarithm; y = contribution (%) of the dairy unit to overall farm income; Cm = membership in group Cm (yes/no, i.e. 1/0); WP = cultivated winter pasture area (ha animal^{-1}); HL = presence of hired labour (yes/no, i.e. 1/0); SP = cultivated summer pasture area (ha animal^{-1}). animal = cattle, irrespective of age and physiological status.

with previous observations by Magalhaes (2009) who reported a decrease of women's participation in dairy production with an increasing economic importance of this activity associated with the strengthening and modernisation of dairy cooperatives and markets.

4.2   Factors determining milk production

The fodder maize area was most influential for individual cow productivity and total daily milk output. The three dairy cooperatives in the study region (see below) are working towards improved feeding strategies; they sell quality maize seeds to their members at an affordable price and provide training courses and individual farm support for making maize silage. The latter is more profitable than the still widely practiced feeding of bulk maize, given the higher nutritive value of silage and the resulting higher milk production (O'Mara *et al.*, 1998; Wander, 2001).

Considering the good quality of natural pastures in the region during summer (Overbeck *et al.*, 2007), the use of cultivated winter pastures is more strongly advised by the cooperatives than the use of cultivated summer pastures. This advice aims at reducing the feeding costs and guaranteeing a good milk production during winter when biomass production on natural pastures is reduced due to climatic conditions. The positive effect of cultivated winter pasture manifested in its (statistical) importance for the contribution of the dairy unit to overall family income.

On farms selling milk to a private company, factors influencing individual cow productivity and contribution of milk sales to farm income were cultivated summer pasture area per cattle head and amount of concentrate feeds used. In the whole study region, cultivated summer pasture is mainly used by specialised dairy farmers (Weber, 2004), while less specialised producers rely on summer grazing of natural pastures (Nero, 2004), which was also the case in the present study. The importance of concentrate feeding and summer pasture cultivation on farms selling milk to a private company points to an intensified animal nutrition on farms that produce higher amounts of milk.

4.3   Milk marketing to cooperatives and private companies

In our analysis, the factors explaining farmers' decision to choose dairy cooperatives as their milk marketing channel were related to their investments into the dairy unit during the past ten years and their feeding strategies, mainly the area of pasture per cattle head, but also depended on the area cultivated with subsistence crops. Among others, these factors point to the attractiveness of the cooperatives' extension programs towards improving dairy units. These results are in accordance with Bourdieu (2005) and Magalhaes (2007) who claimed that dairy cooperatives are essential for the sustainable development of the associated local farmers.

Dairy cooperatives provide extension and on-farm advice, milk production equipment and feed, offer incentives for formal education, foster knowledge exchange, marketing courses and introduce new feed crops or feed conservation methods (Rajendran & Mohanty, 2004). The collective action of cooperatives on commodity markets is an important factor for the economic viability of their (small-scale) members, not only because these organisations are able to buy inputs and sell outputs at better price than the individual, but they also help their members to adapt to new production and marketing patters and standards, and to face regional or (inter)national market competition (Farina, 2002).

In the study region, COSULATI (Cooperativa Sul-Rio-Grandense de Laticinios Ltda.) is the most important cooperative. Located in Pelota and founded in 1932, it has around 20,000 members, employees and collaborators. Of the interviewed farmers, 34 were COSULATI members. In Sao Lourenco do Sul, COOPAR (Cooperativa Mista de Pequenos Agricultores da Regiao Sul) was founded in 1992 and at the time of study had 2050 members, of whom 32 were participating in our study. Thirty-two interviewed farmers in Cangucu were members of COOPAL (Cooperativa de Pequenos Agricultores Produtores de Leite da Regiao Sul). Founded in 1999 with initially 200 members, this cooperative had 650 members in 2010. The 63 interviewees who sold their milk to a private company were all dealing with Brazil Foods (BRF). Existing since 2009, a branch of this company is located in Sao Lourenco do Sul.

The dominance of dairy cooperatives in the South of Rio Grande do Sul State contradicts the predictions of Farina (2002), according to whom the milk market in southern Rio Grande do Sul should by now be in the hands of private companies, because only these were thought to be able to respond to the changes on the milk market since the beginning of the 1990s (Magalhaes, 2007). Whereas the private company BRF encouraged its suppliers through financial incentives to produce $> 100 \, \mathrm{L \, day^{-1}}$ of milk, by granting discounts on farm inputs, the three cooperatives motivated their members to deliver $> 50 \, \mathrm{L \, day^{-1}}$ but still enabled farmers with variable milk quantities to stay in business. By selling their raw milk directly to consumers in their neighbourhoods, Cm farmers reduce the involvement of marketing intermediaries. This gives them the opportunity to tailor their production to the consumer demand for raw milk and to sustain doorstep milk delivery systems.

4.4 Future development of dairy farming in southern Rio Grande do Sul

Since most of the interviewed farmers were classified as Cm and half of them sold their milk to cooperatives, they face a high risk of being expelled from the milk market when private companies are expanding and the stake of cooperatives dwindles (Costales et al., 2008; dos Santos et al., 2008; Novo et al., 2010). Government policies such as special credit schemes for private companies in order to increase nation-wide productivity of the dairy sector are indirectly supporting the move towards specialised dairy farms with modern technology (Fundação Banco do Brasil, 2011). Our multiple regression analyses indicate that M and CM farmers will be able to deal with both private companies and milk cooperatives, since they can adapt their milk production system to either milk market. Given their more professional herd and feeding management and higher milk production, M farmers are superior to CM farmers and thus the optimal choice for private companies. Due to modest milk production and less professional production system, most Cm farmers will be bound to sell their milk to cooperatives. As long as cooperatives continue to collect milk of all farmers in order to increase total milk volume, these farmers will still have a place in the market. Yet, for both dairy cooperatives and small-scale producers it is high time to comply with the ongoing processes of mechanisation and modernisation in milk production and marketing, in order to stay in business. Since crop and livestock integration is an important characteristic of Brazilian small-scale agriculture (IBGE, 2012) and was in fact practiced on most of the farms surveyed in this study, the current situation of many dairy units is worrisome: Cm farmers who just market surplus milk reap a low income from their dairy unit and do thus not invest in its modernisation. This diverges from the overall development of the Brazilian milk market that requires very professional management and marketing (Farina, 2002). Therefore, even though the doomsday scenario of abandonment of small-scale milk production (Farina, 2002) has so far not materialised in the South of Rio Grande do Sul, cooperatives should be aware that laissez faire of their members with respect to extensive and somewhat incidental management of the dairy unit may soon threaten these members' milk production and with this probably also the cooperatives' further existence.

5 Conclusions

In contrast to predictions from the early years 2000, market exclusion of small-scale dairy producers and reduced activities of dairy cooperatives were not the most critical points in southern Rio Grande do Sul State. Instead, fewer specialised larger scale milk producers and an increasing number of small-scale mixed producers, along with an adaptation of dairy cooperatives to a professionalising and globalizing dairy sector were observed. Even though specialised dairy farms offer higher and more efficient milk production than mixed farms with regular or even only occasional milk marketing, the latter two types represented the majority of milk producers in the study region. Considering the importance of mixed small-scale farms, milk cooperatives are important for supporting their milk production and marketing, and thus for the regional development. The current modernisation, diversification and enlargement of the cooperatives' milk processing plants must therefore be paralleled by considerable improvements in the fields of animal nutrition, herd management and milk marketing on the mixed farms in order to sustain their longer-term productivity and economic survival. Given the technical assistance programs already in place, dairy cooperatives are key players in this process. For the sake of sustainable regional development and employment, government initiatives that support the cooperatives' endeavours would be most welcome.

Acknowledgements

We highly appreciate the participation of the farmers and milk cooperatives from Pelotas, Sao Lourenco do Sul and Cangucu in the present study. We also thank Meike Wollni for her guidance with respect to the economic part of data analysis. The first author received four-year research scholarship from Catholic Academic Exchange Service (KAAD), Bonn, Germany.

References

Alonso, J. A. F. & Bandeira, P. S. (1994). *Crescimento econômico da região sul no Rio Grande do Sul: causas e perspectivas. 1. ed..* Fundação de Economia e Estatística, Siegfried Emanuel Heuser, Porto Alegre, RS.

Archer, K. J., Lemeshow, S. & Hosmer, D. M. (2006). Goodness-of-fit tests for logistic regression models when data are collected using a complex sample design. *Computational Statistics and Data Analysis*, 51, 4450–4464.

Bourdieu, P. (2005). *The social structures of the economy*. Polity Press, Cambridge.

Cortez-Arriola, J., Rossing, W. A., Massiotti, R. D. A., Scholberg, J. M., Groot, J. C. & Tittonell, P. (2015). Leverages for on-farm innovation from farm typologies? An illustration for family-based dairy farms in north-west Michoacán, Mexico. *Agricultural Systems*, 135, 66–76.

Costales, C. A., Pica-Ciamarra, U. & Otte, J. (2008). Social consequences for mixed crop livestock production systems in development countries. *In:* Livestock in a Changing Landscape: Drivers, Consequences and Responses. Vol. 1, pp. 249–267, FAO, Rome, Italy.

Daskalopoulou, I. & Petrou, A. (2002). Utilising a farm typology to identify potential adopters of alternative farming activities in Greek agriculture. *Journal of Rural Studies*, 18 (1), 95–103.

Defesa Civil do Rio Grande Do Sul (2011). Indice pluviometrico e dados do estado. Available at: http://www2.defesacivil.rs.gov.br/estatistica/pluviometro_consulta.asp (accessed on: 23 May 2013).

DESER (2009). O leite da agricultura familiar. Department of Rural Socioeconomic Studies (DESER), Curitiba – PR, Brazil. Available at: http://www.deser.org.br/search_results.asp?criterio=leite (accessed on: 13 September 2015).

Emtage, N. H. & Harrison, S. (2006). Landholder typologies used in development of natural resource management programs in Australia. *Australian Journal of Environmental Management*, 13, 79–94.

FAOSTAT (2015). FAO statistical databases. FAO, Rome, Italy. Available at: http://faostat.fao.org/site (accessed on: 4 January 2016).

Farina, E. M. M. Q. (2002). Consolidation, multinationalisation, and competition in Brazil: Impacts on horticulture and dairy products systems. *Development Policy Review*, 20, 441–457.

FEE (2015). Em 2013, o Rio Grande do Sul liderou o crescimento do Produto Interno Bruto (PIB) na nova série, entretanto perdeu a quarta posição entre as maiores economias do País. Published on 19 November 2015. Fundação de Economia e Estatística (FEE), Porto Alegre – RS. Available at: http://www.fee.rs.gov.br/indicadores/pib-rs/estadual/destaques/

Fundação Banco do Brasil (2011). Bovinocultura

de Leite – Parte 1. Desenvolvimento Regional Sustentável. Série Cadernos de Propostas para Atuação em Cadeias Produtivas. pp. 5–34, Brasília.

Garcia, A. (2005). *Discovering statistics using SPSS.* Sage Publications Ltda. London, UK.

Gerosa, S. & Skoet, J. (2012). Milk availability: Trends in production and demand and medium-term outlook. ESA Working paper No. 12-01. Agricultural Development Economics (ESA), FAO, Rome, Italy. Available at: http://www.fao.org/docrep/015/an450e/an450e00.pdf (accessed on: 12 September 2017).

Guilhoto, J. J. M., Silveira, F. G., Ichihara, S. M. & Azzoni, C. R. (2006). Os componentes do complexo pecuário e patronal do Brasil. *Revista de Economia e Sociologia Rural,* 44 (3), 355–383.

Huynh, T. H., Franke, C., Piorr, A., Lange, A. & Zasada, I. (2014). Target groups of rural development policies: Development of a survey-based farm typology for analysing self-perception statements of farmers. *Outlook on Agriculture,* 43 (2), 75–83.

IBGE (2010). Censo de População de 2010 Rio de Janeiro: IBGE. Instituto Brasileiro de Geografia e Estatística (IBGE). Available at: http://www.ibge.gov.br/censo2010/ (accessed on: 2 March 2013).

IBGE (2012). Producao Agropecuaria no Brasil. Instituto Brasileiro de Geografia e Estatística (IBGE). Available at: http://www.ibge.gov.br/home/estatistica/indicadores/agropecuaria/producaoagropecuaria/abate-leite-couro-ovos_201204_publ_completa.pdf. (accessed on: 12 May 2013).

IBGE (2014). Contas regionais do Brasil 2012. Instituto Brasileiro de Geografia e Estatística (IBGE). Available at: http://www.ibge.gov.br/english/estatistica/economia/contasregionais/2012/default_pdf.shtm (accessed on: 24 November 2014).

IBM Corp (2010). *IBM SPSS Statistics for Windows, Version 19.0.* IBM Corporation, Armonk, NY.

Magalhaes, R. S. (2007). Habilidades sociais no Mercado de leite. *Revista de Administracao de Empresas,* 47, 15–25.

Magalhaes, R. S. A. (2009). A "masculinização" da produção de leite. *Revista de Economia e Sociologia Rural,* 47 (1), 275–300.

Matthey, H., Fabiosa, J. F. & Fuller, F. H. (2004). Brazil: The future of modern agriculture. MATRIC Briefing Papers, 8. Iowa State University,

USA. Available at: http://lib.dr.iastate.edu/matric_briefingpapers/8 (accessed on: 12 July 2011).

MDA (2009). Ministerio do Desenvolvimento Agrario (MDA): Características da agricultura familiar. Available at: http://www.mda.gov.br

MDA (2013). Ministerio do Desenvolvimento Agrario (MDA): Producao de Leite. Available at: http://www.mda.gov.br/o/892998

Novo, A., Kees, J., Slingerland, M. & Giller, K. (2010). Biofuel, dairy and beef production in Brazil: Competing claims on land use in Sao Paulo state. *Journal of Peasant Studies,* 37, 769–792.

OCB (2013). Notícias. Organização das Cooperativas do Brasil (OCB). Available at: http://www.ocb.org.br/site/brasil_cooperativo/index.asp (accessed on: 12 May 2013).

O'Mara, F. P., Fitzgerald, J. J., Murphy, J. J. & Rath, M. (1998). The effect on milk production of replacing grass silage with maize silage in the diet of dairy cows. *Livestock Production Science,* 55 (1), 79–87.

Overbeck, G. E., Muller, S. C., Fidelis, A., Pfadenhauer, J., Pillar, V. D., Blanco, C. C., Boldrini, I. I., Both, R. & Forneck, E. D. (2007). Brazil's neglected biome: The South Brazilian Campos. *Perspectives in Plant Ecology, Evolution and Systematics,* 9 (2), 101–116.

Radeny, M., van den Berg, M. & Schipper, R. (2012). Rural poverty dynamics in Kenya: Structural declines and stochastic escapes. *World Development,* 40 (8), 1577–1593.

Rajendran, K. & Mohanty, S. (2004). Dairy cooperatives and milk marketing in India: Constrains and opportunities. *Journal of Food Distribution Research,* 35 (2), 34–41.

Sacco dos Anjos, F. A. (1995). *Agricultura familiar em transformacao: O caso dos colonos-operários de Masaranduba, Santa Catarina.* Editora da UFPEL, Pelotas.

dos Santos, M. V., Rennó, F. P., Prada e Silva, L. F. & Laranja da Fonseca, L. F. (2008). Importância sócio-econômica do leite. *Revista CFMF,* 14 (44), 9–15.

Somda, J., Kamuanga, M. & Tollens, E. (2005). Characteristics and economic viability of milk production in the smallholder farming systems in The Gambia. *Agricultural Systems,* 85, 42–58.

Sraïri, M. T., Hasni Alaoui, I., Hamama, A. & Faye, B. (2005). Relations entre pratiques d'élevage et qualité globale du lait de vache en étables suburbaines au

Maroc. *Revue de Médecine Vétérinaire*, 156,155–162.

Tavernier, E. M. & Tolomeu, V. (2004). Farm typology and sustainable agriculture: does size matter? *Journal of Sustainable Agriculture*, 24 (2),33–46.

Toleubayev, K., Jansen, K. & van Huis, A. (2010). Knowledge and agrarian de-colectivisation in Kazakstan. *The Journal of Peasant Studies*, 37 (2),353–377.

Vasconcelos Dantas, V., Pedroso Oaigen, R., Souza dos Santos, M. A., Spacek Godoy, B., da Silva, F., Pinto Corrêa, R., Nogueira Domingues, F. & Soares Simon Marques, C. (2016). Characteristics of cattle breeders and dairy production in the southeastern and northeastern mesoregions of Pará state, Brazil. *Semina: Ciências Agrárias*, 37 (3),1475–1488.

Vilela, D., Bressan, M. & Cunha, A. S. (eds.) (2001). *Cadeia de lácteos no Brasil: restrições ao seu desenvolvimento*. Juiz de Fora: Embrapa Gado de Leite, MCT/CNPq, Brasília.

Wagner, S. A., Gehlen, I. & M., W. J. (2004). Padrao tecnológico em unidades de producao familiar de leite no Rio Grande do Sul relacionado com diferentes tipologias. *Ciencia Rural Santa Maria*, 34 (5),1579–1584.

Wander, A. E. (2001). Multifarm mechanization of small farms in the Centro-Serra region of the Brazilian state Rio Grande do Sul. *Revista REDES Santa Cruz do Sul*, 6 (2),41–53.

Weber, M. (2004). Metodologia das Ciencias Sociais (Original title: Gesammelte Aufsätze zur Wissenschaftslehre), 5ª edição. Cortez Editora, São Paulo, Brazil.

World Bank (2011). Money converter. Available at: http://www.worldbank.org (accessed on: 12 September 2011).

Zilli, J. B. & Candaten, J. (2016). Main impacts on value of milk production in different regions from Rio Grande Do Sul, Brazil. *Journal of Agricultural Science*, 8 (9),184–198.

PERMISSIONS

All chapters in this book were first published in JARTS, by Kassel University Press; hereby published with permission under the Creative Commons Attribution License or equivalent. Every chapter published in this book has been scrutinized by our experts. Their significance has been extensively debated. The topics covered herein carry significant findings which will fuel the growth of the discipline. They may even be implemented as practical applications or may be referred to as a beginning point for another development.

The contributors of this book come from diverse backgrounds, making this book a truly international effort. This book will bring forth new frontiers with its revolutionizing research information and detailed analysis of the nascent developments around the world.

We would like to thank all the contributing authors for lending their expertise to make the book truly unique. They have played a crucial role in the development of this book. Without their invaluable contributions this book wouldn't have been possible. They have made vital efforts to compile up to date information on the varied aspects of this subject to make this book a valuable addition to the collection of many professionals and students.

This book was conceptualized with the vision of imparting up-to-date information and advanced data in this field. To ensure the same, a matchless editorial board was set up. Every individual on the board went through rigorous rounds of assessment to prove their worth. After which they invested a large part of their time researching and compiling the most relevant data for our readers.

The editorial board has been involved in producing this book since its inception. They have spent rigorous hours researching and exploring the diverse topics which have resulted in the successful publishing of this book. They have passed on their knowledge of decades through this book. To expedite this challenging task, the publisher supported the team at every step. A small team of assistant editors was also appointed to further simplify the editing procedure and attain best results for the readers.

Apart from the editorial board, the designing team has also invested a significant amount of their time in understanding the subject and creating the most relevant covers. They scrutinized every image to scout for the most suitable representation of the subject and create an appropriate cover for the book.

The publishing team has been an ardent support to the editorial, designing and production team. Their endless efforts to recruit the best for this project, has resulted in the accomplishment of this book. They are a veteran in the field of academics and their pool of knowledge is as vast as their experience in printing. Their expertise and guidance has proved useful at every step. Their uncompromising quality standards have made this book an exceptional effort. Their encouragement from time to time has been an inspiration for everyone.

The publisher and the editorial board hope that this book will prove to be a valuable piece of knowledge for researchers, students, practitioners and scholars across the globe.

LIST OF CONTRIBUTORS

Mst. Esmat Ara Begum, Stefanos A. Nastis and Evangelos Papanagiotou
Department of Agricultural Economics, School of Agriculture, Aristotle University of Thessaloniki, Greece

Hossein Sadeghi and Laleh Rostami
Department of Natural Resources and Environmental Engineering, College of Agriculture, Shiraz University, 71441-65186, Shiraz, Iran

Musa Hasen Ahmed
School of Agricultural Economics and Agribusiness, Haramaya University, Dire Dawa, Ethiopia

Richard Alepa and Rajashekhar Rao B. K
Department of Agriculture, The Papua New Guinea University of Technology, Lae 411, Morobe Province, Papua New Guinea

Ashinie Bugale, Miguel Aguila and Joachim Müller
Institute of Agricultural Engineering (440e), University of Hohenheim, Stuttgart, Germany

Wolfram Spreer
Department of Highland Agriculture and Natural Resources, Faculty of Agriculture, Chiang Mai University, Thailand

Setegn Gebeyehu
International Rice Research Institute, IRRI -WARDA Office, Dar Es Salaam, Tanzania

Lovince Asimwe, Abiliza Kimamboa, Germana Laswai and Louis Mtenga
Department of Animal Science and Production, Sokoine University of Agriculture, Morogoro, Tanzania

Martin Weisbjerg
Department of Animal Science, AU Foulum, Aarhus University, Tjele, Denmark

Jorgen Madsen
Department of Larger Animal Sciences, University of Copenhagen, Frederiksberg, Denmark

John Safarid
Institute of Rural Development Planning, Dodoma, Tanzania

Melanie Willich and Andreas Buerkert
Organic Plant Production and Agroecosystems Research in the Tropics and Subtropics, Universität Kassel, Witzenhausen, Germany

Anne Kathrin Schiborra and Laura Quaranta
Animal Husbandry in the Tropics and Subtropics, Universität Kassel, Witzenhausen and Georg-August-Universität Göttingen, Göttingen, Germany

Christine Bosch, Manfred Zeller and Domenica Deffner
Hans-Ruthenberg Institute, University of Hohenheim, 70593 Stuttgart, Germany

Birthe K. Paul and Wanjiku L. Chiuri
International Center for Tropical Agriculture (CIAT), Nairobi, Kenya

Brigitte L. Maass
Department for Crop Sciences, University of Göttingen, Germany
International Center for Tropical Agriculture (CIAT), Nairobi, Kenya

Fabrice L. Muhimuzi and Gaston S. Amzati
Université Evangélique en Afrique (UEA), Bukavu, DR Congo

Samy B. Bacigale
International Institute of Tropical Agriculture (IITA), Bukavu, DR Congo

Benjamin M. M. Wimba
Institut National pour l'Etude et la Recherche Agronomiques (INERA), Bukavu, DR Congo

Lizah Khairani
Graduate School of Agriculture, University of Miyazaki, Miyazaki, Japan (present: Faculty of Animal Husbandry, University of Padjadjaran, Indonesia)

Vichet Sorn
General Directorate of Agriculture, Ministry of Agriculture, Forestry and Fisheries, Phnom Penh, Cambodia

Renny Fatmyah Utamy
Interdisciplinary Graduate School of Agriculture and Engineering, University of Miyazaki, Miyazaki, Japan

Yasuyuki Ishii, Sachiko Idota and Kiichi Fukuyama
Faculty of Agriculture, University of Miyazaki, Japan

Pyseth Meas
Department of International Cooperation, Ministry of Agriculture, Forestry and Fisheries, Phnom Penh, Cambodia

Tara Pin
Chea Sim University of Kamchaymear, Prey Veng Province, Cambodia

Martin Gummert
International Rice Research Institute, Metro Manila, Philippines

Sophie Graefea
Georg-August-Universität Göttingen, Tropical Silviculture and Forest Ecology, Göttingen, Germany

Inge Armbrecht
Universidad del Valle, Departamento de Biología, Cali, Colombia

Lucía Gaitán
International Center for Tropical Agriculture (CIAT), Pham Van Dong, Tu Liem, Hanoi, Vietnam

Gholamreza Sanjari
Research Institute of Forests and Rangelands, Tehran, Iran

Hossein Ghadiri and Bofu Yu
Griffith School of Environment; Griffith University, Australia

Yodai Okuyama
RECS International Inc., Chiyoda-ku, Tokyo, 102-0075, Japan

Atsushi Maruyama, Michiko Takagaki and Masao Kikuchib
Graduate School of Horticulture, Chiba University, Matsudo, Chiba, 271-8510, Japan

Aberra Melesse, Sandip Banerjee, Aster Abebe and Amsalu Sisay
School of Animal and Range Sciences, Hawassa University, Ethiopia

Degnet H/Meskel
Wachamo University, Hossana, Ethiopia

Siaka Seriba Diarra, Malakai Koroilagilagi, Simeli Tamani, Latu Maluhola, Sila Isitolo, Jiaoti Batibasila, Tevita Vaea, Vasenai Rota and Ulusagogo Lupea
School of Agriculture and Food Technology, Alafua Campus, University of the South Pacific, Apia, Samoa

Abiodun O. Claudius-Cole
Department of Crop Protection and Environmental Biology, University of Ibadan, Ibadan, Nigeria

Lawrence Kenyon
AVRDC – The World Vegetable Center, Shanhua, Tainan, Taiwan

Daniel L. Coyne and Abiodun O. Claudius Cole
International Institute of Tropical Agriculture (IITA), Oyo Road, PMB 5320, Ibadan (Oyo State), Nigeria

Olufemi Ernest Ojo
Department of Veterinary Microbiology and Parasitology, College of Veterinary Medicine, Federal University of Agriculture Abeokuta, Abeokuta, Nigeria

Olajoju Jokotola Awoyomi and Morenike Atinuke Dipeolu
Department of Veterinary Public Health and Reproduction, College of Veterinary Medicine, Federal University of Agriculture Abeokuta, Abeokuta, Nigeria

Eniola Fabusoro
Department of Agricultural Extension and Rural Development, College of Agricultural Management and Rural Development, Federal University of Agriculture Abeokuta, Abeokuta, Nigeria

Youssouf Toukourou, Dassouki Sidi Issifou, Ibrahim Traore Alkoiret and Armand Paraïso
Department of Animal Production, Faculty of Agronomy, University of Parakou, Parakou, Benin Republic

Guy Appolinaire Mensah
National Agricultural Research Institute of Benin (NARIB), Agricultural Research Centre Agonkanmey, 01 Cotonou, Benin Republic

Aline dos Santos Neutzling, Luc Hippolyte Dossa and Eva Schlechta
Animal Husbandry in the Tropics and Subtropics, University of Kassel and Georg-August-Universität Göttingen, Steinstrasse 19, 37213 Witzenhausen, Germany

Luc Hippolyte Dossa
Faculté des Sciences Agronomiques, Université d'Abomey-Calavi, 03 BP 2819 Cotonou Jéricho, République du Bénin

Index